PHYSICAL PROPERTIES OF FOODS AND FOOD PROCESSING SYSTEMS

PHYSICAL PROPERTIES OF FOODS AND FOOD PROCESSING SYSTEMS

M. J. LEWIS
Department of Food Science and Technology
University of Reading, UK

WOODHEAD PUBLISHING LIMITED
Cambridge England

Published by Woodhead Publishing Limited, Abington Hall, Abington
Cambridge CB1 6AH, England
www.woodhead-publishing.com

First published 1987 Ellis Horwood Limited
Reprinted and issued in paperback 1990
Reprinted 1996 Woodhead Publishing Limited
Reprinted 2002, 2006

© 1996, Woodhead Publishing Limited
The author has asserted his moral rights.

This book contains information obtained from authentic and highly regarded sources. Reprinted material is quoted with permission, and sources are indicated. Reasonable efforts have been made to publish reliable data and information, but the author and the publisher cannot assume responsibility for the validity of all materials. Neither the author nor the publisher, nor anyone else associated with this publication, shall be liable for any loss, damage or liability directly or indirectly caused or alleged to be caused by this book.

Neither this book nor any part may be reproduced or transmitted in any form or by any means, electronic or mechanical, including photocopying, microfilming and recording, or by any information storage or retrieval system, without permission in writing from the publisher.

The consent of Woodhead Publishing Limited does not extend to copying for general distribution, for promotion, for creating new works, or for resale. Specific permission must be obtained in writing from Woodhead Publishing Limited for such copying.

Trademark notice: Product or corporate names may be trademarks or registered trademarks, and are used only for identification and explanation, without intent to infringe.

British Library Cataloguing in Publication Data
A catalogue record for this book is available from the British Library.

ISBN -13: 978-1-85573-272-8
ISBN -10: 1-85573-272-6

Table of Contents

Preface 13

Acknowledgements 15

1 Units and Dimensions
   1.1 Introduction 17
   1.2 Fundamental units 18
   1.3 Mass [M] (kg) 18
      1.3.1 Mass balances 19
         1.3.1.1 Example of mass balance 19
         1.3.1.2 Mass balances (hourly basis) 20
   1.4 Length [L] (m) 22
   1.5 Time [T] (s) 22
   1.6 Temperature [Θ] (K) 22
      1.6.1 Food-processing temperatures 26
   1.7 Other fundamental units 27
      1.7.1 Electric current (A) 27
      1.7.2 Luminous intensity (cd) 27
      1.7.3 Amount of substance (mol) 28
   1.8 Prefixes in common use 28
   1.9 Derived units 28
   1.10 Area $[L^2]$ (m^2) 29
   1.11 Volume $[L^3]$ (m^3) 31
      1.11.1 Surface-area-to-volume ratio 32
   1.12 Density $[ML^3]$ $(kg\ m^{-3})$ 32
      1.12.1 Specific gravity 32
   1.13 Velocity $[LT^{-1}]$ $(m\ s^{-1})$ 33
      1.13.1 Angular velocity 34
   1.14 Momentum $[MLT^{-1}]$ $(kg\ m\ s^{-1})$ 34
   1.15 Acceleration $[LT^{-2}]$ $(m\ s^{-2})$ 35
   1.16 Force $[MLT^{-2}]$ $(kg\ m\ s^{-2}$ or N) 36
      1.16.1 Centrifugal force 38
   1.17 Pressure $[ML^{-1}T^{-1}]$ $(kg\ m^{-1}\ s^{-2}$ or $N\ m^{-2})$ 38
      1.17.1 Vacuum measurement 40
      1.17.2 Pressures used in food-processing operations 40

1.18 Work $[ML^2T^{-2}]$ (kg m^2 s^{-2} or J) 42
1.19 Power $[ML^2T^{-3}]$ (kg m^2 s^{-3} or W) 43
    1.19.1 Pumping power 44
1.20 Energy $[ML^2T^{-2}]$ (J) 45
1.21 Summary of the main fundamental and derived units 45
1.22 Dimensional analysis 45
1.23 Concentration 47
1.24 Symbols 49
    1.24.1 Greek symbols 49
    1.24.2 Dimensionless groups 50

2 Density and Specific Gravity
2.1 Introduction 51
2.2 Solid density 53
2.3 Bulk density 55
    2.3.1 Relationship between porosity, bulk density and solid density 57
2.4 Liquid density and specific gravity 58
    2.4.1 Density bottles 59
    2.4.2 Hydrometers and hydrometer scales 60
        2.4.2.1 Hydrometer scales 60
    2.4.3 Liquid density values 62
    2.4.4 Density of milk 64
2.5 Gases and vapours 65
2.6 Density of aerated products: over-run 66
2.7 Symbols 68
    2.7.1 Greek symbols 68

3 Properties of Fluids, Hydrostatics and Dynamics
3.1 Introduction 69
3.2 Hydrostatics 69
    3.2.1 Pressure measurement 70
    3.2.2 Vacuum measurement 73
3.3 Archimedes' principle 74
3.4 Factors affecting frictional losses 74
3.5 Streamline and turbulent flow 75
    3.5.1 Further distinctions between streamline and turbulent flow 77
        3.5.1.1 Streamline flow 77
        3.5.1.2 Turbulent flow 78
3.6 Reynolds number in agitated vessels 78
3.7 The continuity equation 79
3.8 Bernoulli's equation 80
3.9 Pressure drop as a function of shear stress at a pipe wall 82
3.10 Frictional losses 82
    3.10.1 Frictional losses in straight pipes 82
    3.10.2 Frictional losses in other fittings 85
    3.10.2 Total system losses 86
3.11 Relative motion between a fluid and a single particle 86
3.12 Fluid flow through packed and fluidized beds 89
    3.12.1 Fluidization 92
3.13 Fluid flow measurement 92
    3.13.1 Variable-head measurement 92
        3.13.1.1 Pitot tube 94
    3.13.2 Variable-area meters 95
3.14 Fluid transportation and pumping 96

| | 3.14.1 Positive-displacement pumps 96
| | 3.14.2 Centrifugal pumps..............................97
| | 3.14.3 Other characteristics of pumps......................99
| | 3.14.4 Flow control..................................100
| | 3.14.5 Special types of pump..........................101
| | 3.15 Vacuum operations..................................102
| | 3.16 Aseptic operations..................................102
| | 3.17 Average residence time...............................102
| | 3.18 Distribution of residence times........................103
| | 3.19 Continuous stirred tank reactor 105
| | 3.20 Flowability of powders...............................106
| | 3.21 Symbols 106
| | 3.21.1 Greek symbols...............................107
| | 3.21.2 Dimensionless groups.........................107

4 Viscosity
4.1 Introduction....................................108
4.2 Ideal solids and liquids 109
4.3 Shear stress and shear rate 109
4.4 Newtonian fluids and dynamic viscosity 110
    4.4.1 Temperature effects.........................112
4.5 Kinematic viscosity 112
4.6 Relative and specific viscosities.....................113
4.7 Non-Newtonian behaviour.........................114
4.8 Time-independent fluids 115
    4.8.1 Plastic fluids..............................117
4.9 Time-dependent fluids 117
4.10 The power law equation 119
4.11 Methods for determining viscosity..................121
    4.11.1 Streamline flow methods....................121
    4.11.2 Capillary flow viscometers..................123
    4.11.3 Falling-sphere viscometers..................125
4.12 Rotational viscometers...........................127
    4.12.1 Concentric-cylinder viscometers..............128
    4.12.2 Cone-and-plate viscometers.................129
    4.12.3 Single-spindle viscometers..................129
4.13 Viscometer selection 130
4.14 Viscosity data..................................130
    4.14.1 Milk and milk products.....................131
    4.14.2 Oils and fats.............................132
    4.14.3 Sugar solutions...........................133
    4.14.4 Hydrocolloids............................133
4.15 Some sensory aspects 135
4.16 Symbols 135

5 Solid Rheology and Texture
5.1 Introduction....................................137
5.2 The perception of texture 137
5.3 Texture assessment by sensory methods 138
5.4 Texture evaluation by instrumental methods..........139
    5.4.1 Fundamental methods.......................141
    5.4.2 Imitative methods..........................141
    5.4.3 Empirical methods 141
    5.4.4 Chemical and microscopic methods............142
5.5 Fundamental properties 143

  5.5.1 Young's modulus 145
  5.5.2 Shear modulus 147
  5.5.3 Poisson's ratio 147
  5.5.4 Bulk modulus 148
  5.5.5 Relationship between these properties 148
 5.6 Viscoelastic behaviour 149
  5.6.1 Deformation–time relationships 150
  5.6.2 Stress relaxation experiments 150
  5.6.3 Weissenberg effect 152
  5.6.4 Oscillatory methods 153
  5.6.5 Empirical methods 154
   5.6.5.1 The Brabender system 154
   5.6.5.2 Extensometer or extensograph 155
 5.7 Gelation 155
 5.8 Model systems 157
 5.9 Objective texture measurement: empirical testing 158
  5.9.1 Penetrometer 158
  5.9.2 Extrusion and extruders 159
  5.9.3 General Foods texturometer 161
  5.9.4 Instron universal testing machine 161
  5.9.5 Other instrumental methods 162
 5.10 Size reduction and grinding 162
 5.11 Expression 165
 5.12 Symbols 165
  5.12.1 Greek symbols 166

6 Surface Properties
 6.1 Introduction 167
 6.2 Surface tension 169
 6.3 Surface activity 171
 6.4 Temperature effects 173
 6.5 Methods for measuring surface tension 173
  6.5.1 Capillary rise 174
  6.5.2 Bubble methods 176
  6.5.3 Drop-weight techniques 178
  6.5.4 Direct measurement of capillary pull 179
 6.6 Interfacial tension 180
 6.7 Work of adhesion and cohesion 182
 6.8 Emulsions 184
 6.9 Young's equation (solid–liquid equilibrium) 188
 6.10 Detergency 188
 6.11 Foaming 191
 6.12 Wettability and solubility 194
 6.13 Stabilization (dispersion and colour) 195
 6.14 Other unit operations 196
 6.15 Symbols 198
  6.15.1 Greek symbols 199
  6.15.2 Dimensionless groups 199

7 Introduction to Thermodynamic and Thermal Properties
 7.1 Introduction 200
 7.2 Conservation and conversion of energy 202
 7.3 Thermal energy and thermal units 203
 7.4 Thermodynamic terms 205
  7.4.1 Systems and surroundings 205

Table of Contents

    7.4.2 Adiabatic and isothermal processes 206
    7.4.3 Reversible and irreversible processes 206
   7.5 The first law of thermodynamics 207
   7.6 Enthalpy 207
   7.7 Entropy and the second law of thermodynamics 208
   7.8 Diagrammatic representation of thermodynamic changes 210
   7.9 The Carnot cycle 211
    7.9.1 The reverse Carnot cycle 212
   7.10 Heat or energy blances 213
   7.11 Energy value of food 214
   7.12 Energy conservation in food processing 216
   7.13 Symbols 218
    7.13.1 Greek symbols 219

8 Sensible and Latent Heat Changes
   8.1 Introduction 220
   8.2 Specific heat 220
   8.3 Relationship between specific heat and composition 222
   8.4 Specific heat of gases and vapours 224
   8.5 Determination of specific heats of materials (experimental) 226
    8.5.1 Method of mixtures 226
    8.5.2 Method of cooling 227
    8.5.3 Electrical methods 229
   8.6 Latent heat 229
   8.7 Behaviour of water in foods during freezing 232
   8.8 Latent heat values for foods (fusion) 233
   8.9 Enthalpy–composition data 234
   8.10 Oils and fats: solid–liquid transitions 237
   8.11 Differential thermal analysis and differential scanning calorimetry ... 240
   8.12 Dilatation 243
   8.13 Symbols 244
    8.13.1 Greek symbols 245

9 Heat Transfer Mechanisms
   9.1 Introduction 246
   9.2 Heat transfer by conduction 246
   9.3 Steady- and unsteady-state heat transfer 247
   9.4 Thermal conductivity 248
   9.5 Heat transfer through a composite wall 250
   9.6 Thermal conductivity of foods 252
    9.6.1 Compositional factors 254
    9.6.2 Temperature effects 255
    9.6.3 Pressure effects 255
   9.7 Determination of thermal conductivity 256
   9.8 Thermal diffusivity 257
   9.9 Particulate and granular material 258
   9.10 Heat transfer by convection (introduction) 258
   9.11 Heat film coefficient 259
    9.11.1 Evaluation of heat film coefficients 261
   9.12 Combination of heat transfer by conduction and convection 262
    9.12.1 Overall heat transfer coefficient 263
    9.12.2 Limiting resistances 266
    9.12.3 Biot number 268
    9.12.4 Example of heat exchanger design 268
   9.13 Application to heat exchangers 269

Table of Contents

- 9.13.1 Regeneration efficiency ... 270
- 9.13.2 Continuous heat exchangers ... 271
- 9.14 Direct steam injection ... 272
- 9.15 Fouling ... 274
- 9.16 Evaporator design ... 275
- 9.17 Heat transfer by radiation ... 277
- 9.17.1 Characteristics of electromagnetic radiation ... 277
- 9.18 Radiation emitted from heated surfaces ... 279
- 9.19 Stefan's law ... 280
- 9.20 Infrared radiation ... 282
- 9.21 Radio-frequency waves ... 282
- 9.21.1 Microwave and dielectric heating ... 282
- 9.21.2 Absorption of microwave energy ... 283
- 9.22 Irradiation ... 287
- 9.23 Symbols ... 290
- 9.23.1 Greek symbols ... 290
- 9.23.2 Dimensionless groups ... 291

10 Unsteady-state Heat Transfer
- 10.1 Introduction ... 292
- 10.2 Heat transfer to a well-mixed liquid ... 292
- 10.3 Unsteady-state heat transfer by conduction ... 294
- 10.4 Heat transfer involving conduction and convection ... 295
- 10.5 Thermal processing ... 301
- 10.5.1 D and Z values ... 303
- 10.7 F_0 evaluation ... 304
- 10.7 Commercial sterility and UHT processes ... 306
- 10.8 Heat penetration into canned foods (f_h and f_c values) ... 306
- 10.9 Freezing and thawing times ... 309
- 10.10 Refrigeration methods ... 312
- 10.11 Plate freezers ... 313
- 10.12 Cold-air freezing ... 314
- 10.13 Immersion freezing ... 315
- 10.14 Cryogenic freezing ... 316
- 10.15 Vacuum cooling and freezing ... 318
- 10.16 Chilling ... 319
- 10.17 Controlled-atmosphere storage and heat of respiration ... 320
- 10.18 Symbols ... 320
- 10.18.1 Greek symbols ... 323
- 10.18.2 Dimensionless groups ... 323

11 Properties of Gases and Vapours
- 11.1 Introduction ... 324
- 11.2 General properties of gases and vapours ... 324
- 11.2.1 Distinction between gases and vapours ... 326
- 11.3 Properties of saturated vapours ... 327
- 11.4 Properties of saturated water vapour (steam tables) ... 329
- 11.5 Wet vapours ... 332
- 11.6 Superheated vapours ... 334
- 11.7 Thermodynamic charts ... 336
- 11.8 Diagrammatic representation of some thermodynamic processes ... 338
- 11.8.1 Compression ... 338
- 11.8.2 Cooling and condensation at a constant pressure ... 339
- 11.8.3 Throttling expansion ... 340
- 11.8.4 Evaporation at a constant pressure ... 340

Table of Contents

- 11.9 Vapour compression refrigeration cycle ... 340
- 11.10 Introduction to air–water systems ... 344
  - 11.10.1 Absolute and relative humidity ... 344
  - 11.10.2 Dew-point temperature ... 346
  - 11.10.3 Wet-bulb temperature ... 347
  - 11.10.4 Adiabatic saturation temperature ... 347
- 11.11 Humidity charts ... 349
- 11.12 Determination of other properties from humidity charts ... 351
- 11.13 Example of interpretation of charts ... 351
- 11.14 Mixing of air streams ... 354
- 11.15 Water in food ... 355
- 11.16 Sorption isotherms ... 357
- 11.17 Water activity in food ... 359
  - 11.17.1 Hysteresis ... 361
  - 11.17.2 Frozen foods ... 362
- 11.18 Water activity–moisture relationships ... 363
- 11.19 Symbols ... 364
  - 11.19.1 Subscripts ... 364

12 Electrical Properties
- 12.1 Introduction ... 366
- 12.2 Electrical units ... 366
- 12.3 Electrical resistance and Ohms' law ... 368
  - 12.3.1 Electrical conductance ... 371
- 12.4 Electrical energy ... 372
- 12.5 Magnetic effects associated with an electric current ... 373
- 12.6 Measurement of electrical variables ... 374
  - 12.6.1 Potentiometer ... 375
  - 12.6.2 Electrical resistance measurement ... 376
- 12.7 Resistivity and specific conductance of foods ... 377
- 12.8 Electrical sensing elements ... 380
  - 12.8.1 Thermocouples ... 380
  - 12.8.2 Resistance thermometers ... 384
  - 12.8.3 Semiconductors (thermistors) ... 385
  - 12.8.4 Strain gauge transducer ... 386
  - 12.8.5 Humidity measurement ... 387
  - 12.8.6 Other sensors ... 387
- 12.9 Process control and automation ... 387
- 12.10 Alternating current ... 389
  - 12.10.1 Introduction ... 389
  - 12.10.2 Production and characteristics of an alternating EMF ... 389
  - 12.10.3 Voltage, current and power measurement ... 391
  - 12.10.4 Electrical heating ... 392
- 12.11 AC circuits ... 392
  - 12.11.1 Inductance ... 393
  - 12.11.2 Capacitance ... 394
  - 12.11.3 AC circuits containing resistors, inductors and capacitors ... 395
  - 12.11.4 Q factor ... 400
  - 12.11.5 Measurement of capacitance ... 400
- 12.12 Dielectric properties ... 402
  - 12.12.1 Dielectric constant ... 402
  - 12.12.2 Dielectric loss factor ... 402
- 12.13 Dielectric properties of foods ... 403
- 12.14 Power factor ... 407
- 12.15 Transformer action ... 408

Table of Contents

    12.16 Three-phase supply 408
    12.17 Electric motors 410
    12.18 Symbols 411
        12.18.1 Greek symbols 412
        12.18.2 Subscripts 412

13 Diffusion and Mass Transfer
    13.1 Introduction 413
    13.2 Diffusion 415
    13.3 Fick's law 415
    13.4 Gaseous diffusion 416
        13.4.1 Equimolecular counter-diffusion 416
        13.4.2 Diffusion of a gas through a stagnant layer 417
        13.4.3 Experimental determination of diffusivity 419
    13.5 Diffusivity in liquids 420
    13.6 Solid diffusion 421
    13.7 Two-film theory 423
        13.7.1 Overall mass transfer coefficient 424
        13.7.2 Relationship between overall mass transfer coefficients and film coefficients 425
        13.7.3 Determination of mass film coefficients 426
    13.8 Unsteady-state mass transfer 427
    13.9 Simultaneous heat and mass transfer 428
        13.9.1 Hot-air drying 428
        13.9.2 Spray drying 432
        13.9.3 Freeze drying (lyophilization) 434
    13.10 Packaging materials 437
    13.11 Membrane processes 440
    13.12 Symbols 443
        13.12.1 Greek symbols 444
        13.12.2 Dimensionless groups 444
        13.12.3 Subscripts 445

Bibliography and references 446

Index 459

Preface

This book has been written in order to fill a need for a text dealing with the physical properties of foods as well as those physical principles involved in food-processing operations.

While many of those connected with or interested in food or the food industry, including students of food science and technology, may well be versed in advanced chemistry or biology, a smaller number seem to be equally qualified and confident in physics or mathematics. For this reason the aim has been to produce a text for those with an ordinary understanding of physics and mathematics and rudimentary knowledge of the principles of differentiation, integration logarithms and the exponential function.

Although serving as an introduction to the physical properties of foods and the physics involved in food processing, ample references provide pathways to more advanced treatments and authoritative reviews on the subjects.

A further objective of the book is to provide numerical values of the properties of foods such as are necessary for solving food-processing calculations or selecting the most appropriate equipment. It will assist in answering the kinds of questions asked by practising technologists or even inquisitive students, such as the following. What is the density of Golden Delicious apples? What is the porosity of rape seed? What is the specific heat of a pork pie? What is the thermal conductivity of beetroot? What is the hardness of spaghetti? What is the viscosity of a custard? What is the spreadability of margarine compared with butter? What is the monomolecular layer moisture value for bananas? What is the water activity of Christmas cake? What is the electrical conductivity of cheese whey? What is the dielectric loss factor of mashed potato? What is the diffusion rate of sulphur dioxide into fresh vegetables? To this end, a wide variety of references are cited to provide published values and equations which relate to the compositional properties of foods as well as to environmental conditions such as temperature, pressure or humidity. Simple experimental methods are also described, to form the basis of informative practical exercises on food materials and to provide answers, where data are not readily available.

While answers to these questions assume importance in food-processing operations and quality control, it is important to realize that the principles and properties described within the text are by no means unique to food. Many of these principles will be applicable to most applied biology subjects, biotechnology and chemical engineering, including soil studies, pharmaceutical products, agricultural produce, fermentation products and enzyme preparations. Therefore, students and practitioners working in these areas may find this book useful.

Finally, it is hoped that this book may stimulate you, the reader, to take a greater interest in the physical properties of the diverse range of foods currently available, and to integrate this with your chemical, biochemical and microbiological knowledge, in order to improve your general appreciation of the field of food studies.

M. J. Lewis

This book is dedicated to
my mother Nancy and my late father Jack

Acknowledgements

Permission to use material from the following sources is gratefully acknowledged.

Fig. 3.5, George Newnes Ltd, from Turnbull et al. (1962).
Fig. 3.14, Pergamon Press, from Coulson and Richardson (1977).
Fig. 8.6, D. Reidel, from Rha (1975a).
Fig. 9.16, Elsevier Applied Science Publishers, from Lewis (1986a).
Fig. 9.17, The APV Company Ltd, Crawley.
Fig. 9.21, Ellis Horwood Ltd, from Milson and Kirk (1980).
Fig. 10.2, John Wiley & Sons, from Henderson and Perry (1955).
Figs 10.4, 10.5 and 10.6, American Society of Heating, Refrigerating and Air-Conditioning Engineers, from American Society of Heating, Refrigerating and Air-Conditioning Engineers (1985).
Fig. 10.11, The Editor, *Refrigeration, Air Conditioning and Heat Recovery*, from Ede (1949).
Fig. 11.13, Pergamon Press, from Coulson and Richardson (1977).
Fig. 12.24, The Editor, *IEEE Transactions*, from Bengtsson and Ohlsson (1974).
Fig. 12.25, The Editor, *Journal of Microwave Power*, from Bengtsson and Risman (1971).
Table 5.1, Academic Press, from Bourne (1982).
Tables 6.11 and 6.12, Butterworths, from Shaw (1970).

Much of the material presented in this book has been disseminated to students in the Department of Food Science, University of Reading and during lecturing visits to universities in Tanzania and Zimbabwe. I am grateful for the comments and criticism from many of these students over that time period, which have helped to improve the presentation. However, to achieve a comprehensive coverage of the subject, I have had to resort to material which has had no previous public airing and I would like to express my thanks to Dr Ann Walker and Dr David Thomson from the Department of Food Science, University of Reading, for their assistance with some of these topics. I am also grateful to Dr Reg Scott for all his good advice and interest shown in the development of the Department of Food Science since his retirement in 1975 and for his valued suggestions for improving the text.

Finally, I would like to acknowledge the special friendship, support encouragement and patience of Gay Hawley, throughout the preparation of this text.

M. J. Lewis

1

Units and dimensions

1.1 INTRODUCTION

Confusion often arises from the diverse system of weights and measures at present operating in both the UK and overseas. It is possible to buy petrol in gallons (gal) or litres (l), beer in pints, vegetables in pounds (lb) and butter in grams (g). Temperatures are still measured in both degrees Fahrenheit (°F) and degrees Celsius (°C); imagine our surprise if the weather forecaster announced temperatures in kelvins (K). It has been common usage for pressure to be measured in pounds-force per square inch (lbf in^{-2}) (often referred to colloquially as psi) e.g. when checking car tyres, but more recently units such as bars and kilograms-force per square centimetre (kgf cm^{-2}) are becoming more widespread. Electrical energy is measured in kilowatt hours (kW h) units, our gas in therms and the power of our motor vehicles in brake horsepower (hp).

Although most science subjects in schools are now taught using the International System of Units (SI), it often comes as a shock to students embarking on courses of higher education to be confronted with the centimetre gram second (cgs) or the Imperial system of units (ft lb s). Most textbooks and articles in earlier scientific journals published in the English language are written using Imperial units; it was not until the 1970s that contributions in food science textbooks began to use SI units. Although American textbooks and papers still quote mainly Imperial units, this is now also changing. The food industry is also steeped in tradition and it will take a considerable time period for all instruments, instruction manuals and personnel to be converted to SI units.

In theory, in the UK, the change from Imperial units to the SI system should have been completed by the early 1980s; in practice, this conversion is nowhere near complete and we now have to contend with a mixed system of units. It is necessary for students of applied science to be familiar with the two metric system of units (i.e. cgs and SI) and the Imperial system, in order to derive the maximum benefit from the available literature. This first chapter is written with the first objective in mind. However, throughout the book the major items of theory will be presented in SI units, and relevant conversion factors will be given, where necessary.

1.2 FUNDAMENTAL UNITS

The fundamental dimensions for the main system of measurements are mass [M], length [L], time [T] and temperature [θ]. The fundamental units in the main systems (together with the corresponding abbreviations in parentheses) are summarized in Table 1.1.

Table 1.1 — Fundamental units for the three main systems of measurement.

Property	SI system	cgs system	British (Imperial) system
Mass	kilogram (kg)	gram (g)	pound (lb)
Length	metre (m)	centimetre (cm)	foot (ft)
Time	second (s)	second (s)	second (s) hour (s)
Temperature	kelvin (K) or degree Celsius (°C)	kelvin (K) degree Celsius (°C)	degree Fahrenheit (°F)

Electric current, luminous intensity and the amount of substance (in units of moles (mol)) are also included amongst the fundamentals the SI system. The Imperial system also includes force F as one of the fundamentals, whereas in the SI system it is included amongst the derived units. The fundamental units will now be discussed in more detail.

1.3 MASS [M] (kg)

Mass is defined as the amount of matter in a body. The international prototype kilogram is a simple cylinder of platinum–iridium alloy, with height equal to diameter, which is kept by the International Bureau of Weights and Measures at Sèvres, near Paris. The mass of this is taken to represent one kilogram (1 kg).

It is important to distinguish between mass and weight. Strictly speaking, weight is defined as the force acting on an object as a result of gravity. Consequently the weight of an object will change as the gravitational force changes, whereas the mass will remain constant. It is often assumed that the acceleration due to gravity is constant at all points on the surface of the Earth and that an object will weigh the same everywhere, for most intents and purposes, this is a reasonable assumption. In practice the mass of an object is determined by comparing the force it exerts with that exerted by a known mass. Consequently we often refer to the weight of an object when we actually mean its mass.

Weight, mass or portion control is extremely important in all packaging and filling operations, where the weight is declared on the label.

1.3.1 Mass balances

In batch processing operations, such as mixing, blending, evaporation and dehydration, the law of conservation of mass applies. This can be used in the form of mass balances for evaluating these processes.

In a batch mixing process it is possible to perform a total mass balance and a mass balance on each of the components, e.g. fat or protein. Sometimes, as in evaporation, all components are grouped together as total solids.

The mass balance states that

input = output

Total and component balances will be illustrated by the following example.

1.3.1.1 Example of mass balance

It is required to produce 100 kg of low fat cream (18% fat) from double cream containing 48% fat and milk containing 3.5% fat; all concentrations are given in weight per weight (w/w). (strictly speaking, mass per mass). How much double cream and milk are required? With all such problems it is helpful to represent the process diagrammatically (see Fig. 1.1).

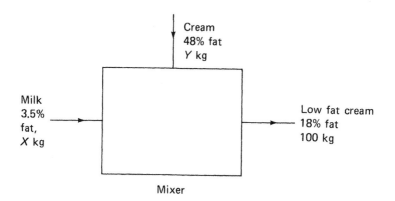

Fig. 1.1 — Mass balance in a cream standardization unit.

Let the mass of the milk and the mass of the double cream required be X kg and Y kg, respectively. Then the total balance is

$$\underset{\text{(input)}}{X+Y} = \underset{\text{(output)}}{100} \quad (1.1)$$

The component balance on the fat is

$$0.035X + 0.48Y = 100 \times 0.18 \qquad (1.2)$$

Substituting X from equation (1.1) into equation (1.2) gives

$$0.035(100 - Y) + 0.48Y = 18$$
$$3.5 - 0.035Y + 0.48Y = 18$$

Therefore,

$$Y = 32.58 \text{ kg}$$
$$X = 67.42 \text{ kg}$$

The blending operation requires 67.4 kg of milk and 32.6 kg of double cream. Standardization is the name given to the process where the fat content of milk products is adjusted by the addition of cream or skim-milk.

In a continuous process, the same principles apply but an additional term is introduced to account for the fact that some material may accumulate. The equation becomes

$$\text{input} - \text{output} = \text{accumulation}$$

However, many continuous food-processing operations take place under steady-state conditions, once they have settled down and can be analysed as such. A steady state is achieved when there is no accumulation of material, i.e.

$$\text{input} = \text{output}$$

Such operations are analysed in the same way as batch operations, normally on a time basis (hourly), as in the following example.

1.3.1.2 Mass balances (hourly basis)

If 100 kg h^{-1} of liquid containing 12% total solids is to be concentrated to produce a liquid containing 32% total solids, how much water is removed each hour?

Again the process can be represented diagramatically (Fig. 1.2).

Let mass of water removed be m and mass of concentrate produced be C. Therefore the total balance is

$$100 = m + C$$

and the solids balance is:

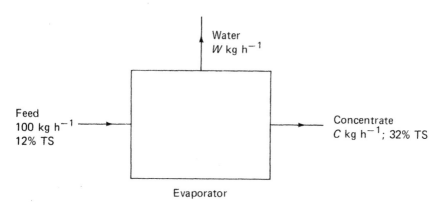

Fig. 1.2 — Mass balance in an evaporation plant.

$$100 \times 0.12 = C \times 0.32$$

It is assumed that the water leaving the evaporator contains no solid. Thus,

$$C = 37.5 \text{ kg}, \quad m = 62.5 \text{ kg}$$

Water needs to be removed at the rate of 62.5 kg h^{-1}. This fixes the evaporative capacity of the equipment. Such calculations involving total solids are extremely useful in evaporation and dehydration processes.

In some cases, chemical or biological reactions take place and an extra term for the production of new components will need to be considered.

input + production − output = accumulation

Mass balances can also be used for evaluating losses occurring during food processing. For example, a large creamery may process 1,000,000 l of milk a day producing butter and skim-milk powder. If the amount and composition of milk processed are known together with the amounts and compositions of butter and skim-milk powder, it is possible to determine the processing losses; most of this will end up in the effluent stream and require extra expensive treatment.

For example, let us evaluate the losses occurring during the conversion of 10^6 l of full cream milk to 40 000 kg of butter and 92 000 kg of skim-milk powder. The input is as follows (note that kg m^{-3} is equivalent to g l^{-1}): milk, 10^6 l; fat, 35 kg m^{-3}; milk solids not fat (MSNF), 90 kg m^{-3}. The output is as follows: butter, 40 000 kg (fat, 84%; MSNF, 1%; water, 15% (w/w)); skim-milk powder, 92 000 kg (fat, 1%; MSNF, 95%, water, 4% (w/w)). The losses occurring are shown in Table 1.2. This table shows that 480 kg of fat and 2200 kg of MSNF are not recovered in these products.

Table 1.2 — Loss during conversion of full cream milk.

	Input (kg)	Output (kg)
Fat	35 000	34 520 (butter, 33 600; skim-milk powder, 920)
MSNF	90 000	87 800 (butter, 400; skim-milk powder, 87 400)

Some of the MSNF will be retained in the butter-milk, which is an additional byproduct. Such accounting procedures rely on being able to measure volumes, volumetric flow rates and concentrations accurately. Further examples of mass balances have been given in Earle (1983), Toledo (1980) and Blackhurst et al. (1974).

1.4 LENGTH [L] (m)

One metre (1 m) was originally the length between two marks on a specially constructed bar of platinum-iridium, kept st Sèvres, near Paris (see section 1.3). In the age of atomic physics it has been defined more precisely; the wavelength of orange light emitted by a discharge lamp containing a pure isotope of krypton (^{86}Kr) at 63 K is 6.058×10^{-7} m. However, such conditions are not so easy to reproduce and, for a manufacturer interested in producing accurate metre rules, it is quite obvious which of the two standards would be most useful. Instruments used for measuring small distances accurately are vernier calipers, micrometers and travelling microscopes.

1.5 TIME [T] (s)

One mean solar second was based on astronomical observations and was equal to 1/86 400 of a mean solar day. It is now based on the duration of the electromagnetic radiation emitted from ^{133}Cs and is the time required for 9 192 631 770 wavelengths to pass a stationary observer, i.e. 1 s = 9 192 631 770 periods.

1.6 TEMPERATURE [θ] (K)

Temperature is defined as the degree of hotness of a body. In a spontaneous change, heat (energy) is always transferred from an object at a high temperature to one at a lower temperature, until thermal equilibrium is achieved (i.e. the temperatures are equal). To set up a scale of temperature it is necessary to use some reproducible fixed points, which are normally the melting point or boiling point of pure substances, and some easily measured property of a substance that changes in a uniform manner, as the temperature changes.

The two scales of temperature most commonly encountered are the

Temperature [θ] (K)

Fahrenheit and Celsius scales. The Fahrenheit system is still widely used and favoured by the 'older generation' of food technologists and American publishing houses.

The fixed points most easily reproduced are the melting and boiling points of pure water at atmospheric pressure (Table 1.3).

Table 1.3 — Melting and boiling points of pure water at atmospheric pressure.

	Melting point of water	Boiling point of water
Celsius scale (°C)	0	100
Fahrenheit scale (°F)	32	212

Temperature conversions can be achieved by the following equations or by reference to Table 1.4:

Table 1.4 — Temperature conversion chart.

°C←°F	Temperature	°C→°F	°C←°F	Temperature	°C→°F
−240	−400	—	10.0	50	122
−188.4	−300	—	12.8	55	131
−128.9	−200	−328	15.6	60	140
−101.1	−150	−238	18.3	65	149
−73.3	−100	−148	21.1	70	158
−67.8	−90	−130	23.9	75	167
−62.2	−80	−112	26.7	80	176
−56.7	−70	−94	29.4	85	185
−51.1	−60	−76	32.2	90	194
−45.6	−50	−58	35	95	203
−40	−40	−40	37.8	100	212
−37.2	−35	−31	40.6	105	221
−34.4	−30	−22	43.3	110	230
−31.7	−25	−13	46.1	115	239
−28.9	−20	−4	48.9	120	248
−26.1	−15	5	54.4	130	266
−23.3	−10	14	60	140	284
−20.6	−5	23	65.6	150	302
−17.8	0	32	71.1	160	320
−15	5	41	76.7	170	338
−12.2	10	50	82.2	180	356
−9.4	15	59	87.8	190	374
−6.7	20	68	93.3	200	392
−3.9	25	77	121.1	250	482
−1.1	30	86	148.9	300	572
1.7	35	95	176.7	350	662
4.4	40	104	204.4	400	752
7.2	45	113			

Conversions: $°C = (°F - 32) \times \frac{5}{9}$; $°F = °C \times \frac{9}{5} + 32$.

$$°C = (°F - 32) \times \frac{5}{9} \quad \text{or} \quad °F = \frac{9}{5} \times °C + 32$$

A temperature of –
- – 40 marks the point where the two scales coincide, i.e. –
  – 40 °C = – 40 °F.

The interval, one degree Celsius (1 degC), is 1/100 times the temperature difference between the boiling point and freezing point of water, whereas the interval, one degree Fahrenheit (1 degF), is 1/180 times this temperature difference. Therefore, it is more precise to record temperatures to ± 1 degF than ± 1 degC.

Conversion of temperature differences is made by the use of the following equations:

$$\text{degC} = \text{degF} \times \frac{5}{9} \quad \text{or} \quad \text{degF} = \text{degC} \times \frac{9}{5}$$

It will be seen in Chapter 9 that heat transfer rates are proportional to temperature difference. There may be many cases where it is necessary to convert temperature differences as well as temperatures. It should be noted that the Z value for an organism, which is a measure of how the heat resistance of an organism changes with temperature, is a temperature difference rather than a temperature. Most heat-resistant spores have a Z value equal to 10 degC (18 degF), i.e. an increase in temperature of 10 degC (18 degF), will decrease the processing time required by a factor of 10 (see section 10.5.1).

Other fixed points which are used are the boiling point of oxygen (– 182.97 °C), the boiling point of sulphur (444.6 °C) and the melting points of antimony (630.5 °C), silver (960.8 °C) and gold (1063.0 °C).

On the Celsius and Fahrenheit scales the numerical value attached to a particular temperature appears to be rather arbitrary and in both scales it is possible to achieve temperatures below zero. However, as temperatures is reduced, a point is reached at which all molecular motion stops and the kinetic energy of the molecule becomes zero. The temperature at this point is known as absolute zero or zero kelvin (0 K); this is the lower fixed point on the absolute scale of temperature. A second 'easily' reproducible fixed point is the triple point of water, the temperature at which ice, liquid water and water vapour are all in equilibrium. This occurs at a temperature of 273.16 K. Therefore the interval 1 K is equal to 1/273.16 of the temperature difference between the triple-point temperature and absolute zero.

On the absolute scale the freezing point and boiling point of water are 273.15 K and 373.15 K, respectively. Thus the interval 1 K is equal to the interval 1 degC. Temperature conversions can be made using the following equation:

$$K = 273.15 + °C$$

Lord Kelvin later showed that the work produced from an ideal heat engine taking heat in at a source temperature θ_1 and rejecting it a sink temperature θ_2 was proportional to the temperature difference and that the efficiency of heat engine working between two fixed temperatures would always be the same, regardless of the working fluid. The efficiency of the heat engine, which is a measure of the conversion of heat to work, is termed the Carnot efficiency CE.

$$CE = \frac{\theta_1 - \theta_2}{\theta_1} \times 100$$

This represents the maximum conversion efficiency of heat to work. Thus the efficiency of such an ideal engine, which depends only on the temperature of the source and the sink, can be used to define the thermodynamic scale of temperature (section 7.9).

Unfortunately, this is not a convenient scale for determining temperatures experimentally. To set up a practical scale, use is made of a property of a material which is easy to measure and which varies with temperature in a simple fashion. Examples of such properties are listed in Table 1.5.

Table 1.5 — Types of thermometer commonly used.

Type of thermometer	Property
Mercury or alcohol thermometer	Expansion of a liquid
Constant-volume gas thermometer	Variation in pressure of a fixed volume of gas
Resistance thermometer and thermistor (see section 12.8)	Variation in electrical resistance
Thermocouple (see section 12.13)	Variation in electromotive force (emf) set up between two different metals

Measurement of these properties at two fixed points is often sufficient to establish a scale of temperature. For example, let the heights h of mercury and electrical resistances R at the ice point, steam point and unknown temperature be given as follows: at the ice point (0°C), h_0 and R_0; at the steam point (100°C), h_{100} and R_{100}; at the unknown temperature (θ°C), h_θ and R_θ.

Then the temperature θ on each of the scales can be determined from

$$\theta = \frac{h_\theta - h_0}{h_{100} - h_0} \times 100 \quad \text{or} \quad \theta = \frac{R_\theta - R_0}{R_{100} - R_0} \times 100$$

However, there is no reason why the different types of thermometer should record exactly the same temperature when immersed in the same fluid. For this reason the International Scale of Temperature states which thermometers should be used for different temperature ranges. Other types of thermometer have been discussed by Jones (1974a).

1.6.1 Food-processing temperatures

The most common thermometers use in food-processing operations are mercury-in-steel thermometers, resistance thermometers and thermocouples. Mercury-in-glass thermometers are not used because they are easily broken and the resulting mercury is extremely toxic. A wide range of temperatures are encountered, ranging from $-196\,°C$, when using liquid nitrogen for freezing, to $1300\,°C$ when using direct flame techniques for the sterilization of canned products. However, the usual range is between $-40\,°C$ and $250\,°C$. Table 1.6 shows typical temperatures for a range of

Table 1.6 — Typical temperatures used for processing and storing foods.

$-40\,°C$ to $-20\,°C$	Freon 12 direct refrigeration techniques, air temperatures in blast and fluidized-bed freezers. Primary storage of frozen foods
$-20\,°C$ to $-15\,°C$	Domestic deep-freeze cabinets. Temperatures at which most of the water in foods is frozen
$-5\,°C$ to $-1\,°C$	Freezing points of most foods
$-1\,°C$ to $+10\,°C$	Chilled storage and transportation
Ambient to $50\,°C$	Milling operations, freeze drying, irradiation reverse osmosis, filtration, separation
$50\,°C$ to $100\,°C$	Vacuum evaporation, ultrafiltration, pasteurization, homogenization, hot-air drying, sterilization of acid products
$110\,°C$ to $125\,°C$	Sterilization of low-acid foods; reference temperature for thermal processing ($121\,°C$ or $250\,°F$)
$140\,°C$ to $150\,°C$	Roller drying, long-life products (ultrahigh temperature (UHT))
$150\,°C$ to $250\,°C$	Spray drying, extrusion cooking, baking

food-processing operations.

It can be seen that the food processor has a wide range of temperatures to deal with. Furthermore, accurate temperature control is essential in the canning of low-acid products (a temperature error of 2 degC can lead to considerable under processing (see the lethality table in section 10.6) and in the storage of chilled and frozen produce to ensure optimum retention of quality. Recommendations for chilled storage of perishable products are given in an International Institution of Refrigeration publication (1979) and in section 10.16. In large-scale equipment, such as retorts and ovens, it is important to ensure that there are no temperature fluctuations and, when measuring physical properties of food such as viscosity and surface tension, it is essential to have accurate temperature control.

It is very important to check the accuracy of thermometers at regular intervals both at the fixed points and over the temperature range in which they are most often used. It may also be necessary to record temperatures during certain processing operations, e.g. pasteurization, canning and to keep records for inspection by the local health authorities. Electrical thermometers are extremely useful for this purpose.

The food processer still works mainly in temperatures on the Fahrenheit and Celsius scales. However, when certain equations such as the ideal gas equations ($pV = RT$) or the Arrhenius equation ($k = A \exp(-E/RT)$) are used, it is essential that the temperature be substituted in absolute temperatures, expressed in kelvin or degrees Rankine (°R):

$$0\,K = -273.15\,°C = -459.67\,°F \quad \text{and} \quad °R = °F + 459.67$$

(note that °R is how the absolute temperature is expressed on the Fahrenheit scale). Further details have been provided by Gruenwedel and Whitaker (1984).

1.7 OTHER FUNDAMENTALS

The fundamental units in the SI system are completed with the addition of electric current, luminous intensity, and the amount of substance (units of mole) (mol). These will now be defined.

1.7.1 Electric current (A)
Electric current is a measure of the flow of electrons. One ampere (1 A) is that flow of electrons which, when flowing down two long parallel conductors, of negligible cross-sectional area, placed 1 metre apart in a vacuum produces between the two wires a force of 2×10^{-7} N per metre of the length (see section 12.2).

1.7.2 Luminous intensity (cd)
One candela (1 cd) is the luminous intensity, in the perpendicular direction, of a surface of $1/600\,000\,m^2$ of a perfect radiator (black body) at the temperature of freezing platinum (1772 °C) under a pressure of one standard atmosphere (1 atm) or 1.013 bar.

1.7.3 Amount of substance (mol)

One mole (1 mol) is the amount of substance of a system which contains as many elementary entities as there are atoms in 12×10^{-3} kg of ^{12}C. When the mole is used, the elementary entity must be specified and may be atoms, molecules, ions, electrons, other particles or specified groups of such particles.

Kaye and Laby (1973) make some interesting comments about the use of these definitions of the given fundamental dimensions for setting up practical scales of measurement.

1.8 PREFIXES IN COMMON USE

The prefixes given in Table 1.7 are commonly used with the metric system of units, but not with the Imperial system.

Table 1.7 — Prefixes in common use before units.

Prefix	Symbol	Multiple	Prefix	Symbol	Multiple
exa	E	10^{18}	deci	d	10^{-1}
peta	P	10^{15}	centi	c	10^{-2}
tera	T	10^{12}	milli	m	10^{-3}
giga	G	10^{9}	micro	μ	10^{-6}
mega	M	10^{6}	nano	n	10^{-9}
kilo	k	10^{3}	pico	p	10^{-12}
hecto	h	10^{2}	femto	f	10^{-15}
deca	da	10^{1}	atto	a	10^{-18}

Compound prefixes (e.g. mμ) are not allowed.

1.9 DERIVED UNITS

Measurements such as volume $[L^3]$, density $[ML^{-3}]$, velocity $[LT^{-1}]$ and pressure $[ML^{-1}T^{-1}]$ are termed derived units, because they are made up of combinations of the fundamental units. Derived units can be expressed in two alternative forms, e.g. m/s or m s^{-1}. Both forms of presentation are commonly encountered; in this book, the second form will be used. Let us consider as an example of a derived unit the velocity of an object which is measured by dividing the distance covered by the time. The dimensions are $[LT^{-1}]$, which remain constant regardless of the system of units being used.

The respective units of velocity in the different systems are m s^{-1} (or cm s^{-1}) and ft s^{-1}.

All equations must be *dimensionally* consistent, i.e. the dimensions on both sides of the equation must be the same. If an equation is not dimensionally consistent, it cannot be correct.

The important derived units will now be covered.

1.10 AREA $[L^2]$ (m^2)

Area is defined as the product of two lengths. Therefore the dimensions of area are $[L^2]$ and in the SI system the unit is the square metre (m^2), where

$$1\,m^2 = 10^4\,cm^2 = 10.76\,ft^2$$

Common areas encountered are as follows:

$$\text{surface area of a circle} = \pi r^2 = \frac{\pi D^2}{4}$$

$$\text{surface area of a sphere} = 4\pi r^2 = \pi D^2$$

where r and D are the radius and the diameter, respectively.

Thus the total surface area of a can of radius r and length h is equal to $2\pi rh + 2\pi r^2 = 2\pi r(r+h)$. One system for can size measurement still commonly encountered is to express the diameter and height each in terms of three figures: the first is the value in inches, and the second and third are the values in sixteenths of an inch. Thus a 200, 307 can would have a diameter equal to 2 in and a height equal to $3\frac{7}{16}$ in. Table 1.8 lists some common can sizes at present used in the UK, with equivalent metric dimensions.

In many physical and chemical processes the rate of reaction is proportional to the surface area. Therefore it is often desirable to maximize the surface area. This is so in drying operations and in aerobic fermentations. Food, in either a solid or a liquid form, is prepared in strips or cubes or in the form of a spray to increase the drying rate, and air is broken into bubbles to increase the oxygen transfer rate. For homogenization of milk the fat globule is reduced in size to prevent separation from taking place; in this case the increased surface area may be a disadvantage as oxidation reaction rates will increase.

The approximate surface areas for a variety of fruit have been quoted by Mohsenin (1970), i.e. apples, 17.2–25.2 in^2; plums, 5.4–7.0 in^2; pears, 22.2–23.0 in^2.

In the design of food-processing equipment, it is necessary to be able to predict the surface area required for such pieces of equipment as heat exchangers, evaporators, driers and filtration equipment.

The size and shape of argicultural produce are used for grading purposes and for separating and sifting material. Sieves are commonly used for this purpose.

The particle size distribution of milled or ground commodities such as

Table 1.8 — Sizes of some common UK round open-top cans.

	Diameter (see text for meaning)	Diameter (mm)	Height (see text for meaning)	Height (mm)	Capacity (fl oz)	Capacity (ml)	Examples
Baby food	202	52	213¼	69	4.5	128	Baby food, fruit juices
Picnic	211	65	301	74	8.3	236	Fruit, vegetables
A1	211	65	400	99	11.1	315	Soups, vegetables
UT Tall	300	73	408¾	115	15.7	446	Beans, pet food, meat
A2½	401	99	411	119	29.7	843	Pet foods, peas
A6	603		600		92.5	2627	
A10	603		700		109.0	3096	Catering packs

Adapted from Hersom and Hulland (1980).

flour, coffee, oil seeds and other materials needs to be known and controlled for ensuring that subsequent processing and handling is optimized.

Some of these factors have been discussed in more detail by Mohsenin (1970) and Arthey (1975).

1.11 VOLUME $[L^3]$ (m³)

Volume is the product of three lengths, with dimensions $[L^3]$. The SI unit of volume is the cubic metre (m³). However, this is a very large unit of volume and is often more conveniently subdivided into litres (l) (dm³) and ml (or cubic centimetres (cm³)):

$$1 \, m^3 = 10^3 \, l \, (dm^3) = 10^6 \, ml \, (cm^3)$$

In Imperial units, volumes are measured in gallons (gal) where 4.54 l = 1 gal. Note that the British (Imperial) gallon (gal) is *larger* than the American gallon (US gal):

$$1 \, US \, gal = 0.8327 \, gal$$

The output of breweries is still often quoted in barrels, where one standard barrel is equal to 36 gal. Barrels can be subdivided into kilderkins (18 gal) and firkins (9 gal) (4 firkins = 1 barrel); a larger measure containing 63 gal is called a hogshead.

The capacity of food-processing equipment which handles fluids is often expressed as a volumetric flow rate, in terms of gallons per hour (gal h^{-1}) or litres per hour (l h^{-1}). Thus, pilot plant pasteurizers may process milk at the rate of 200 l h^{-1}, whereas the larger-scale pasteurizers in dairies or breweries may be capable of handling up to 50 000 l h^{-1}. In this text, volumetric flow rate will be given the symbol Q with units of m³ s^{-1}, unless otherwise stated.

1.11.1 Surface-area-to-volume ratio

The sphere is the shape that offers the minimum surface area for a given volume:

$$\text{volume of a sphere} = \frac{4}{3}\pi r^3 \quad \text{or} \quad \frac{1}{6}\pi D^3$$

When foods are heated in packages (cans, bottles or flexible pouches), it is desirable to maximize the surface-area-to-volume ratio; this will maximize the heat transfer rate and reduce the time required for the temperature at the slowest heating point (normally the centre of the package) to reach the desired processing temperature. This will improve the quality of the product.

The volume of a can, in terms of the radius r and height h, equals $\pi r^2 h$. Therefore,

$$\text{surface-area-to-volume ratio} = \frac{2(r+h)}{rh}$$

If the surface-area-to-volume ratio of the following two cans are compared, it can be seen that the ratio decreases as the can size increases: for a 211,301 can (capacity 235 ml) the surface-area-to-volume ratio is $2.14\,\text{in}^2\,\text{in}^{-3}$, whereas for a 300, 410 can (capacity 457 ml) it is $1.765\,\text{in}^2\,\text{in}^{-3}$. Therefore, longer processing times are required to process larger cans; this can lead to a reduction in product quality. One important advantage of flexible pouches over cans is their high surface-area-to-volume ratio.

Dispersing material into small droplets will also give a large increase in the surface-area-to-volume ratio. For example, one spherical fat globule of volume $1\,\text{cm}^3$ has a surface area of $4.83\,\text{cm}^2$. If this is broken up into a thousand globules of equal size, the new total surface area is equal to $48.3\,\text{cm}^2$, which gives an increase of $43.47\,\text{cm}^2$.

Measurement and control of volumetric flow rate are very important in many operations, particularly for metering purposes and control of residence times in thermal processes.

1.12 DENSITY $[\text{ML}^{-3}]$ (kg m^{-3})

Density is defined as the mass of a substance divided by its volume:

$$\text{density} = \frac{\text{mass}}{\text{volume}}$$

The dimensions of density are $[\text{ML}^{-3}]$ and the units are kilograms per cubic metre (kg m^{-3}). Most people are familiar with the fact that water has a density of $1\,\text{g ml}^{-1}$:

$$1\,\text{g ml}^{-1} = \frac{10^{-3}\,\text{kg}}{10^{-6}\,\text{m}^3} = 10^3\,\text{kg m}^{-3}$$

The density of water in Imperial units is $62.3\,\text{lb ft}^{-3}$.

1.12.1 Specific gravity

Specific gravity is defined as the mass of a substance divided by the mas of an equal volume of water therefore specific gravity is a *dimensionless quantity*, i.e. a quantity that has no units.

Liquids, pastes, suspensions and solids are generally considered to be incompressible, i.e. the density is hardly affected by moderate changes in temperature and pressure. Gases and vapours are compressible.

The density of food materials (liquids and solids) and of vapours, commonly used in food-processing operations, will be discussed in more detail in Chapter 2.

1.13 VELOCITY $[LT^{-1}]$ $(m s^{-1})$

The average velocity of an object is defined as distance covered divided by time taken to cover that distance. The dimensions are $[LT^{-1}]$n and the SI units are metres per second $(m s^{-1})$. If an object starts from a particular point and the distance s moved from that point is plotted against time t (see Fig. 1.3(a)); the instantaneous velocity at any time t can be obtained from

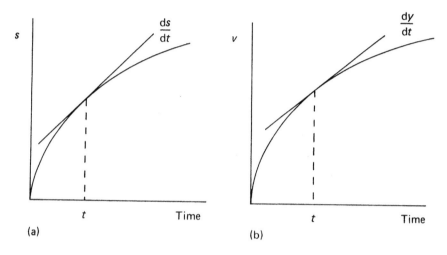

Fig. 1.3 — Introduction to the derivative notation: (a) graph of distance s travelled against time t; (b) graph of velocity v against time t.

the gradient of the tangent to the curve at that particular point. This is represented by the derivative notation ds/dt.

Velocity is a vector quantity, i.e. it has both a magnitude and a direction. Therefore, strictly speaking, to specify velocity correctly, the magnitude (or speed) and the direction should be quoted.

In fluid flow applications it is important to know the *average velocity* of a fluid in a pipe or a channel. The average velocity v is determined from the equation

$$\text{average velocity } (m s^{-1}) = \frac{\text{volumetric flow rate } Q \ (m^3 s^{-1})}{\text{cross-sectional area } A \ (m^2)}$$

For a pipe or tube of circular cross-section and diameter D,

$$v = \frac{4Q}{\pi D^2}; \text{ (note the dimensional consistency)}$$

The fluid is imagined as flowing along the pipe as a plug or piston and this situation is known as plug or piston flow; the velocity at all points across the tube is the same (Fig. 1.4).

Average velocity is a useful concept. However, in practice, there is a velocity distribution across the tube or channel, with the maximum velocity being attained at the centre (see section 3.18). However, in most fluid flow problems, average velocity is used; this will be the case here, unless otherwise stated.

High fluid velocities are required when cleaning and sanitizing food processing equipment and pipelines, using in-place-cleaning techniques.

1.13.1 Angular velocity

If an object moves in a circle of radius r (Fig. 1.5(a)), the velocity of the object can be defined in terms of either an angular velocity or a tip speed.

In terms of angular velocity, it is necessary to define a radian (rad). 1 rad is the angle subtended by an arc of a circle equal to the radius of the circle. Thus a revolution (rev) can be defined:

$$2\pi \text{ rad} = 1 \text{ rev} = 360°$$

$$1 \text{ rad} = 57.3°$$

The angular velocity N or ω is the angle subtended per unit time; it can be expressed in revolutions per minute (rev min^{-1}), revolutions per second (rev s^{-1}) for N or radians per second (rad s^{-1}) for ω. Therefore,

$$\omega = 2\pi N$$

To serve as an example, an angular speed of 600 rev min^{-1} can be expressed as $N = 10$ rev s^{-1} or $\omega = 2\pi \times 10$ rad s^{-1}.

The tip speed or linear velocity v_t of an object rotating with an angular velocity ω situated a distance r from the centre is given by

$$\text{tip speed} = \omega r = 2\pi r N \quad \text{m s}^{-1}$$

The dimensions of rotational speed (angular velocity) are $[T^{-1}]$ (as revolution and radian are both dimensionless).

An object moving round a circular path at a constant speed is *not* moving at a constant velocity as its direction is continuously changing.

1.14 MOMENTUM $[MLT^{-1}]$ (kg m s^{-1})

Moving objects are said to possess momentum. Linear momentum is defined as the product of the mass and the velocity. Thus an object hitting a wall and rebounding from that wall will change its momentum even if its speed has not changed, simply because the direction has changed. Whenever there is a change of momentum, this gives rise to an impressed force (see section 1.16).

Fig. 1.4 — Average velocity of fluid in a pipe under flow conditions.

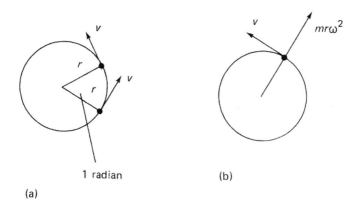

Fig. 1.5 — Circular motion: (a) measurement of angular velocity; (b) the centrifugal force acting on an object.

The law of conservation of linear momentum states that, in a perfectly elastic collision between two objects, i.e. no mechanical energy is lost, the total momentum before the collision is equal to the total momentum after the collision.

1.15 ACCELERATION $[LT^{-2}]$ (m s^{-2})

If the velocity of an object is plotted against time (see Fig. 1.3(b)), the acceleration at any time t is given by the gradient of the tangent to the curve at that time; the derivative notation dv/dt is used. Thus, acceleration is the rate of change in velocity and has dimensions of $[LT^{-2}]$. An object moving with a constant velocity has no acceleration, and the velocity obtained under these conditions is often referred to as the *terminal velocity*.

The most common acceleration encountered is that due to gravity. The acceleration due to gravity which is given the symbol g at Greenwich is 9.81 m s^{-2}. However, this value is not constant over the Earth's surface, as the Earth is not a perfect sphere. On other planets, the acceleration due to its gavity would be different, being about $\frac{1}{6}g$ on the Moon. The acceleration due to gravity on a planet is a function of the diameter of the planet. The variation in g with latitude and height has been given by Jones (1974a).

1.16 FORCE [MLT^{-2}] ($kg\,m\,s^{-2}$ or N)

We are all familiar with concepts of pushing, pulling, biting, chewing, breaking and hammering. In all these situations we are subjecting an object to a force; under the influence of that force the object will deform. The study of the deformation of objects under the influence of applied forces is termed *rheology*.

Solids and liquids are subjected to complex forces, both during processing and while being chewed and digested. The rheological behaviour of the food is extremely important, as it will affect the type of processing equipment selected and the evaluation of the texture of that food (e.g. its viscosity, smoothness, hardness, chewiness and crispness). Therefore it is necessary to be able to quantify these forces and to measure the resulting deformation.

Newton's second law of motion states that, when an object changes momentum, it is subjected to a force F and that the magnitude of the force is proportional to the rate of change in momentum. This can be expressed mathematically as

$$\text{force} \propto \text{rate of change in momentum}$$

i.e.

$$\text{force} \propto \frac{\mathrm{d}}{\mathrm{d}t}(mv)$$

Thus, if the mass of the object remains constant as it does in the cases considered in this book,

$$\text{force} \propto m\frac{\mathrm{d}v}{\mathrm{d}t} = km\frac{\mathrm{d}v}{\mathrm{d}t} \quad (1.3)$$

In the SI system the unit of force is the newton (N). 1 N is defined as that force which gives a mass of 1 kg an acceleration of one metre per second per second ($1\,\mathrm{m\,s^{-2}}$).

Substituting this in equation (1.3) gives

$$1\,\mathrm{N} = k \times 1\,\mathrm{kg} \times 1\,\mathrm{m\,s^{-2}}$$

Therefore,

$$k = 1$$

Thus, force is defined as mass multiplied by the acceleration and the dimensions are [MLT^{-2}]. 1 N could be written as $1\,\mathrm{kg\,m\,s^{-2}}$, but it is obviously much simpler to use the symbol N instead of $\mathrm{kg\,m\,s^{-2}}$.

It is often difficult to picture what some of these physical quantities mean in real terms; imagine being hit on the head with an apple (mass, 0.1 kg)

Sec. 1.16] Force [MLT^{-2}] (kg m s^{-2} or N)

falling from a tree ($g = 9.81\,\mathrm{m\,s^{-2}}$). The force that it strikes you with would be approximately 1 N. Alternatively, the force exerted by an apple held in the hand is approximately 1 N.

The units of force are summarized in Table 1.9. From this table it can be

Table 1.9 — Units of force.

System of units	Force	=	Mass	×	Acceleration
SI	1 newton (N)	=	1 kg	×	1 m s^{-2}
cgs	1 dyne (dyn)	=	1 g	×	1 cm s^{-2}
Imperial	1 poundal (pdl)	=	1 lb	×	1 ft s^{-2}

seen that

$$1\,\mathrm{dyn} = 1\,\mathrm{g\,cm\,s^{-2}} = 10^{-3}\,\mathrm{kg}\,10^{-2}\,\mathrm{m\,s^{-2}} = 10^{-5}\,\mathrm{N}$$

Thus the dyne is a very small unit of force.

Units of force also in common use are the kilogram-force (kgf) (which is called the kilopond (kp) in some European countries) and the pound-force (lbf); these units are abbreviated in this way to distinguish them from the units of mass, the kilogram (kg) and the pound (lb).

One kilogram-force (1 kgf) is the force acting on a mass of 1 kg due to gravity. Thus,

$$1\,\mathrm{kgf} = 1\,\mathrm{kg} \times 9.81\,\mathrm{m\,s^{-2}} = 9.81\,\mathrm{N}$$

Similarly,

$$1\,\mathrm{lbf} = 32.2\,\mathrm{pdl}$$

The force acting on a body due to gravity is called the *weight* (mg). Thus an object will not have a constant weight over the surface of the Earth because the Earth is not a perfect sphere.

The pound-force and kilogram-force are commonly used for pressure measurement. Often the symbol f is omitted; this is a sloppy practice and may well lead to confusion.

One major difference between the British (Imperial) system and the SI system is that force is a fundamental dimension on the British system and a derived unit on the SI system. This can lead to confusion since the use of the British system warrants the use of gravitational conversion factor g_c which keeps appearing in the equations. The same material or equations, particularly with mechanics or fluid flow, may appear to be different when presented side by side in the different units. The gravitational constant g_c has been discussed by Toledo (1980). In my opinion these of SI units is much simpler and more satisfactory. A force of 10 lbf would be equivalent to the force acting on a mass 10/2.205 kg, i.e. 10/2.205 × 9.81 = 44.49 N.

When there are not net forces acting on an object, that object is said to be at equilibrium. This may mean that the object is either at rest or moving with a constant velocity.

1.16.1 Centrifugal force

An object moving round a circle at a constant tip speed is constantly changing its velocity because of its change in direction. The acceleration that the object is subjected to is $r\omega^2$ or v_t^2/r. This gives rise to an inward force, known as the centripetal force. To balance this force (to keep the object in the same relative position), there is an outward force acting on the object, called the centrifugal force (Fig. 1.5(b)):

$$\text{centrifugal force} = mr\omega^2 = \frac{mv_t^2}{r}$$

Centrifugal forces are used in high-speed separators or clarifiers for removal of particulate matter or the separation of immiscible liquids of different densities (cream from milk or solvents from water). Basically, a centrifugal force is applied to speed up the separation that would inevitably take place under the influence of gravity. The efficiency of a separator can be defined in terms of the number of gravitational forces or 'g' forces, where

$$\text{number of '}g\text{' forces} = \frac{\text{centrifugal force}}{\text{gravitational force}} = \frac{mr\omega^2}{mg} = \frac{r\omega^2}{g}$$

Effectively, it gives a measure of how much quicker the separation is compared with one influenced only by gravity. It is affected by the diameter of the centrifuge and its rotational speed.

A centrifuge with a rotor arm of radius 4 in rotating at 2000 rev min^{-1} would be operating at an angular velocity ω of 209.46 rad s^{-1}, would have a tip speed v_t of 21.3 m s^{-1} and would give 454 'g' forces.

1.17 PRESSURE [ML^{-1}T^{-2}] (kg m^{-1} s^{-2} or N m^{-2})

The units of pressure probably cause more confusion than any other units do. The definition of absolute pressure is simple enough:

$$\text{absolute pressure} = \frac{\text{magnitude } F \text{ of force}}{\text{surface area } A \text{ over which it is applied}}$$

The dimensions of pressure are therefore [MLT^{-2}/L^2] or [ML^{-1}T^{-2}]. In the SI system the unit of pressure is the newton per square metre (N m^{-2}). This is also known as the pascal (Pa). The pascal is a small unit of pressure. In many applications the bar and mega pascal (MPa) are often used where

Sec. 1.17]   Pressure $[ML^{-1}T^{-2}]$ $(kg\,m^{-1}s^{-2}$ or $Nm^{-2})$

$$1\,\text{bar} = 10^5\,\text{Pa}\,(N\,m^{-2}) = 0.1\,\text{MPa}$$

We shall see later that 1 bar is approximately equal to 1 atm.

Thus, absolute pressure could also be measured in $lbf\,in^{-2}$, or $kgf\,cm^{-2}$ ($kp\,cm^{-2}$). The appropriate conversions would have to be made: 1 atm = 14.69 $lbf\,in^{-2}$ = 1.013 bar = 0.76 m Hg = 1.033 $kgf\,cm^{-2}$.

We are also used to measuring pressure in terms of a height or head of fluid. For example, we know that a pressure of 1 atm will support a column of mercury 760 mm (0.76 m) in height when a simple barometer is set up (atmospheric pressure is subject to day-to-day fluctuations). Expressed in these terms, the dimensions of pressure appear to be length [L]; this is certainly inconsistent with the dimensions of absolute pressure $[ML^{-1}T^{-2}]$.

There is a direct relationship between the absolute pressure $(N\,m^{-2})$ and the height of head of fluid (m) capable of being supported by that pressure, which is expressed as follows:

$$p = \rho g h$$

where p is the absolute pressure $(N\,m^{-2}$ (Pa)), ρ is the fluid density $(kg\,m^{-3})$, g is the acceleration due to gravity $(= 9.81\,m\,s^{-2})$ and h is the head or height of fluid supported (m).

This can be applied to any fluid. 1 atm would support a column of water equal to

$$\frac{1.013 \times 10^5}{9.81 \times 10^3}\,\text{m in height, i.e. } 10.33\,\text{m}$$

In fluid flow operations it is common to talk about the pressure developed by a pump or pressure losses in pipelines in terms of a head of fluid.

A further complication exists because many gauges measure pressures above or below atmospheric pressure. Fig. 1.6 shows the names given to the

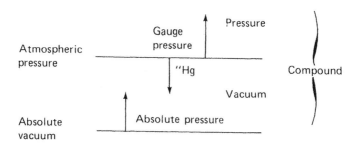

Fig. 1.6 — Relationship between gauge pressure and absolute pressure.

gauges used in the different regions. Pressure gauges will measure pressures above atmospheric pressure, vacuum guagues will measure pressures below atmospheric pressure, and compound gauges will measure both; these are occasionally used on steam lines or for measuring internal pressures of canned foods. Thus a gauge pressure reading of 5 lbf in^{-2} would correspond to an absolute pressure of $14.69 + 5$ lbf in^{-2} (g) = 19.69 lbf in^{-2} (a). Note that the symbols (a) and (g) denote absolute and gauge pressures, respectively.

The relationship between absolute and gauge pressure is as follows:

$$\text{absolute pressure} = \text{gauge pressure} + \text{atmospheric pressure}$$

The distinction is not always made clear in the literature. Obviously, at high pressures (e.g. 3000 lbf in^{-2} (g)) the difference would be small but, at low pressures (e.g. 15 lbf in^{-2}), it is important to know whether this is a gauge or an absolute pressure. Failure to distinguish could make a difference in the temperature of saturated steam of approximately 21 °C (see section 11.4).

1.17.1 Vacuum measurement

In the Imperial system, atmospheric pressure is designated as zero inches of mercury (0 inHg). As the pressure is reduced, the vacuum reading increases. A reading of 15 inHg designates 15 inHg below atmospheric pressure and is a higher absolute pressure than a vacuum reading of 20 inHg. 29.92 (or 30) inHg would correspond to a complete vacuum (zero) absolute pressure. One situation where this system is still widely encountered is in vacuum measurement in canned foods.

In the metric system, vacuum is measured in pascals (Pa), torrs (Torr) or microns (micrometers) or mercury (μmHg) and is taken as being above absolute zero pressure:

$$1 \text{ Torr} = 1 \text{ mmHg} = 10^3 \text{ μmHg } (10^{-6} \text{ mHg}) = 133.3 \text{ Pa}$$

Thus, if a vacuum or pressure measurement of 4.6 Torr is recorded, this indicates that the pressure is 4.6 mmHg above absolute zero pressure. This could also be recorded as 133.3×4.6 Pa or 613.18 Pa.

Inches of mercury (inHg) can be converted to an absolute pressure (Pa) by the following equation (see also Table 1.10):

$$\text{absolute pressure (Pa)} = (29.92 - X)\frac{2.54}{76} \times 1.013 \times 10^5$$

$$= (29.92 - X) \times 3.3856 \times 10^3$$

where X (inHg) is the vacuum reading.

Sec. 1.17] Pressure $[ML^{-1}T^{-2}]$ ($kg\,m^{-1}s^{-2}$ or Nm^{-2})

Table 1.10 — Conversion from in Hg to mmHg and kPa.

Pressure (inHg)	Pressure (mmHg)	Pressure (kPa)	Pressure (inHg)	Pressure (mmHg)	Pressure (kPa)
1	734.6	97.91	16	353.6	47.13
2	709.2	94.53	17	328.2	43.75
3	683.8	91.14	18	302.8	40.36
4	658.4	87.76	19	277.4	36.97
5	633.0	84.37	20	252.0	33.59
6	607.6	80.99	21	226.6	30.20
7	582.2	77.60	22	201.2	26.82
8	556.8	74.22	23	175.8	23.43
9	531.4	70.83	24	150.4	20.05
10	506.0	67.44	25	125.0	16.67
11	480.6	64.06	26	99.6	13.28
12	455.2	60.67	27	74.2	9.89
13	429.8	57.29	28	48.8	6.51
14	404.4	53.09	29	23.4	3.12
15	379.0	50.52	29.5	10.7	1.43

1.17.2 Pressures used in food-processing operations

As with temperature, the food processor has to deal with a wide range of operating pressures. Some processes are listed in Table 1.11.

Table 1.11 — Pressures found during food-processing operations.

Operation	Pressure	Operation	Pressure
Homogenization	30–300 bar (g)	Hot-air drying	
Reverse osmosis	40–80 bar (g)	Spray drying	Atmospheric
Steam generation		Evaporation	pressure or
Compressed air	6–10 bar (g)	Canning acid products	slightly above
Ultrafiltration		Blast freezing	
UHT processes	4–8 bar (g)	Vacuum drying	
		Vacuum evaporation	1–100 kPa
Canning and bottling low-acid foods	1 bar	Vacuum filtration	
		Freeze drying	0.1 Torr to atmospheric pressure, normally below 4.6 Torr

Several of these processes are pressure activated, particularly the separation processes. In other processes, particularly when boiling and condensing fluids are dealt with, fixing the pressure will determine the operating temperature. This is important in evaporation, in canning and in the operation of refrigeration equipment. In a process such as homogenization, the viscosity and mouth feel of the product will depend very much on the homogenization pressure, e.g. single cream (18% fat), needs quite high pressures (3000 lbf in^{-2} (g)) to obtain a reasonable viscosity whereas double cream (48% fat) requires a much lower pressure (500 lbf in^{-2} (g)). Pressures significantly higher than this may cause double cream to solidify in the container. The pressure required in a pumping operation will also affect the pump selection.

Therefore, it can be seen that the measurement and control of pressure are very important in food-processing operations. Pressure measurement is covered in more detail in section 3.2.

1.18 WORK [ML^{2}T^{-2}] (kg m^2 s^{-2} or J)

'I love work: I could sit and watch it for hours'; so goes an old expression. Mechanical work is concerned with moving forces through distances. The definition of the amount of work done is the product of the force and the distance. Thus, work has dimensions of [ML^{2}T^{-2}]. In the SI system the unit of work is the joule (J); this is equivalent to a newton metre (N m) or kilograms metre^2 per second per second (kg m^{-2} s). The units of work are summarized in Table 1.12.

Table 1.12 — Units of work.

System of units	Work	=	Force	×	Distance
SI	1 Joule (J)		1 N	×	1 m
cgs	1 erg	=	1 dyn	×	1 cm
Imperial	1 ft lbf	=	1 lbf	×	1 ft

Thus the work done in lifting 1 kg of sugar a height of 2 m is 2 × 1 × 9.81 = 19.6J.

When a gas is compressed, work is done on the gas by the surroundings. Conversely, when a gas expands, work is done on the surroundings by the gas. In either case the total work done can be evaluated.

Fig. 1.7 illustrates an expansion process, plotted on a pressure–volume diagram. The gas originally at the condition p_1, V_1 is explained to the final condition p_2, V_2. If we consider a very small change (dV) in volume, the work done in changing this is equal to dV.

The total work done in going from (1) to (2) is given by

$$\int_{V_1}^{V_2} \rho dV$$

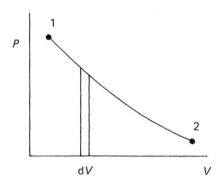

Fig. 1.7—Compressional work: (a) compression or expansion of a gas in a cylinder; (b) work done by a gas during an expansion process.

This is equivalent to the total area under the curve. If the units of pressure are $N m^{-2}$ and the units of volume are m^3, then the work done by the gas will be in joules. It is important to know the work done, so that the motor (electric, steam, petrol, etc.) to drive a compressor or a blower can be correctly sized. Compressors are used for delivering relatively small volumes of air at fairly high pressures. Typical uses would be for providing compressed air for automatic control operations or for attaining low temperatures using a refrigeration cycle. Blowers or fans tend to provide large volumes of air at relatively low pressures. This is used for freezing and dehydration processes and for pneumatic conveying of particulate matter.

Electrical work is concerned with either passing a current through a conductor or moving a charge through a potential difference (see section 12.4). Again the units are measured in joules.

1.19 POWER $[ML^2T^{-3}]$ ($kg\, m^2\, s^{-3}$ or W)

Power is defined as the rate of doing work. The power rating of an object is the work done divided by the time taken. In the SI system. one watt (1 W) is the power exerted when one joule (1 J) of work is performed in 1 s. Therefore,

$$\text{watt} = \frac{\text{joule}}{\text{second}} \quad \text{or} \quad W = J s^{-1}$$

These are often used interchangeably. For example, thermal conductivity units may be expressed in terms of $J s^{-1} m^{-1} K^{-1}$ or $W m^{-1} K^{-1}$.

In the British system the unit of power is the horsepower (hp). This is equivalent to $550 \,\text{ft}\,\text{lbf}\,\text{s}^{-1}$. The conversion is as follows:

$$1\,\text{hp} = 745.7\,\text{W}$$

Electrical appliances are also rated in W or kW. We pay for our electricity in terms of kilowatt hours (kW h) or units. 1 kW h is the amount of work that we obtain when an object rated 1 kW is run for 1 h:

$$1\,\text{kW}\,\text{h} = 1000\,\text{J}\,\text{s}^{-1} \times 3600\,\text{s} = 3.6 \times 10^6\,\text{J} \quad \text{or} \quad 3.6\,\text{MJ}$$

Thus, 1 unit of electricity, currently priced at 5.1p, will buy you enough work (theoretically) to lift approximately 180 000 bags of sugar (1 kg) a height of 2 m. Who could complain about the cost of electricity when expressed in those terms?

The work done is expressed as follows:

$$\text{total work done (J)} = \text{power rating}\,(\text{J}\,\text{s}^{-1}) \times \text{time (s)}$$

1.19.1 Pumping power

When a fluid is transported along a pipe, an amount of energy is needed to overcome the frictional forces. Pumps are usually used for this purpose. The fluid is taken in at the inlet (suction) side and discharged at a higher pressure (Fig. 1.8). If the inlet pressure and outlet pressure are p_i ($\text{N}\,\text{m}^{-2}$) and p_o

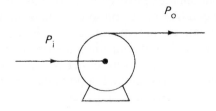

Fig. 1.8 — Power utilization during a pumping operation.

($\text{N}\,\text{m}^{-2}$), respectively, the head H developed by the pump is

$$H = \frac{p_o - p_i}{\rho g}$$

where ρ is the fluid density.

The theoretical power P_0 (W) delivered to the fluid is equal to

P_0 (W) = mass flow rate (kg s^{-1}) × head developed (m) × acceleration due to gravity (m s^{-2})

Thus, if the flow rate is given in terms of volumetric flow rate Q (m^3 s^{-1}), the theoretical power: P_0 is given by

$$P_0 = \frac{Q_\rho(p_o - p_i)}{\rho g} g = Q(p_o - p_i)$$

1.20 ENERGY [ML^{2}T^{-2}] (J)

Energy is defined as the capacity to do work. Any substance or person which is said to possess energy can do useful mechanical work, i.e. move forces through distances. The units of energy are similar to those of work, and energy contents of chemicals or foods are expressed in joules or kilocalories. Units of thermal energy and energy transfer processes will be discussed in section 7.2. Electrical energy and units in section 12.2.

1.21 SUMMARY OF THE MAIN FUNDAMENTAL AND DERIVED UNITS

The main fundamental and derived units are summarized in Table 1.13.
    Other physical properties, which will be considered in later chapters, are listed in Table 1.14. A more detailed set of metric and other conversion tables has been compiled by Lewis (1973).

1.22 DIMENSIONAL ANALYSIS

In many food and chemical engineering situations the technique of dimensional analysis has been used to analyse the problem; it is particularly useful when the factors that affect the process are known but the mathematical equations describing the process are not. For example, the factors affecting the rate of heat transfer (measured by the heat film coefficient), to a fluid flowing through a pipe are as follows:

$$h = \phi(D, k, \mu, v, c, \rho)$$

where ϕ denotes 'a function of', h (W m^{-2} K^{-1}) is the heat film coefficient, D (m) is the pipe diameter, k (W m^{-1} K^{-1}) is the thermal conductivity of fluid, μ (N s m^{-2}) is the dynamic viscosity of fluid, v (m s^{-1}) is the average velocity of fluid, c (J kg^{-1} K^{-1}) is the specific heat of fluid and ρ (kg m^{-3}) is the density of fluid.

    Dimensional analysis can be used to arrange these factors into dimensionless groups. The number of independent dimensionless groups is determined by the Buckingham pi (π) theorem; this states that the number of

Table 1.13 — Summary of the main fundamental and derived units.

Property	Dimensions	SI unit	Imperial unit	Conversion factors
Mass	[M]	kilogram (kg)	pound (lb)	1 lb = 0.4356 kg
Length	[L]	metre (m)	foot (ft)	1 ft = 0.3048 m
Time	[T]	second (s)	second (s)	
Temperature	[Θ]	kelvin (K)	degree Fahrenheit (°F)	
Area	[L^2]	m^2	ft^2	1 ft^2 = 9.29 × 10^{-2} m^2
Volume	[L^3]	m^3	ft^3	1 ft^3 = 2.832 × 10^{-2} m^3
			gal (US gal in USA)	1 gal = 4.546 l 1 US gal = 3.785 l
Volumetric flow rate	[L^{3}T^{-1}]	m^3 s^{-1}	ft^3 s^{-1} gal h^{-1}	1 ft s^{-1} = 2.832 × 10^{-2} m^3 s^{-1} 1 gal h^{-1} = 1.263 × 10^6 m^3 s^{-1}
Density	[ML^{-3}]	kg m^{-3}	lb ft^{-3}	1 lb ft^{-3} = 16.02 kg m^{-3}
Velocity	[LT^{-1}]	m s^{-1}	ft s^{-1}	1 ft s^{-1} = 0.3048 m s^{-1}
Acceleration	[LT^{-2}]	m s^{-2}	ft s^{-2}	1 ft s^{-2} = 3.048 × 10^{-1} m s^{-2}
Force	[MLT^{-2}]	newton (N)	lbf	1 lbf = 4.448 N
Pressure	[ML^{-1}T^{-2}]	pascal (Pa)	lbf in^{-2}	1 lbf in^{-2} = 6.895 × 10^3 Pa
Work	[ML^{2}T^{-2}]	joule (J)	British thermal unit (Btu)	1 Btu = 1.055 × 10^3 J
Power	[ML^{2}T^{-3}]	watt (W)	horsepower (hp)	1 hp = 745.7 W
Energy	[ML^{2}T^{-2}]	joule (J)	British thermal unit (Btu)	

Table 1.14 — Dimensions and conversions for some important physical properties.

Property	Dimensions	SI units	conversion
Dynamic viscosity	[ML^{-1}T^{-1}]	N s m^{-2} (Pa s)	1 lb ft^{-1} s^{-1} = 1.488 Pa s
Kinematic viscosity	[L^{2}T^{-1}]	m^2 s^{-1}	1 ft^2 s^{-1} = 92.9 × 10^{-3} m^2 s^{-1}
Surface tension	[MT^{-2}]	N m^{-1}	1 dyn cm^{-1} = 10^{-3} N m^{-1}
Specific heat	[LT^{-2}Θ^{-1}]	J kg^{-1} K^{-1}	1 Btu lb^{-1} °F^{-1} = 4.187 × 10^3 J kg^{-1} K^{-1}
Thermal conductivity	[MLT^{-3}Θ^{-1}]	W m^{-1} K^{-1}	1 Btu ft^{-1} h^{-1} °F^{-1} = 1.731 W m^{-1} K^{-1}
Thermal diffusivity	[L^{2}T^{-1}]	m^2 s^{-1}	1 ft^2 s^{-1} = 92.9 × 10^{-3} m^2 s^{-1}
Latent heat	[L^{2}T^{-2}]	J kg^{-1}	1 Btu lb^{-1} = 2.326 × 10^3 J kg^{-1}
Heat film coefficient	[MT^{-3}Θ^{-1}]	W m^{-2} K^{-1}	1 Btu h^{-1} ft^{-2} °F^{-1} = 5.678 W m^{-2} K^{-1}
Overall heat transfer coefficient	[MT^{-3}Θ^{-1}]	W m^{-2} K^{-1}	

[a] Surface tension is rarely recorded in Imperial units.

groups is equal to the difference between the number of factors (seven in this example) and the number of basic dimensions (four).

The dimensionless groups which are derived are

$$\text{Nusselt number} \quad Nu = \frac{hD}{k} = \frac{(Wm^{-2}K^{-1})(m)}{(Wm^{-1}K^{-1})}$$

$$\text{Reynold's number} = Re = \frac{vDp}{\mu} = \frac{(ms^{-1})(m)(kg\,m^{-3})}{(kg\,m^{-1}s^{-1})}$$

$$\text{Prandtl number} = Pr = \frac{(J\,kg^{-1}K^{-1})(kg\,m^{-1}s^{-1})}{(J\,s^{-1}m^{-1}K^{-1})}$$

This can be expressed in the form

$$\frac{hD}{k} = \phi \left(\frac{vDp}{\mu}\right)^a \left(\frac{c\mu}{k}\right)^b$$

This type of relationship can then be investigated experimentally to determine how the Nusselt number containing the heat film coefficient will be influenced by the Reynolds number and Prandtl number. For fully turbulent flow in a tube the following empirical relationship has been found:

$$Nu = 0.023\,(Re)^{0.8}\,(Pr)^{0.4}$$

This allows the heat film coefficient to be evaluated for a wide range of flow conditions and fluids. Dimensional analysis is also extremely important for the scaling-up engineering operations. The reader is referred to texts by Loncin and Merson (1979), American Society of Heating, Refrigerating and Air-conditioning Engineers (1981) and Weast (1982) for a more detailed discussion of these problems.

1.23 Concentration

Concentration can be expressed in many ways, the most common being weight per unit weight (w/w) and weight per unit volume (w/v) (strictly speaking, these should be mass per unit mass and mass per unit volume respectively). Therefore the concentration (w/w) is the mass of the substance divided by the total mass of the food. It is commonly used for solids and for liquids when the quantity is expressed in terms of mass rather than volume; it is usually expressed as a percentage. Thus a food containing 15% total solids (w/w) will contain 15 kg of solid matter in every 100 kg of food (solid or liquid). Sugar concentrations (°Brix) are measured as kilograms of sugar per 100 kg of sugar solution (see section 2.4).

The concentration (w/v) is the mass of solute dissolved in unit volume of the solution and is expressed in kilograms per cubic metre ($kg\,m^{-3}$) or grams per litre ($g\,l^{-1}$), where

$$1\,\text{kg}\,\text{m}^{-3} = \frac{10^3\,\text{g}}{10^3\,\text{l}} = 1\,\text{g}\,\text{l}^{-1}$$

The density of the solution will be required to convert from concentration (w/w) to concentration (w/v) and vice versa. The molar concentration or molarity (in units of M) is the concentration of solution in grams per litre divided by the molecular weight of the solute.

The mole fraction of a component in a mixture is the number of moles of the substance compared with the total number of moles in the system and thus has no units.

These different ways of expressing concentration are reviewed for a solution of sucrose made by adding 20 kg of sucrose to 80 kg of water. The density of the resulting solution is $1083\,\text{kg}\,\text{m}^{-3}$.

If the total mass of solution is 100 kg, therefore

$$\text{concentration (w/w)} = \frac{20}{100} \times 100$$
$$= 20\%$$

Similarly, of the total mass of solution is 100 kg and the density is $1083\,\text{kg}\,\text{m}^{-3}$, therefore

$$\text{volume} = \frac{\text{mass}}{\text{density}} = \frac{100}{1083} = 0.0923\,\text{m}^3$$

and

$$\text{concentration (w/v)} = \frac{\text{mass of solid}}{\text{total volume}} = \frac{20}{0.0923}$$
$$= 216.7\,\text{kg}\,\text{m}^{-3}\,(\text{g}\,\text{l}^{-1})$$

To express the concentrations as a molarity, we have

$$\text{molarity of sucrose solution} = \frac{\text{concentration (g}\,\text{l}^{-1})}{\text{molecular weight}} = \frac{216.7}{342}$$
$$= 0.634\,\text{M}$$

To express the concentration as a mole fraction, we have

$$\text{number of moles of sucrose} = \frac{20}{342}$$
$$= 0.0584$$

and

$$\text{number of moles of water} = \frac{80}{18} = 4.444$$

Therefore,

$$\text{mole fraction of sucrose} = \frac{\text{number of moles of sucrose}}{\text{total number of moles}} = \frac{0.0584}{4.5024} = 0.0130$$

1.24 SYMBOLS

A	surface area, collision frequency (Arrhenius equation)
CE	Carnot efficiency
c	specific heat
D	diameter
F	force
g	acceleration due to gravity
g_c	gravitational constant
h	height or heat film coefficient (in the Nusselt number)
H	head developed by a pump
k	thermal conductivity or reaction velocity constant (in the Arrhenius equation)
[L]	dimensions of length
m	mass
[M]	dimensions of mass
N	rotational speed
p	pressure
P_0	power
Q	volumetric flow rate
r	radius
R	electrical resistance or gas constant (in the gas and Arrhenius equations)
s	distance
t	time
T	temperature (in the ideal gas and Arrhenius equations)
[T]	dimensions of time
v	average velocity
v_t	tip speed
V	volume
X	vacuum reading

1.24.1 Greek symbols

μ	viscosity
$[\theta]$	dimensions of temperature
ω	angular velocity

1.24.2 Dimensionless groups
Nu   Nusselt number
Pr   Prandtl number
Re   Reynolds number

2
Density and Specific Gravity

2.1 INTRODUCTION

This chapter covers the property of density in all its aspects for solid, liquid and gaseous systems and shows how and why it is used in food-processing systems. The density of liquids is fairly straightforward, but solids in particulate form, such as peas or powders, have a bulk density as well as a solid density to consider. Gases and vapours, unlike solids and liquids, are considered to be compressible and some foods such as whipped cream and ice-cream have air incorporated into them during their preparation, the amount of air being measured by the overrun. Examples and applications of density measurement and its importance in food-processing applications are given throughout this chapter.

The density of a substance is equal to the mass of the substance divided by the volume that it occupies:

$$\text{density} = \frac{\text{mass}}{\text{volume}}$$

Density has dimensions of $[ML^{-3}]$. In the SI system of units it is measured in kilograms per cubic metre ($kg\ m^{-3}$). It is usually denoted by the Greek symbol rho ρ. Water has its maximum density of $1000\ kg\ m^{-3}$ at 4 °C:

$$10^3\ kg\ m^{-3} = \frac{10^3 \times 10^3\ g}{10^6\ ml} = 1\ g\ ml^{-1}$$

In the Imperial system, density is measured in pounds per cubic foot ($lb\ ft^{-3}$), where

$$1\ lb\ ft^{-3} = 16.02\ kg\ m^{-3}$$

The densities of some common solids and liquids are given in Tables 2.1 and 2.2.

Table 2.1 — Densities of some common solid materials.

Material	Density (kg/m³)	Temperature (°C)
Aluminium	2640	0
Cast iron	7210	0
Copper	8900	0
Mild steel	7840	18
Stainless steel	7950	20
Brick	1760	20
Concrete	2000	20
Glass soda	2240	20
Wood	200	30

Adapted from the data of Earle (1983).

Table 2.2 — Densities of some selected fluids.

Fluid	Density (kg/m³)	Temperature (°C)
Acetone	792[a]	20
Carbon tetrachloride	1595[a]	20
Glycerol	1260[a]	0
Mercury	13600[a]	—
Milk	1028–1035[a]	—
Acetic acid	1050[b]	20
Olive oil	910[b]	20
Rape seed oil	900[b]	20
Tallow	900[b]	65
Cream (20% fat)	1010[b]	3

[a]From the data of Weast (1982).
[b]From the data of Earle (1983).
Additional data for a wide range of solids has been given by Weast (1982).

In most engineering problems, solids and liquids are assumed to be incompressible, i.e. the density is hardly affected by moderate changes in temperature and pressure. In reality, the densities of water and other substances do change with temperature. Density differences in heated fluids provide the driving force for natural convection (see section 9.10).

In almost all cases, density decreases as temperature increases. Table shows denisty changes for water, alcohol and a variety of vegetable oils, over the temperature range from -20 °C to 80 °C.

Table 2.3 — Effect of temperature on the densities of some fluids.

Temperature (°C)	Density (kg/m³) of the following fluids						
	Water	Ethyl alcohol[a]	Maize oil[b]	Sunflower oil[b]	Sesame oil[b]	Soya oil[b]	Cotton oil[b]
−20	993.5		947	944	946	947	949
−10	998.1		940	937	939	941	942
0	999.9	806.3	933	930	932	934	935
4	1000.0	802.9					
10	999.7	792.9	927	923	925	927	928
20	998.2	789.5	920	916	918	920	921
40	992.2		906	903	905	907	908
60	983.2		893	899	891	893	894
80	971.8		879	876	878	879	881

[a]From the data of Weast (1982).
[b]From the data of Tschubik and Maslow (1973).

The specific gravities of a wide range of oils derived from plant and animal sources have been tabulated by Weast (1982).

2.2 SOLID DENSITY

For particulate matter (such as peas, beans, grain, flour and powders), milk, coffee and starch, one may be interested in the density of individual particles or units or the density of the bulk of the material which includes the void volume between the individual units.

The *solid* or *particle* density will refer to the density of an individual unit. This unit may or may not contain internal pores. The solid density is defined as the mass of particles divided by the volume of particles and will take into account the presence of such pores.

The densities of solid constituents, disregarding any internal pores, have been summarized by Peleg (1983) and are shown in Table 2.4.

Most fresh fruit and vegetables contain between 75% and 95% water. Therefore, many food densities should not be far removed from the value for water, namely 1000 kg m^{-3}, provided that they do not contain too much air.

Table 2.4 — Densities of solid constituents.

Constituent	Density (kg m^{-3})	Constituent	Density (kg m^{-3})
Glucose	1560	Fat	900–950
Sucrose	1590	Salt	2160
Starch	1500	Citric acid	1540
Cellulose	1270–1610	Water	1000
Protein	1400		

In theory, if the composition of the food is known, the density ρ_f can be estimated from

$$\rho_f = \frac{1}{m_1/\rho_1 + m_2/\rho_2 + m_3/\rho_3 + \ldots + m_n/\rho_n}$$

where ρ_f is the density of the food, m_1 to m_n are the mass fractions of constituents 1 to n and ρ_1 to ρ_n are the densities of constituents 1 to n (n is the number of constituents).

For example, for an apple containing 84.4% moisture, 14.55% sugar, 0.6% fat and 0.2% protein (densities are in kg m^{-3}; see Table 2.4),

$$\rho_f = \frac{1}{0.844/1000 + 0.1455/1590 + 0.006/925 + 0.002/1400}$$

$$= \frac{10^{-3}}{0.840 + 0.0920 + 0.006 + 0.001}$$

$$= 1064 \text{ kg m}^{-3} \ (1.064 \text{ g ml}^{-1})$$

However, there appears to be some discrepancy here, because apples will usually float in water. Mohsenin (1970) quotes a value of 846 kg m^{-3} at 29 °C. Therefore, considerable quantities of air must be trapped within the pores. This air is removed during the blanching process.

If the densities and volume fractions are known, the density can be evaluated from

$$\rho_f = V_1\rho_1 + V_2\rho_2 + V_3\rho_3 + \ldots + V_n\rho_n$$

where V_1 to V_n are the volume fractions of constituents 1 to n and ρ_1 to ρ_n are the densities of constituents 1 to n.

Water at 0 °C has a density of 999 kg m^{-3}, whereas the value for ice at 0 °C is 916 kg m^{-3}. As the temperature of ice is reduced, the density increases. The values are recorded in Table 2.5. The solid densities for food materials are not well documented. Milson and Kirk (1980) present values for different types of food. Some of these are recorded in Table 2.5. Solid densities are recorded for a variety of cereals in Table 2.6.

Table 2.5 — Densities (or the specific gravity SG) of some foods.

Food	Density (kg m^{-3})	Food	Density (kg m^{-3})
Fresh fruit	865–1067	Fresh fish	1056
Fresh vegetables	801–1095	Frozen fish	967
Frozen fruit	625–801	Meat	1.07 (SG)
Frozen vegetables	561–977	Ice (0 °C)	916
		Ice ($-$10 °C)	933
		Ice ($-$20 °C)	948

The densities of frozen fruit and vegetables are less than those of their fresh counterparts, whereas the density of frozen fish is much higher than that of fresh fish. No explanation for this discrepancy is offered. Obviously, foods undergo some drastic decrease in density as the water freezes.

Solid densities can be determined by the principle of flotation, using liquids of known densities (see section 2.4.3). The temperature of the liquid should be equal to that of the solid.

Most dry solids, or powders, without internal pores will have solid densities between 1400 kg m^{-3} and 1500 kg m^{-3}, as the solid densities for most food constituents are fairly similar.

Solid densities are important in separation processes (e.g. sedimentation and centrifugation) and in pneumatic and hydraulic transport of powders and particulates. Processing conditions, particularly during dehydration and agglomeration, can significantly affect the extent and nature of pore formation, and hence the solid density of the material.

2.3 BULK DENSITY

When mixing, transporting, storing and packaging particulate matter, e.g. peas and flour, it is important to know the properties of the bulk material. When such solids are poured into a container, the total volume occupied will contain a substantial proportion of air. The *porosity* ε of the packed material is that fraction of the total volume which is occupied by the air, i.e.

$$\varepsilon = \frac{\text{volume of air}}{\text{total volume}}$$

The porosity will be affected by the geometry, size and surface properties of the material. In addition, if this container is tapped, the total volume and hence the porosity will decrease, until eventually the system reaches an equilibrium volume. The density of the bulk material under these conditions is generally known as the *bulk density*.

The bulk density of the material will therefore depend upon a number of factors, namely the solid density, the geometry, the size, the surface properties and the method of measurement. Normally, bulk density is determined by placing a known weight of powder (20 g (British Standards Institution) or 50 g (continental) see Society of Dairy Technology (1980)) into a measuring cylinder, tapping the cylinder a fixed number of times and determining the resulting bulk volume:

$$\text{bulk density} = \frac{\text{mass}}{\text{bulk volume}}$$

However, the recommended procedures employ slightly different conditions, and so the values in the literature should be treated with caution. Table 2.6 shows some 'approximate' bulk density values for a wide range of food materials in powdered form. Table 2.7 shows some bulk density values

Table 2.6 — Bulk densities of a variety of powders.

Powder	Bulk density (kg m^{-3})	Powder	Bulk density (kg m^{-3})
Oats[a]	513	Milk[b]	610
Wheat[a]	785	Salt (granulated)[b]	960
Flour[a]	449	Sugar (granulated)[b]	800
Cocoa[b]	480	Sugar (powdered)[b]	480
Coffee (instant)[b]	330	Wheat flour[b]	480
Coffee (ground and roasted)[b]	330	Yeast (bakers)[b]	520
Corn starch[b]	560	Egg (whole)[b]	340

[a]From the data of Milson and Kirk (1980).
[b]From the data of Peleg (1983).

for fruit and vegetables; Table 2.8 lists solid densities, bulk densities and moisture contents for a selection of cereals. A range of values is presented for each cereal as different varieties were measured.

Table 2.7 — Bulk densities of some fruit and vegetables.

Fruit or vegetable	Bulk density (kg m^{-3})	Fruit or vegetable	Bulk density (kg m^{-3})
Apples	544–608	Oranges	768
Carrots	640	Peaches	608
Grapes	368	Onions	640–736
Lemons	768	Tomatoes	672

Adapted from the data of Mohsenin (1970).

The bulk density of spray-dried products is affected by the solids content of the feed, immediately prior to drying, and the inlet and outlet air

Table 2.8 — Moisture contents, solid densities and bulk densities of different cereals.

Cereal	Moisture content	Solid density (kg m^{-3})	Bulk density (kg m^{-3})
Barley	7.5–8.2	1374–1415	565–650
Oats	8.5–8.8	1350–1378	358–511
Rice	8.6–9.2	1358–1386	561–591
Wheat	6.2–8.5	1409–1430	790–819

Adapted from the data of Mohsenin (1970).

temperatures. Some data for a variety of milk powders are given in Table 2.9.

Table 2.9 — Selected properties of spray-dried milk powders.

Mean porosity	46.6%
Range of porosity	42.9–50.9%
Range of particle density	0.770–1.225 g ml^{-1}
Range of powder (bulk) density	0.35–0.65 g ml^{-1}

Whole milk powder	Powder density (g ml^{-1})	Particle density (g ml^{-1})	Porosity (%)
11% solids; 90 °C outlet	0.39	0.802	51
34% solids; 90 °C outlet	0.55	1.017	46
47% solids; 90 °C outlet	0.66	1.150	42
47% solids; 113 °C outlet	0.46	0.913	50
Straight through, instantized	0.63	1.100	43
Re-wet instantized	0.33	1.100	70

Adapted from the data of Mettler (1980).

In addition the method of atomization will affect the bulk density. Early designs of atomizer wheel produced powders of 0.5–0.55 g ml^{-1}, whereas later designs typified by the varied wheel produced bulk densities of 0.55–0.65 g ml^{-1}. Later designs have used steam to occlude air from the fluid. Jet nozzles can produce powders with bulk densities as high as 0.83 g ml^{-1}.

2.3.1 Relationship between porosity, bulk density and solid density
This relationship is given by

$$\text{porosity } \varepsilon = \frac{\text{volume of air}}{\text{volume of bulk sample}}$$
$$= \frac{\text{volume of bulk sample} - \text{true solid volume}}{\text{volume of bulk sample}}$$

$$= 1 - \frac{\text{solid volume}}{\text{bulk volume}}$$

Note that the mass of solid and the mass of bulk are equal. Therefore

$$\text{porosity} = 1 - \frac{\text{bulk density}}{\text{solid density}}$$

$$= 1 - \frac{\rho_b}{\rho_s}$$

$$= \frac{\rho_s - \rho_b}{\rho_s}$$

Note that porosity can be expressed as a fraction or as a percentage. This equation can be used for solids with or without internal pores.

2.4 LIQUID DENSITY AND SPECIFIC GRAVITY

Water has its maximum density of 1000 kg m^{-3} at 4 °C. As the temperature increases above 4 °C, the density will decrease (see Table 2.3). The addition of all solids, with the exception of fat, to water will increase its density. Density measurement can be used for pure substance as an indication of total solids.

However, it is often more convenient to measure the specific gravity SG of a liquid, where:

$$SG = \frac{\text{mass of liquid}}{\text{mass of equal volume of water}}$$

$$= \frac{\text{density } \rho_L \text{ of liquid}}{\text{density } \rho_w \text{ of water}}$$

(Note that this can also be used for solids as well as liquids)

Specific gravity is dimensionless. The specific gravity of a fluid changes less than the density, as temperature changes. For example, maize oil has densities of 927 kg m^{-3} and 893 kg m^{-3} at 10 °C and 60 °C, respectively. However, the specific gravity only decreases from 0.927 to 0.908. When specific gravities are quoted, it is normally at a particular temperature. If the specific gravity of a material is known at a temperature T °C, its density at T °C will be given by

$$\rho_L = (SG)_T \times \rho_w$$

where ρ_L is the density of liquid at T °C, $(SG)_T$ is the specific gravity at T °C and ρ_w is the density of water at T °C (obtained from tables).

Sec. 2.4]   Liquid density and specific gravity

A shorthand notation often used in brewing is in terms of the original gravity OG. Thus a wort with an original gravity of 42 would have a starting specific gravity before fermentation of 1.042.

Specific gravities are conveniently measured, using density bottles, pycnometers or hydrometers.

2.4.1 Density bottles

A density bottle (Fig. 2.1(a)) can be used to determine the specific gravity of an unknown liquid and a particulate solid provided that the solid is insoluble in the liquid. Care should be taken to ensure that all the air is removed from the bottle when the liquid is added to the solid.

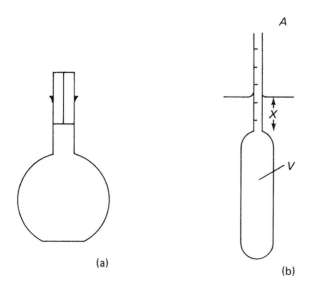

Fig. 2.1 — Equipment used for measuring density: (a) a specific gravity bottle; (b) a hydrometer.

The following readings are taken: the weight w_1 of the specific gravity bottle empty; the weight w_2 of the bottle full of water; the weight w_3 of the bottle full of liquid; the weight w_4 of the bottle plus solid (approximately one-third full); the weight w_5 of the bottle plus solid plus liquid to fill.

The specific gravity of the liquid is equal to

$$\frac{w_3 - w_1}{w_2 - w_1}$$

Now the weight of the solid is $w_4 - w_1$, and the weight of liquid having some volume of solid is $w_3 - w_1 - (w_5 - w_4)$. Therefore the specific gravity of the solid is equal to

$$\frac{w_4 - w_1}{w_3 - w_1 - (w_5 - w_4)} \frac{w_3 - w_1}{w_2 - w_1}$$

i.e.

$$\frac{\text{weight of solid}}{\text{weight of equal volume liquid}} \times \text{specific gravity of liquid}$$

Toluene has been recommended as a suitable solvent for determining the specific gravity of food materials (Mohsenin, 1970).

2.4.2 Hydrometers and hydrometer scales

The constant-weight hydrometer works on the principle that a floating body displaces its own weight of fluid. A diagram of a hydrometer is shown in Fig. 2.1(b). The instrument is placed in the fluid and the density of the fluid is read from the scale on the stem. Let the volume to the base of stem be V, the cross-sectional area of stem be A and the weight of the hydrometer be W. When immersed in a liquid of density ρ, the length of the stem immersed is x. Thus the volume of liquid displaced is $Ax + V$. The weight of liquid displaced equals $\rho(Ax + V)$, which by the principle of flotation equals W. Therefore,

$$\rho = \frac{W}{Ax + V}$$

Hydrometers are easy to use and are available in a range of sizes, e.g. 1.00–1.100 and 1.100–1.200, for different applications.

2.4.2.1 Hydrometer scales

The following are some examples of special hydrometers or hydrometer scales.

An *alcoholometer* is used for testing alcoholic solutions, the scale shows the amount of alcohol by volume (0–100%).

The *Brix saccharometer* shows directly the percentage sucrose by weight in the solution, at the temperature indicated on the instrument. Degrees Brix is therefore percentage of sugar w/w and is still in common use (Table 2.10).

The *density hydrometer*, sometimes known as a lactometer, is used for determining the density of milk. The instrument has scale divisions ranging from 25 (1025 kg m^{-3}) to 35 (1035 kg m^{-3}). If the fat content is known, then by reference to appropriate tables the hydrometer readings can be used to give MSNF (see section 2.4.4). Such hydrometers, which offer a narrow range of measurement and are sensitive to small changes in density, are constructed by having a large bulb volume V and a narrow stem, i.e. a small

Liquid density and specific gravity

Table 2.10 — Relationship between specific gravity, the Brix scale and the Baumé scale for sucrose.

Brix reading (° Brix)	Concentration (g/l)	Specific gravity at 20 °C	Baumé reading (° Baumé)
0	0	1.0000	0.00
5	50.9	1.0120	2.79
10	103.8	1.0400	5.57
15	158.9	1.0610	8.34
20	216.2	1.0829	11.10
25		1.1055	13.84
30	338.1	1.1290	16.57
35		1.1533	19.28
40	470.6	1.1785	21.97
45		1.2046	24.63
50	614.8	1.2317	27.28
55		1.2598	29.90
60	771.9	1.2887	32.49
65		1.3187	35.04
70	943.1	1.3496	37.56
75		1.3814	40.03
80	1129.4	1.4142	42.47

Adapted from the data of Spencer and Meade (1957).

cross-sectional area A.

A *Twaddell hydrometer* has the scale so arranged that the reading multiplied by five and added to 1000 gives the specific gravity with reference to water as 1000. It is only used for liquid more dense than water (Table 2.11).

The Baumé scale: there are two scales, one for use with fluids heavier than water and one with lighter fluids. In the former, 0 °C corresponds to a specific gravity of 1.000 and 66 °C corresponds to a specific gravity of 1.842. The following equation gives the means of conversion on this scale:

$$SG = \frac{m_B}{m_B - d}$$

where m_B is a constant equal to 145 in the USA and equal to 146.78 on the new scale, and d is the Baumé reading.

Conversions from density (g ml^{-1}) to degrees (Baumé (°Baumé) and degrees Twaddell (°Twaddell) are shown in Table 2.11.

An *oleometer* is used for vegetable and sperm oils; the scale is from 50 to 0 and corresponds to a specific gravity of from 0.870 to 0.970.

Table 2.11 — Conversion from density values to Baumé and Twaddell scales.

Density (g ml^{-1})	Baumé reading (° Baumé)	Twaddell reading (° Twaddell)
1.00	0.00	0
1.05	6.91	10
1.10	13.18	20
1.15	18.91	30
1.20	24.17	40
1.25	29.00	50
1.30	33.46	60
1.35	37.59	70
1.40	41.43	80
1.45	45.00	90
1.50	48.33	100
1.55	51.45	110
1.60	54.38	120

Compiled from the data of Weast (1982).

2.4.3 Liquid density values

The specific gravities of sucrose solutions if different strengths are shown in Table 2.10. Similar tables are given for glucose syrups by Norrish (1967). Table 2.12 shows the specific gravity and glycerol. Tables with much more detail have been presented by Weast (1982) for a wide range of aqueous solutions.

Table 2.12 — Specific gravities of some aqueous solutions (at 20 C).

Amount (wt.%) in aqueous solution	Fructose	Lactose	Maltose	Ethanol	Ethylene glycol	Glycerol
5	1.012	1.020	1.020	0.991	1.006	1.012
10	1.040	1.039	1.040	0.984	1.013	1.023
15	1.062	1.062	1.060	0.977	—	—
20	1.084	1.077(18%)	1.082	0.970	1.023	1.048
30	1.130		1.129	0.956	—	—
40	1.177		1.179	0.937	1.053	1.100
50	1.232		1.233	0.916	—	—
60	1.288		1.288	0.893	1.078	1.155
70	1.347			0.869	—	—
80				0.845		1.211
90				0.819		—
100				0.791		1.263

Compiled from the data of Weast (1982).

Table 2.13 shows the specific gravities of sodium chloride and calcium chloride brines; these are used as direct immersion refrigerants (see section

10.13). Also shown are the freezing points and specific heats of these solutions (see Chapter 8).

Table 2.13 — Specific gravities, freezing points and specific heats of aqueous solutions of sodium chloride and calcium chloride.

Concentration (w/w)	Specific gravity[a]	Freezing point (°F)	Specific heat (kcal kg^{-1} K^{-1})
Sodium chloride			
1	1.007	31.8	0.992
2	1.015	29.3	0.984
4	1.030	26.6	0.968
6	1.045	23.9	0.946
8	1.061	21.2	0.919
10	1.076	18.7	0.892
12	1.091	16.0	0.874
15	1.115	12.2	0.855
20	1.155	6.1	0.829
24	1.187	1.2	0.795
26	1.204	−1.1	0.771
Calcium chloride			
1	1.007	31.1	0.99
2	1.015	30.4	0.97
4	1.032	28.6	0.94
6	1.049	26.6	0.91
8	1.067	24.3	0.88
10	1.085	21.3	0.86
12	1.103	18.1	0.83
15	1.131	12.2	0.795
20	1.179	−1.4	0.730
24	1.218	−17.1	0.69
26	1.239	−27.0	0.68
28	1.261	−39.2	0.67
30	1.283	−54.4	0.65

[a] At 39 °F for sodium chloride, and at 66 °F for calcium chloride.
Compiled from the data of Hall et al. (1971).
Similar data have been presented by Weast (1982).

Information about the relation of density or specific gravity to concentration can be used to make solutions of different densities for determining the densities of solid foods, using the principle of flotation. The fluid density in which the solid appears neither to sink nor to float is noted.

The mean density values and total solids contents are given for a variety of fruit juices in Table 2.14.

Table 2.14 — Densities and solids contents of some fruit juices.

Fruit juice	Mean density (g/ml^{-1})	Mean total solids
Orange	1.042	10.8
Grapefruit	1.040	10.4
Lemon	1.035	10.0
Lime	1.035	9.3
Apple	1.060	13.0
Blackcurrant	1.055	13.5

Adapted from the data of Egan *et al.* (1981).

2.4.4 Density of milk

The density of bovine milk usually falls within the range 1025–1035 kg m^{-3}. the densities of the respectively solid constituents are regarded as fat (930 kg m^{-3}), water (1000 kg m^{-3}) MSNF (1614 kg m^{-3}). *British Standard* 734 (British Standards Institution, 1937, 1959) gives information on density hydrometers for use in milk. Tables are supplied for determining the total solids of milk, knowing the specific gravity and fat content. Temperature correction tables are also presented. Fat contents range between 1% and 10% and the total solids determination is based on the following equations: according to *British Standard* 734 (British Standards Institution, 1937),

$$C_T = 0.25D + 1.21F + 0.66$$

and, according to *British Standard* 734 (British Standards Institution (1959),

$$= 0.25D + 1.22F + 0.72$$

where C_T is the total solids concentration (w/w), $D = 1000(SG - 1)$ (SG is the specific gravity) and F is the fat percentage.

Thus, milk at 26 °C with a fat content of 3.5% and a specific gravity of 1.032 would be corrected to a value of 1.0322 at 20 °C and have a total solids of 13.05, according to *British Standard* 734 (British Standards Institution, 1959). The value would be slightly lower, using the earlier equation. Total solids are normally expressed to the nearest 0.05%. Obviously the 1959 tables are preferred but several people are still of the opinion that they tend to overestimate the total solids and that the 1937 formula is more accurate

(Egan et al., 1981). Lewis (1986b) reviews how the density of milk and cream are affected by temperature changes.

Compositional and other factors, such as solid-to-liquid fat ratios and degree of hydration of protein, which affect the density of milk, evaporated milk and cream have been discussed by Webb et al. (1974).

2.5 GASES AND VAPOURS

Gases and vapours are compressible, and densities are affected by changes in temperature and pressure. Under moderate conditions, most gases obey the ideal gas equation, which can be written as

$$pV_m = RT$$

where p (N m^{-2}) is the pressure, V_m (m^3 kmol^{-1}) is the molar volume, i.e. the volume occupied by 1 kmol, $R = 8.314$ kJ kmol^{-1} K^{-1}) is the gas constant and T (K) is the temperature.

Note that the molecular weight of any gas expressed in kilograms (1 kmol) occupies 22.4 m^3 at 273 K and 1 atm. For example, 29 kg of air occupies a volume of 22.4 m^3 at 273 K and 1 atm. Therefore,

$$\text{density of air} = \frac{\text{mass}}{\text{volume}}$$

$$= \frac{29}{22.4}$$

$$= 1.29 \text{ kg m}^{-3}$$

At 100 °C and 1 atm,

$$\frac{V_1}{T_1} = \frac{V_2}{T_2}$$

Therefore,

$$\text{new volume} = 22.4 \times \frac{373}{273} = 30.605 \text{ m}^3$$

$$\text{new density} = \frac{\text{mass}}{\text{volume}}$$

$$= \frac{29}{30.605}$$

$$= 0.947 \text{ kg m}^{-3}$$

The densities of some common gases are given in Table 2.15. The values agree very well with those calculated using the ideal gas equation. The densities of moist air at different barometric pressures, water vapor pressure and temperatures have been given in some detail by Weast (1982). These values may be extremely useful for dehydration calculations.

Table 2.15 — Densities of some common gases.

Gas	Density (kg m^{-3}) at the following temperatures	
	0 °C	100 °C
Air	1.29	0.94
Carbon dioxide	1.98	1.46
Nitrogen	1.30	

Taken from the data of Earle (1983).

With thermodynamic fluids such as steam and refrigerants, reference is often made to the *specific volume* V_g. This is the volume occupied by unit mass of vapour, which is the inverse of density. Saturated water vapour at 1.013 bar (100 °C) has a specific volume of 1.67 m^3 kg^{-1} whereas at 10 °C it would have a specific volume of 106.43 m^3 kg^{-1}. This shows that vapours have very large specific volumes at reduced pressures; consequently, in operations involving the removal of water vapour at low pressures, such as vacuum evaporation or freeze drying, the vacuum pump needs to be large enough to handle the large volumes produced.

2.6 DENSITY OF AERATED PRODUCTS: OVERRUN

Some well-known foods are produced by incorporating air into a liquid and producing a foam. In such systems, air is the dispersed phase and the liquid is the continuous phase. The foam is stabilized by surface-active agents which collect at the interface. Examples of foams are cake mixes, whipped creams and desserts, and ice-cream. Obviously the inclusion of air will reduce the density of the product.

The amount of air incorporated is expressed in terms of the over-run, normally as a percentage, where

$$\text{over-run} = \frac{\text{increase in volume}}{\text{original volume}} \times 100$$

$$= \frac{\text{volume of foam} - \text{original volume of liquid}}{\text{volume of liquid}} \times 100$$

For example, with ice-cream, the volume of foam refers to the final volume of the ice-cream, and the volume of liquid to the volume of the original mix.

In practice, over-run is most easily determined by taking a container of fixed volume and weighing it full of the original liquid and full of the final foam. In this case the over-run is determined as follows:

$$\text{over-run} = \frac{\text{weight of original liquid} - \text{weight of same volume of foam}}{\text{weight of same volume of foam}} \times 100$$

The factors that affect over-run in ice-creams have been discussed by Arbuckle (1972) and by Hyde and Rothwell (1973). These include the total solids of the mix and type of freezer used. Generally speaking, the higher the total solids content, the greater is the possible over-run. Some authorities suggest that the over-run should be between two and three times the total solids content. Values for ice-creams generally range between 40% (soft) and 100% (hard) (Webb et al., 1974). Arbuckle (1977) gives typical values for a range of ice-creams and other frozen desserts, and Porter (1975) gives values for hard and soft ice-creams. Some of these values are summarized in Table 2.16. Too much air will result in a snowy fluffy unpalatable product; too little will give a soggy heavy product.

Table 2.16 — Over-runs for frozen desserts.

Dessert	Over-run
Ice-cream (packaged)	70–80
Ice-cream (bulk)	90–100
Sherbert	30–40
Ice	25–20
Soft ice-cream	30–50
Ice milk	50–80
Milk shake	10–15

Adapted from the data of Arbuckle (1977).

For whipping cream, over-runs of 100–120% would be expected (Society of Dairy Technology, 1975). As well as total over-run, it is important to measure the stability of the foam over a period of time.

It is interesting to note that ice-cream is sold by volume rather than by weight. Therefore, it is in the manufacturer's interest to obtain the maximum possible over-run.

2.7 SYMBOLS

A	area
c_T	total solids concentration (w/w)
d	Baumé reading
D	$= 1000\,(SG-1)$, modified density
F	fat content (%)
L	dimensions of length
m	mass fraction
m_B	Baumé constant
M	dimensions of mass
OG	original gravity
p	pressure
R	gas constant
SG	specific gravity
T	temperature
V	volume
V	volume fraction
V_m	molar volume
V_g	specific volume
w	weight
W	hydrometer weight
x	distance

2.7.1 Greek symbols

ε	porosity
ρ	density
ρ_b	bulk density
ρ_f	food density
ρ_s	solid density

3
Properties of fluids, hydrostatics and dynamics

3.1 INTRODUCTION
The properties of fluids both at rest (hydrostatics) and in motion are of relevance in the study of food-processing operations. This chapter concentrates on fluid statics and dynamics whereas Chapter 4 concentrates on the property of viscosity, methods for measuring it, and values for different types of food. Inevitably, there is some degree of overlap between the subject matter. Many foods are fluid in nature e.g. milk, fruit juices, oils emulsions, pastes and suspensions, as are most of the heat transfer fluids, such as hot water, steam, refrigerants and air, used for heating, drying, chilling and freezing operations.

3.2 HYDROSTATICS
When an object is immersed in a fluid, it is subjected to a pressure because of the weight of the fluid acting on top of it. The factors affecting the pressure will be the density of the fluid and the depth of immersion. In Fig. 3.1(a) the

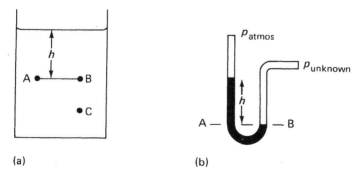

Fig. 3.1 — Fluid statics: (a) variation in pressure with depth; (b) open-tube manometer.

pressure at A is equal to the pressure on the surface plus the pressure due to column of liquid above A. Thus,

$$p_A = p_s + \rho g h$$

(see section 1.17 for definitions of pressure and pressure units) where p_A (N m^{-2}) is the pressure at A, p_s (N m^{-2}) is the surface pressure, ρ (kg m^{-3}) is the fluid density and h (m) is the depth of immersion.

The pressure on the surface can be either atmospheric pressure for an open tank, or above or below atmospheric pressure for a sealed or pressurized tank.

The pressure at B will be the same as that at A, whilst the pressure at C will be greater.

3.2.1 Pressure measurement

A manometer, which is a glass tube shaped in the form of a U and illustrated in Fig. 3.1(b), can be used to determine pressure. It is normally filled with a suitable fluid (mercury, water or carbon tetrachloride). One end is attached to the pressure being measured and the other end is open to the atmosphere. The pressure at A will equal the pressure at B:

$$p_{unknown} = p_{atmos} + \rho g h$$

where h (m) is the height of fluid and ρ (kg m^{-3}) is the fluid density.

Pressures both above and below atmospheric pressure can be measured. For pressures only slightly different from atmospheric pressure, a low-density fluid in the manometer would be suitable whilst, for higher pressures, mercury would be used.

Manometers can also be used to measure differential pressures, such as the pressure drop across a straight length of pipe, a valve, a heat exchanger or a drier (Fig. 3.2). If water is flowing along the pipe, it is essential that the

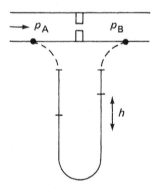

Fig. 3.2 — Manometer used for measuring differential pressures.

manometer fluid and the water do not mix. Carbon tetrachloride is often used with the addition of a suitable dye to make it visible. Other manometer fluids have been listed by Jones (1974a).

In this case the differential pressure $p_A - p_B$ (N m^{-2}) is given by

$$p_A - p_B = gh(\rho_m - \rho)$$

where ρ_m is the density of the manometer fluid and ρ is the density of the fluid in the pipe. If air or some other gas was flowing in the pipe, then the density term becomes negligible.

The measurement of differential pressures can give useful information on flow rates and the performance of food-processing equipment (see section 3.13). For measuring pressures only slightly different from atmospheric pressure or small differential pressures, an inclined manometer is used (Fig. 3.3). This extends the scale a longer distance. A typical

Fig. 3.3 — Inclined-tube manometer.

gradient would be 1 in 20. The scale is drawn along the length of the tube and is often calibrated directly in pressure units, e.g. pascals (Pa). This is extremely useful for measuring the very low differential pressures found at very low liquid velocities and in air flow problems. Such gauges normally have a range between 0.25 mm and 40 mm (water gauge).

In food processing equipment, gauges are normally used for measuring pressure. One very important consideration is their hygienic design. The cheapest and most common form of pressure-measuring element is the Bourdon tube. This is a hollow tube of eliptical cross-section, shaped as in Fig. 3.4. The pressure to be measured is applied to the inside of the tube. The differential pressure between the inside and the outside tends to strengthen the tube. This movement is amplified by a lever-and-spring mechanism and the pressure is indicated by a pointer on a calibrated dial. Fig. 3.4(a) shows how a Bourdon tube may be attached for measurement of pressures of gasses and liquids in pipes and flow channels. The Bourdon tube is very suitable for gases and clean fluids and for measuring fluids not directly in contact with foods, i.e. chilled water, hot water, steam and compressed air. However, as can be seen from the diagram, the fluid being measured will

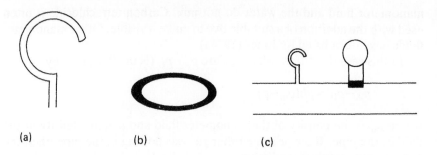

Fig. 3.4 — (a) Workings of a Bourdon-type gauge; (b) cross-section of tube; (c) Bourdon gauge and diaphragm gauge mounted on a pipe.

enter the Bourdon tube, making it not at all suitable for use with food materials. For example, milk would be extremely difficult to remove using cleaning-in-place methods; any milk left in the tube would quickly go sour and act as a source of contamination for subsequent batches. This illustrates a very important aspect in the design of all measuring instruments and fittings such as valves, bends and tees for food-processing equipment. Great care should be taken to ensure that there are no dead spaces and that all parts of the instrument or fittings can be adequately cleaned by acid and alkali detergents and by sanitizing agents.

The most suitable type of gauge for use on food lines are diaphragm gauges, the diaphragm being the force-detecting element; when these are fitted in a pipe line, the diaphragm lies flush with the pipe and the instrument is easily cleaned (Fig. 3.4(c)). The movement of the centre of the diaphragm is transmitted by a ball-and-socket joint and a link to a pointer as in the Bourdon gauge. These are more expensive than Bourdon gauges, but they are the only gauges suitable for use with food, other biodegradable materials and corrosive cleaning or sanitizing fluids.

The pressure exerted by the atmosphere fluctuates and, in cases where it is necessary to know accurately what the pressure is, a barometer can be used. The simplest form of barometer is constructed by filling a tube, which is sealed at one end, with mercury and inverting the tube with the open end below the surface of mercury and fixing the tube vertically; the height of the mercury will give atmospheric pressure.

The densities of dry air at pressures other than 1 atm have been given by Weast (1982). Other types of barometer, using diaphragms or bellows as the pressure-detecting elements, have been described by Jones (1974a). Similar instruments are also available for pressure ranges above and below atmospheric pressure.

A pressure transducer is a device which converts a force into an electrical signal. The force acting on a diaphragm is transmitted to a strain gauge (see section 12.8.4). The resulting electrical signal can then be used for measurement, recording or control functions. These are becoming extremely popular sensors for pressure measurement. Pressure measurement in turn can be used for measuring levels in tanks and silos by taking the pressure measurement at the base of the tank.

3.2.2 Vacuum measurement

Occasionally, it is necessary to measure very low absolute pressures, particularly in high-vacuum applications.

The most common type of gauge used is based on a prototype designed by McLeod in 1874. A sample of the gas is trapped in a vessel of reasonably high volume and compressed into a low volume, thereby increasing its pressure (Boyle's law) and making it easier to read. The detailed workings have been given by Jones (1974a).

A more convenient form of this gauge, which is suitable for measuring pressures between 10 Torr and 10^{-3} Torr, is shown in Fig. 3.5. The gauge,

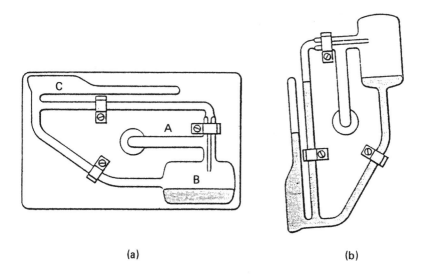

Fig. 3.5 — The McLeod gauge. (From Turnbull *et al.* (1962), with permission.)

which is mounted on a metal panel, swivels about a horizontal axis. The tube A bends at right angles to pass horizontally through the panel and is joined to the vacuum system being measured. When tube A is in the position shown in Fig. 3.5, which is not quite horizontal, all the mercury drains into bulb B and the remainder of the volume of the gauge is at the same pressure as that being measured. To determine the pressure reading, the gauge is rotated to the vertical position, as shown in the diagram, and the gas in bulb C is trapped and compressed by the rising mercury column. The height of the mercury in this bulb is related to the orginal pressure.

One disadvantage is that the gauge will not give a continuous pressure reading. Precautions may also have to be taken to prevent mercury from entering the vacuum chamber. McLeod gauges have also been discussed by Turnbull *et al.* (1962). Such gauges are extremely useful for checking the pressures during freeze-drying processes.

At very low pressures, the thermal conductivity of a gas decreases as its pressure is reduced. Use is made of this principle in a Pirani gauge, which is employed to measure low pressures. A glass or metal enclosure contains a metal filament with a high temperature coefficient of resistance. For a fixed voltage across the filament, the temperature and hence the electrical resistance will depend upon the rate at which heat is conducted away by the surrounding gas, i.e. its thermal conductivity and pressure.

3.3 ARCHIMEDES' PRINCIPLE

When an object is immersed in a fluid, it appears to lose weight. This is because the bottom of the object is subjected to a greater force than the top of the object. This apparent loss in weight is termed the upthrust. Archimedes showed that the upthrust was equal to the weight of fluid displaced.

Upthrust can be considered a force which acts vertically upwards, through the centre of gravity of the displaced fluid. Upthrust is a very important concept when dealing with the flow of mixtures of solids and liquids.

The rest of this chapter is devoted to various aspects of fluid dynamics.

3.4 FACTORS AFFECTING FRICTIONAL LOSSES

When a fluid is flowing along a straight horizontal pipe (Fig. 3.6(a)),

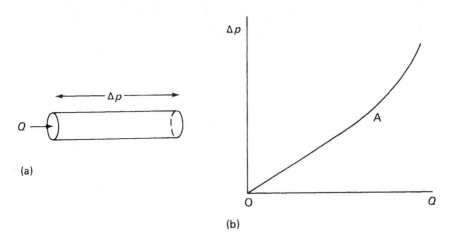

Fig. 3.6 — (a) Fluid flow in a straight pipe; (b) the relationship between the pressure drop and flow rate for fluid flow in a straight pipe.

frictional losses occur which result in a loss of pressure energy. The volumetric flow rate is slowly increased from zero and the corresponding pressure drop is measured; the relationship between the pressure drop and the flow rate is shown in Fig. 3.6(b). Over the range OA, there is a direct linear relationship. However, at point A, the nature of the flow changes and the relationship is no longer linear. Other factors affecting the pressure drop in the region OA will be the viscosity and density of the fluid, and the length and diameter of the pipe.

In any fluid flow situation, e.g. a fluid in a pipe or between parallel plates or an air stream passing over a stationary sphere, there exists a boundary layer adjacent to the surface (Fig. 3.7). Within this boundary layer the

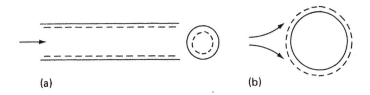

Fig. 3.7 — The boundary layer existing (a) at the wall of a pipe and (b) over the surface of a sphere.

property of the fluid is directly affected by the presence of the solid object (note that the velocity of the fluid in immediate contact with the surface is considered to be zero). The boundary layer can be thought of as a stagnant film or layer; this film offers considerable resistance to both heat and mass transfer to and from the surface. The region outside the boundary layer is termed the bulk fluid region; here the fluid is said to be well mixed. This boundary layer concept is very important in the consideration of heat and mass transfer operations.

3.5 STREAMLINE AND TURBULENT FLOW

When a dye is injected into a fluid flowing along a tube, there are two types of flow pattern that can be observed. At low flow rates, the dye will flow along the tube as a single stream, and apart from molecular diffusion there is no mixing or distribution of the dye with the bulk of the fluid. This is termed streamline flow. As the flow rate increases, the flow will become more turbulent.

Reynolds, in a series of experiments, examined some of the factors that determined whether flow would be streamline of turbulent, in a given situation. The following properties were found to be important: fluid velocity, density, viscosity and the tube diameter.

Using a technique known as dimensional analysis, he arranged these factors into a dimensionless group, which now bears his name and is referred to as the Reynolds number.

The Reynolds number Re is given by the following expression:

$$Re = \frac{vD\rho}{\mu} \frac{(ms^{-1})(m)(kg\,m^{-3})}{(kg\,m^{-1}s^{-1})}$$

where v (m s^{-1}) is the average velocity, D (m) is the pipe diameter, ρ (kg m^{-3}) is the fluid density and μ (kg m^{-1} s^{-1}) is the fluid viscosity. It is dimensionless if care is taken to ensure that the units are expressed consistently in SI, cgs or British units. The Reynolds number can be considered to be a measure of the ratio of the inertial force to the viscous force acting within that fluid. Further analysis showed that, if the value of the Reynolds number was less than 2100, the flow was streamline and if it was greater than 4000, the flow was turbulent. This was true for gases and liquids in all sizes of pipes ranging from narrow-bore capillary tubes through to large-diameter pipes used for transporting gas, oil and water, often over very long distances. Therefore, if the flow conditions and physical properties of the fluid are known, it is possible to determine the flow characteristic prevailing.

The flow regime between 2100 and 4000 is termed a transition region; here the flow could either be streamline or turbulent. It has also been demonstrated that, if streamline flow conditions prevail and the flow rate is very carefully and gradually increased, then Reynolds numbers up to 30 000 can be achieved before the onset of turbulence is observed. However, in this metastable situation (i.e. a Reynolds number greater than 2000, and flow streamline) any slight disturbance would cause the onset of turbulence and streamline flow conditions would not be achieved again until the Reynolds number was reduced below 2000. As an example, consider two fluids of different viscosities (e.g. water and glycerol) flowing along a pipe of diameter (0.1 m) at the same average velocity of 1.0 m s^{-1}. For water (indicated by a subscript w),

$$Re = \frac{vD\rho_w}{\mu_w}$$
$$= \frac{1.0 \times 0.1 \times 10^3}{10^{-3}}$$
$$= 10^5$$

For glycerol (indicated by a subscript g),

$$Re = \frac{vD\rho_g}{\mu_g}$$
$$= \frac{1.0 \times 0.1 \times 1260}{1.47}$$
$$= 85.7$$

Under these conditions, the water would be exhibiting turbulent flow and the glycerol streamline flow behaviour.

The average v is related to the volumetric flow rate Q by the expression

$$v = \frac{4Q}{\pi D^2}$$

If this is substituted into the expression for Reynolds numbers, it becomes

$$\mathrm{Re} = \frac{4Q\rho}{\pi \mu D}$$

In SI units the volumetric flow rate will be measured in cubic metres per second ($m^3\,s^{-1}$).

3.5.1 Further distinctions between streamline and turbulent flow

3.5.1.1 Streamline flow

In streamline flow (Fig. 3.8(a)), there is no diffusion of material in a

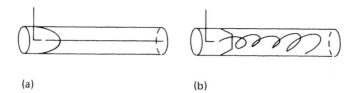

(a)   (b)

Fig. 3.8 — Some characteristics of (a) streamline flow and (b) turbulent flow.

direction at right angles to the bulk flow. The fluid can be imagined as existing in layers, with these layers sliding over each other. The whole fluid acts as a boundary layer; thus, streamline flow represents a true viscous flow condition and is sometimes known as viscous or laminar flow.

If the velocity of the fluid is taken at different distances across the tube, it is found that there is a parabolic distribution. The fluid has its maximum velocity at the centre of the tube and zero velocity at the tube wall. The maximum velocity v_{max} is twice the average velocity. Therefore, there is a wide distribution of velocities, and two particles positioned at different distances from the pipe wall may take considerably different times to pass through the same pipe. This gives rise to a concept of distribution of residence times (see section 3.18). Heat and mass transfer are poor under these conditions. Furthermore, the maximum shear stress is exerted at the wall (wall shear stress), and the shear stress at the centre of the tube is zero.

Streamline flow conditions can be described mathematically by a theoretical relationship between the pressure drop and the volumetric flow rate, as is given by Poiseuilles' equation (see section 3.10). This is the theory behind the determination of viscosity using capillary flow methods.

3.5.1.2 Turbulent flow

When streamline flow conditions exist, there eventually comes a point, as the flow rate increases, where the stream of dye breaks up and good bulk mixing occurs, i.e. the onset of turbulence. Turbulence has been described as an irregular random fluid motion, in a direction transverse to the main flow, which is superimposed on the main flow. This is termed eddying and the process is much more difficult to describe in mathematical terms than is streamline flow. The frictional losses are also much higher than in streamline flow.

Turbulent flow (Fig. 3.8(b)) is a situation in which we find the coexistence of a boundary layer and a well-mixed layer. Near the wall of the tube exists a boundary layer in which the flow is streamline. Away from the wall, this changes and the flow patterns become irregular. As the Reynolds number increases, the flow becomes more turbulent, frictional losses increase and the the boundary layer becomes thinner. Heat and mass transfer processes also become more efficient. The velocity profile across the tube is much flatter and most of the velocity gradient (but not necessarily all) takes place across the boundary layer. In the well mixed region the shear stresses are quite high and the maximum velocity v_{max} is equal to 1.20 times the average velocity. The path of a particle flowing in a tube under turbulent flow conditions will be much more difficult to predict and obviously much more tortuous than in streamline flow. There will be a smaller distribution of residence times in this situation.

Predictions of pressure drops under conditions of turbulent flow are not based on equations that can be derived theoretically but on correlations and equations determined experimentally (see section 3.10). Therefore, in all flow techniques for determining viscosity, it is essential to avoid turbulent flow conditions.

However, convective heat and mass transfer processes can be improved by increasing turbulence, at the expense of increased energy (pumping) costs.

3.6 THE REYNOLDS NUMBER IN AGITATED VESSELS

There is an alternative form for the Reynolds number for equipment for mixing and agitation (Fig. 3.9). For instance, the term for velocity in the conventional Reynolds number has been replaced by tip speed (which is proportional to ND); so the Reynolds number for an agitator of diameter D, rotating at a speed $N(s^{-1})$ on a fluid of viscosity μ and density ρ is equal to

$$\text{Re} = \frac{ND^2\rho}{\mu}$$

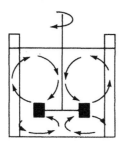

Fig. 3.9 — Turbulence and flow patterns in a stirred tank.

where N is given in revolutions per minute divided by 60 (rev min^{-1}/60). If this value is less than 10, the flow is streamline and, if the value is greater than 10^4, the flow is turbulent. There is a very large transition region in this particular application.

As for pumping applications, this Reynolds number can be used to determine the energy (power) requirements for particular mixing and blending operations. This has been described in more detail by Uhl and Gray (1966) and Earle (1983).

3.7 THE CONTINUITY EQUATION

Consider an incompressible fluid coming to a restriction in a pipe (Fig. 3.10),

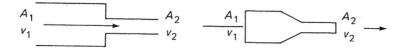

Fig. 3.10 — Fluid flow at a restriction.

where the cross-sectional area for flow decreases from A_1 to A_2. Obviously the volumetric flow rate at point 1 must equal the flow rate at point 2, and therefore

$$v_1 A_1 = v_2 A_2$$

where v_1 and v_2 are the average velocities at points 1 and 2, respectively.

$$v_2 = \frac{v_1 A_1}{A_2}$$

This is known as the continuity equation. Its use is illustrated below.

If fluid flowing along a pipe of diameter 10 cm with an average velocity of $2 \, \text{m s}^{-1}$ comes to a restriction of diameter 5 cm, what is the new fluid velocity?

$$v_2 = \frac{2 \times \pi 10^2/4}{\pi \times 5^2/4}$$

$$= 8 \, \text{m s}^{-1}$$

We shall see later that there is a pressure drop over such a restriction.

3.8 BERNOULLI'S EQUATION

In a pipeline network (Fig. 3.11(a)), it is possible to perform an energy

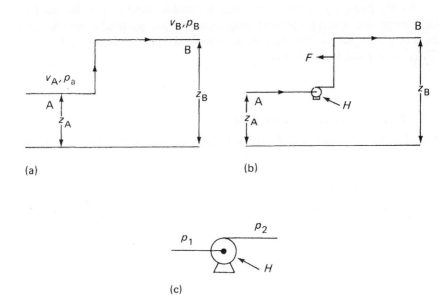

Fig. 3.11 — Application of Bernoulli's equation: (a) on the assumption of no friction and the absence of a pump; (b) including friction and a pump; (c) head developed by pump.

balance between any two points on the network. If we consider points A and B. Bernoulli's principle states that the total energy at A is equal to the total energy at B. First we shall consider the simple case, where there are no frictional losses and no work put in between A and B (i.e. no pump situated between A and B). The total energy at any point is made up of the sum of the potential energy, kinetic energy and pressure energy. One way of expressing these energies is in terms of a head of fluid. Thus,

energy at A = potential energy + kinetic energy + + pressure energy

$$\text{energy at A} = z_A + \frac{v_A^2}{2g} + \frac{p_A}{\rho g}$$

where z_A (m) is the height of point A above an arbitrary datum line, normally ground level, v_A (m s^{-1}) is the velocity of fluid at A and p_A (N m^{-2}) is the presure at A.

Similarly,

$$\text{energy at B} = z_B + \frac{v_B^2}{2g} + \frac{p_B}{\rho g}$$

Equating the energy terms therefore gives

$$z_A + \frac{v_A^2}{2g} + \frac{p_A}{\rho g} = z_B + \frac{v_B^2}{2g} + \frac{p_B}{\rho g}$$

In a normal flow situation, it is necessary to put in energy to transport a fluid to overcome the frictional losses. Therefore, two additional terms are introduced into the equation (Fig. 3.11(b)):

$$\frac{p_A}{\rho g} + \frac{v_A^2}{2g} + z_A + H = \frac{p_B}{\rho g} + \frac{v_B^2}{2g} + z_B + F$$

where F is the frictional loss between A and B expressed in metres of fluid and H is the pressure head developed by the pump.

The main use of this equation is for the calculation of the pressure head developed by a pump. The pressure head developed is given by (Fig. 3.11(c))

$$H = \frac{p_2 - p_1}{\rho g}$$

where p_2 (N m^{-2}) is the discharge presure and p_1 (N m^{-2}) is the suction pressure.

The theoretical power transmitted to the fluid is obtained from the expression

$$W = Hgm$$

where m (kg s^{-1}) is the mass flow rate (see section 1.19.1 and Fig. 1.8).

3.9 PRESSURE DROP AS A FUNCTION OF SHEAR STRESS AT A PIPE WALL

Consider the flow of a fluid flowing with a constant average velocity v through a cylindrical pipe of lengh L and internal diameter D. A pressure drop Δp exists across the tube because of the frictional (or viscous forces). These frictional forces result in a shear stress R_W over the inside surface of the pipe.

A force balance over the pipe (Fig. 3.12) when there is no slip at the walls

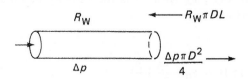

Fig. 3.12 — Frictional losses in a straight pipe (force balance).

gives

$$\frac{\Delta p \pi D^2}{4} = R_w \pi D L$$

Therefore,

$$\Delta p = \frac{4 R_w L}{D}$$

Thus the pressure drop can be expressed in terms of the wall shear stress and the dimensions of the pipe.

3.10 FRICTIONAL LOSSES

3.10.1 Frictional losses in straight pipes

Application of Bernoulli's equation between points A and B on a straight pipe will show that

$$\frac{p_A - p_B}{\rho g} = F$$

Therefore the frictional loss F is related to the pressure drop over the pipe. Under *streamline* flow conditions, $p_1 - p_2$ or Δp is related to the flow rate by the Poiseuille equation; Poiseuille showed mathematically that the pressure drop was related to the volumetric flow rate Q by the following equation:

$$Q = \frac{\Delta p\, \pi D^4}{128 L \mu}$$

Therefore the frictional loss

$$F = \frac{\Delta p}{\rho g}$$
$$= \frac{128 L \mu Q}{\pi D^4 \rho g} \text{ (m)}$$

The situation is not so straightforward under turbulent flow conditions. To overcome the problem a dimensionless basic friction factor ϕ has been defined (Holland, 1973). This basic friction factor is half the fanning friction factor which is used by some researchers. The basic friction factor ϕ is given by

$$\phi = \frac{R_w}{\rho v^2}$$

Experimentally it has been found that the basic friction factor ϕ depends upon the extent of turbulence and the roughness of the pipe. The results have been presented graphically in terms of the basic friction factor plotted against the Reynolds number (Figs. 3.13 and 3.14).

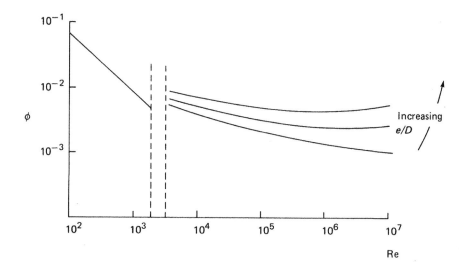

Fig. 3.13 — Friction factor chart (simplified diagram).

$$\frac{\Delta p}{\rho g} = \frac{4R_{\mathrm{w}}L}{D}\frac{1}{\rho g}$$

Noting that $R_{\mathrm{w}} = \phi \rho v^2$, we find that

$$\frac{\Delta p}{\rho g} = 4\phi \rho v^2 \frac{L}{D}\frac{1}{\rho g} = F$$

This can be conveniently arranged as

$$\frac{\Delta p}{\rho g} = 8\phi \frac{L}{D}\frac{v^2}{2g} = F$$

The friction factor chart shows that in the streamline regime (Re < 2000) there is only one line. Under these conditions, it can be shown that $\phi = 8/\mathrm{Re}$.

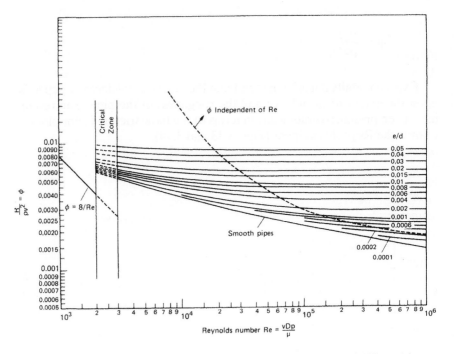

Fig. 3.14 — Friction chart. (From Coulson and Richardson (1977), with permission.)

When the flow is turbulent, there are a series of lines for increasing pipe roughness. Obviously, frictional losses will increase as the pipe roughness increases. The roughness is measured by the dimensionless term e/D where e is the average height of protrusions.

For a smooth pipe, e would have a value of zero. Values of e for various types of pipe are given in Table 3.1.

If charts are not available, the relationship between the basic friction factor and Reynolds number for flow in smooth circular tubes under turbulent flow conditions is given by Blasius' equation. This states that

$$\phi = 0.0396 \, \text{Re}^{-0.25}$$

To obtain the frictional losses in a pipe when the flow is streamline or turbulent, the following procedure can be adopted.

(1) Determine the Reynolds number for the prevailing flow conditions.
(2) Determine the roughness factor e/D.
(3) Read the value of the friction factor from the chart.
(4) Substitute the value in the equation

$$F = 8\phi \frac{L}{D} \frac{v^2}{2g}$$

Consider, as an example, water flowing along a smooth pipe of length 20 m and diameter 2.54×10^{-2} m at different average velocities. Table 3.2 shows the Reynolds numbers, basic friction factors and frictional losses at different velocities.

It can be seen that the frictional losses increase as the flow becomes more turbulent; the relationship is non-linear, when the flow is turbulent. If the same volumetric flow rate is required, as in example B (Table 3.2), using a pipe of twice the diameter, i.e. 5.08 cm, the average velocity would equal 0.5 m s^{-1} and the frictional loss would be only 0.06 m as opposed to 1.61 m. Pumping costs would be considerably reduced, but the capital cost for the pipe would be higher. This gives rise to the concept of an economic pipe diameter which needs to be evaluated in terms of capital costs and current costs; it will therefore be different from time to time and from place to place.

3.10.2 Frictional losses in other fittings

Frictional losses F occur in other fittings, such as corners, bends, exits, restrictions and valves. One can account for these in terms of a head loss by the expression

$$F = K \frac{v^2}{2g}$$

where K is a constant for the particular fitting.

This is a useful expression because the frictional lossess are expressed in terms of number of velocity heads $v^2/2g$.

An alternative system is to express the losses in terms of equivalent pipe diameters. Values for different fittings are given in Table 3.3.

Table 3.1 — Roughness factors e for different pipes

Pipe material	Roughness factor e (m)
Commercial steel or wrought iron	4.5×10^{-5}
Drawn tubing	1.5×10^{-6}
Cast iron	2.6×10^{-4}

Other values have been given by Holland (1973) and Earle (1983).

Table 3.2 — Variation in frictional losses with changes in fluid velocity for water.

Example	Average velocity v (m s^{-1})	Reynolds number Re	Basic friction factor ϕ	Frictional loss F (m)
A	1	2.54×10^4	3.0×10^{-3}	0.48
B	2	5.08×10^4	2.5×10^{-3}	1.61
C	5	1.27×10^5	2.1×10^{-3}	8.70
D	10	2.54×10^5	1.8×10^{-3}	28.90

Viscosity of water, 10^{-3} N s m^{-2}; density of water, 10^3 kg m^{-3}.

Thus, for a fluid flowing round a 90° sharp bend of diameter 5 cm with a velocity of 2 m s^{-1}, the frictional losses could be expressed as $1.2 \times 2^2/9.81$ m, or as being equivalent to that in a straight pipe of 60 times the pipe diameter, i.e. 3 m long.

3.10.3 Total system losses

Once the losses for the individual components have been determined, the total system can be calculated by summing the individual losses.

For example, in the situation in Fig. 3.15(a), the total losses would be made up of those due to the straight sections plus the losses for four bends and three valves.

When total system losses are determined at different flow rates, the resulting graph is known as the system characteristic (Fig. 3.15(b)).

3.11 RELATIVE MOTION BETWEEN A FLUID AND A SINGLE PARTICLE

Many unit operations deal with mixtures of liquids and solids; often the particle size of the solid is quite small or the material is in powdered or granular form. Some examples are shown in Table 3.4. The analysis of these

Sec. 3.11] Relative motion between a fluid and a single particle

Table 3.3 — Frictional losses related to various fittings

Fitting	K	Pipe diameter
Globe valve (fully open)	1.2–6.0	60–300
Gate valve (fully open)	0.15	7
Gate valve (half open)	4	200
Bend 90° standard radius	0.6–0.8	30–40
Bend 90° sharp	1.2	60
Entry from leg of T-piece	1.2	60
Entrance from large tank to pipe (sharp)	0.5	
Entrance from large tank to pipe (rounded)	0.05	

Taken from the data of Earle (1983) and Coulson and Richardson (1977).

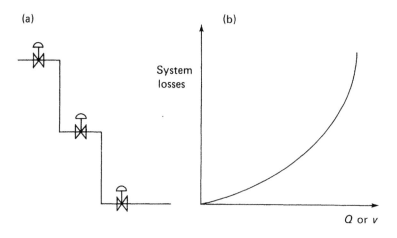

Fig. 3.15 — (a) Evaluation of total system losses; (b) the relationship between total system losses and volumetric flow rate.

Table 3.4 — Unit operations involving solids and liquids

Drying operations	Through-draught drying, fluidized-bed drying, pneumatic drying
Freezing processes	Fluidized-bed freezing
Separation processes	Centrifugation, filtration, cyclone separation
Adsorption processes	Ion exchange processes, bleaching, deodorizing
Conveying	Pumping slurries and suspensions, pneumatic conveying of solids

problems can be simplified initially by considering the forces acting on a single spherical particle.

The forces acting on a spherical solid particle of diameter d and density ρ_s falling through a less dense fluid of density ρ are as follows: acting downwards will be a force due to the weight and given by $\pi d^3 \rho_s g/6$; acting upwards will be a force due to the upthrust, and given by $\pi d^3 \rho g/6$; there is also a force due to the viscous drag, i.e. the fluid exerts a retarding force on the particle.

For a spherical particle, this retarding force F_R is given by the expression

$$F_R = C_d S_p \frac{\rho v^2}{2}$$

where S_p is the projected area of the particle in a plane perpendicular to the direction of flow (for a sphere this equals $\pi d^2/4$) and C_d is a dimensionless drag coefficient. Therefore,

$$F_R = C_d \frac{\pi d^2}{4} \frac{\rho v^2}{2}$$

For a particle moving in a fluid when the upward forces equal the downward forces, the particle is either at rest or moving at its terminal velocity v_t.

Under these conditions,

$$C_d \frac{\pi d^2}{4} \frac{\rho v_t^2}{2} + \frac{\pi d^3}{6} \rho g = \frac{\pi d^3 \rho_s g}{6}$$

This simplifies to

$$v_t = \sqrt{\left[\frac{4d(\rho_s - \rho)g}{3C_d}\right]} \qquad (3.1)$$

The drag coefficient is a function of the particle Reynolds number, defined as

$$(Re)_p = \frac{\rho v_t d}{\mu}$$

where ρ and μ are the density and viscosity of the fluid, respectively. If the particle Reynolds number is less than 0.2, then the flow is streamline and the drag coefficient is given by

$$C_d = \frac{24}{(Re)_p} \qquad (3.2)$$

When C_d is substituted into equation (3.1), the terminal velocity is given by

$$v_t = \frac{d^2(\rho_s - \rho)g}{18\mu} \qquad (3.3)$$

This is known as Stokes' equation. It can be used for estimating terminal velocities or evaluating dynamic viscosities of fluids (see section 4.11.3). When the Reynolds number is between 0.2 and 500, it has been shown that

$$C_d = \frac{24}{(Re)_p}[1 + 0.15(Re)_p^{0.687}] \qquad (3.4)$$

When the Reynolds number is between 500 and 200 000,

$$C_d = 0.44 \qquad (3.5)$$

In most applications involving flow through particles, the particle Reynolds number is lower than 500. Holland (1973) also discusses the situation for non-spherical particles, and between a high concentration of particles, where the particles may well be hindered by other particles and also the wall of the vessel.

3.12 FLUID FLOW THROUGH PACKED AND FLUIDIZED BEDS

When a fluid flows through a packed bed, the material comprising the bed offers a resistance to flow and a pressure drop results. As the fluid enters the bed, the cross-sectional area available for flow decreases and the velocity increases. It is important to distinguish between the velocity v based on the empty column and the actual velocity v_b in the packed bed. If the total volume occupied by the packed bed is V and the porosity is ε, then the volume occupied by the material is $V(1-\varepsilon)$ and the void volume is $V\varepsilon$. If the cross-sectional area available for flow in the empty column is A, then $v = Q/A$ and $v_b = Q/A\varepsilon$. If the surface area per unit volume for the particles is S_0, then the total surface area occupied by the particles is $S_0V(1-\varepsilon)$ and the surface-area-to-volume ratio for the entire bed is $S_0(1-\varepsilon)$. For spherical particles, $S_0 = 6/d$, and, for non-spherical particles with an average particle diameter d, $S_0 = 6/d\psi$, where ψ is a correction factor which takes into account the shape of the particles. (Note that, for spherical particles, $\psi = 1$.)

The surface area S_0 per unit volume and voidage volumes for spheres, cubes, prisms, cylinders, plates, discs and saddles of different sizes have been given by Coulson and Richardson (1978). Values of the shape factor ψ for some non-spherical particles have been tabulated by Perry and Chilton (1973).

The flow of fluid through a packed bed can be analysed in an analogous fashion to the flow through a pipeline, in terms of modified Reynolds numbers and friction factors.

The Reynolds $(Re)_b$ number for a flow in a bed has been defined as

$$(Re)_b \frac{\rho v}{\mu(1-\varepsilon)S_0} \qquad (3.6)$$

A modified friction factor j_f is proposed, where

$$j_f = \frac{\Delta p}{L} \frac{\varepsilon^3}{(1-\varepsilon)S_0 \rho v^2} \qquad (3.7)$$

Under streamline flow conditions, i.e. where $(Re)_b < 2.0$,

$$j_f = \frac{5}{(Re)_b} \qquad (3.8)$$

When these equations are combined, it can be shown that

$$v = \frac{\varepsilon^2 \Delta p}{5 S_0^2 \mu (1-\varepsilon)^2 L} \qquad (3.9)$$

which relates the pressure drop to the flow velocity (based on the empty column).

For spherical particles, where $S_0 = 6/d$, the equation becomes

$$\Delta p = \frac{180 v \mu (1-\varepsilon)^2 L}{\varepsilon^3 d^2} \qquad (3.10)$$

This is known as the Carmen-Kozeny equation; it shows that the factors affecting the pressure drop are the flow velocity, length of bed, fluid viscosity and particle diameter. For example, if the particle size is halved, the pressure drop will increase by four times.

The more generalized form of the equation is

$$\Delta p = K_c \mu L \frac{(1-\varepsilon)^2 S_0^2}{\varepsilon^3} v \qquad (3.11)$$

where K_c is a constant which depends upon the particle shape, porosity and particle size range and lies between 3.5 and 5.5, with its most common value of 5 (Holland, 1973).

When the flow becomes turbulent, the equation proposed by Carmen for the complete range of Reynolds numbers can be used:

Sec. 3.12] Fluid flow through packed and fluidized beds

$$j_f = \frac{5}{(Re)_b} + \frac{0.4}{(Re)_b^{0.1}} \quad (3.12)$$

The friction factor value is then substituted into equation (3.7) to evaluate the pressure drop.

A further increase in the velocity will cause the smallest of the solid particles to be entrained. If the velocity is sufficiently increased, all the particles will be entrained and the particles will be conveyed by the fluid; this is termed pneumatic or hydraulic conveying and is used for transporting particulate matter, slurries and suspensions.

For a spherical particle of diameter d to be entrained, the fluid velocity must exceed the terminal velocity given by equation (3.1). The minimum velocity v_e required to entrain the particles is given by

$$v_e = \sqrt{\left[\frac{4d(\rho_s - \rho)}{3C_d \rho}\right]} \quad (3.13)$$

When the pressure drop across the bed is plotted against the velocity using log–log coordinates, the relationship shown in Fig. 3.16 is found. As

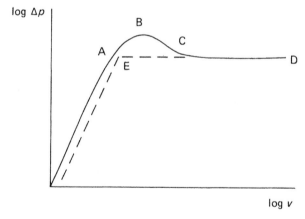

Fig. 3.16 — The relationship between pressure drop and fluid velocity during fluidization.

the velocity increases, the pressure drop increases until a point is reached where the bed becomes fluidized. The velocity corresponding to the fluidization point is determined from equation (3.15) or equation (3.16).

As the velocity is further increased, the pressure drop increases slightly and then falls to a value which is almost independent of the velocity. When the velocity is reduced below the fluidization velocity, the corresponding pressure drop in the re-formed bed is slightly lower than that in the original fixed bed.

There are many examples of processing situations using packed and fluidized beds, particularly in filtration, fluidized-bed freezing and drying and through-draught drying. In all these flow processes the size, shape and density of the particles, as well as the properties of the fluid such as density and velocity, are extremely important.

3.12.1 Fluidization

As the fluid velocity through the bed increases, a point is reached where the upward forces exerted by the fluid on the particles become equal to the apparent weight of particles in the fluid stream, i.e. the weight minus the upthrust. This point marks the start of fluidization, where the particles appear to float in the fluid.

A force balance over the bed indicates that

$$\text{total apparent weight} = Lg(1-\varepsilon)(\rho_s - \rho)$$
$$= \Delta p \tag{3.14}$$

Combining equations 3.11 and 3.14 shows that the fluidization velocity v_F is given by the following:

$$v_F = \frac{(\rho_s - \rho)g}{K_c \mu} \frac{\varepsilon^3}{(1-\varepsilon)S_0^2} \tag{3.15}$$

For spherical particles, this is

$$v_F = \frac{(\rho_s - \rho)g}{\mu} \frac{d^2 \varepsilon^3}{180(1-\varepsilon)} \tag{3.16}$$

where ε is the porosity at the point of fluidization.

Fluidized-bed heat transfer operations such as freezing and drying result in improved heat transfer rates compared with fixed-bed applications because of the larger solid surface areas exposed to the fluid and the improved heat transfer rates at higher velocities (see section 9.11).

Packed-bed fluid flow situations are found in many chromatographic experimental techniques. When particle sizes are small, large pressures are required to achieve reasonable flow rates.

Frictional losses due to pressure drops in these systems may form a substantial part of the total system losses in pumping applications.

3.13 FLUID FLOW MEASUREMENT

3.13.1 Variable-head measurement

When a fluid comes to a restriction in a pipe, there is an increase in kinetic energy (Fig. 3.17). Application of Bernoulli's equation over the restriction shows that there is a reduction in the pressure.

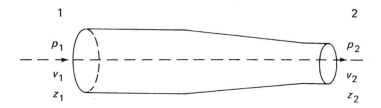

Fig. 3.17 — Fluid flow at a restriction.

$$\frac{p_1}{\rho g} + z_1 + \frac{v_1^2}{2g} = \frac{p_2}{\rho g} + z_2 + \frac{v_2^2}{2g}$$

$z_1 = z_2$ (same level)

Therefore,

$$\frac{p_1 - p_2}{\rho g} = \frac{v_2^2 - v_1^2}{2g}$$

Often $v_2^2 \gg v_1^2$ and v_2 is ignored; then,

$$\frac{p_1 - p_2}{\rho g} = \frac{v_2^2}{2g}$$

i.e. $\quad v_2 = \sqrt{\left[\frac{2(p_1 - p_2)}{\rho}\right]}$

Therefore there is a pressure drop across the restriction that will depend upon v_2.

The volumetric flow rate Q at point 2 is given by

$$Q = v_2 A_2$$

$$= A_2 \sqrt{\left[\frac{2(p_1 - p_2)}{\rho}\right]}$$

The major types of variable-head flow-measuring device are the Venturi meter, orifice plate, flow nozzle and rotameter. These are illustrated in Fig. 3.18.

In all cases, losses due to friction occur. These are accounterd for by a term known as the coefficient C_D of discharge. The highest losses are incurred with the orifice plate ($C_D = 0.60$) and the lowest with the Venturi meter ($C_D = 0.98$). The coefficient of discharge is a measure of the energy which is recovered. Variations in the coefficients of discharge at different flow conditions have been discussed by Perry and Chilton (1973).

The expression for the volumetric flow rate now becomes

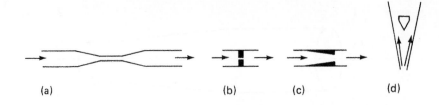

Fig. 3.18 — Representation of (a) a Venturi meter, (b) an orifice plate, (c) a flow nozzle and (d) a rotameter.

$$Q = C_D A_2 \sqrt{\frac{[2(p_1 - p_2)]}{\rho}}$$

If the differential pressure $p_1 - p_2$ is measured using a manometer (section 3.2) and the recorded height is h, then

$$Q = A_2 C_D \sqrt{\left(2gh \frac{\rho_m - \rho}{\rho}\right)}$$

If the velocity v_1 is not ignored from the expression, it can be shown that

$$Q = \frac{A_1 A_2 C_D}{\sqrt{(A_1^2 - A_2^2)}} \sqrt{\left(2gh \frac{\rho_m - \rho}{\rho}\right)}$$

3.13.1.1 Pitot tube

A pitot tube is a useful device for measuring the velocity of a fluid at a point location over the cross-sectional flow area. Thus, it can be used to determine the velocity profile across a tube or at various positions in a hot-air drier or blast freezer.

It consists of two concentric thin tubes which are sealed at one end to form an outer annulus. Small holes are made in outer annulus and the tube is positioned in the flow path parallel to the flow, with the tip facing the oncoming fluid (Fig. 3.19). The centre tube transmits the combined local pressure head and velocity head whilst the outer tube transmits only the pressure head. Therefore the differential pressure gives measure a of the velocity head and kinetic energy of the fluid.

Applications of Bernoulli's equation between points 1 and 2 shows that

$$\frac{p_1 - p_2}{\rho g} = \frac{v_2^2}{2g}$$

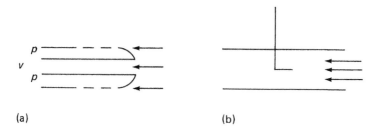

Fig. 3.19 — The construction and positioning of a pitot tube: p, pressure head; v, pressure and velocity head.

When the differential pressure h is measured by a manometer,

$$v = \sqrt{\left(2gh\frac{\rho_m - \rho}{\rho}\right)}$$

Pitot tubes are simple in construction and operation but do suffer from a number of disadvantages. They need very careful alignment and often give a very low differential pressure; they are subject to fouling and only measure the local velocity and not the average velocity. To ascertain the volumetric flow rate, we would need to know the velocity profile across the tube or flow channel and its cross-sectional area. Pitot tubes have been discussed in nore detail by Jones (1974a).

Flow rates in open channels are measured using flumes or weirs (Holland, 1973).

3.13.2 Variable-area meters

In a variable-area meter, often called a rotameter, a float is placed in a precision-made tapered tube and the liquid flows upwards through the tube, between the narrow gap between the edge of the float and the wall of the tube. At a constant flow rate the float will come to equilibrium at a fixed position, when the upward force of the liquid on the float equals the weight of the float. The pressure drop over the restriction is constant but the area of the gap increases as the float rises in the tube (Fig. 3.18(d)). When the flow rate increases, the float rises in the tube and adopts a new equilibrium position. The position of the float is noted by reference to a calibrated scale.

Variable-area meters are extremely useful for measuring flow rates of clean fluids, such as gases and clear liquids. For flow in pipes or for non-transparent liquids, the float and the pipe are made from a non-magnetic material but an extension of the float is made magnetic. The position of the float is followed by another magnet, placed outside the pipe, which moves in accordance with the float and transmits its movement to a pointer (Jones, 1974a).

3.14 FLUID TRANSPORTATION AND PUMPING

When transporting fluids, it is necessary to put in energy to overcome the frictional losses. In selecting a pump for a particular application, many factors have to be taken into account. The most important are summarized as follows

(1) The pressure or head required.
(2) The volumetric flow rate.
(3) The physical properties of the fluid, i.e. density, viscosity, suspended solids, abrasive particles and vapour pressure.
(4) Intermittent or continuous service.
(5) Hygienic considerations.
(6) Susceptibility of product to shear damage.

Pumps are categorized as being of either a *centrifugal* or a *positive-displacement* type. The main distinctions between these are as follows.

For a positive-displacement pump the volume of fluid delivered is largely independent of the discharge pressure. The two major types are the piston pump and the rotary pump. In both cases the volumetric flow rate can be changed by altering the speed of the pump. In contrast with this, a centrifugal pump is one in which the volumetric flow rate is directly affected by the discharge conditions; the flow rate falls as the discharge pressure increases. Some pumps fall in between these broad categories. Other aspects of these pumps will now be discussed.

3.14.1 Positive-displacement pumps

Piston pumps (Fig. 3.20(a)) and rotary pumps (Figs. 3.20(b) and 3.20(c)) both work by forcibly transferring a liquid. With piston pumps, this is achieved by changing the internal volume of the pump. Valves are required on both the suction and the discharge side of the pump. With rotary pumps the fluid is forcibly transferred by rotating gears, lobes, worms or screws; they do not require valves to operate. They are extremely useful for viscous fluids, sometimes containing suspended solids and for generating high pressures. They tend to be fairly expensive and require more frequent servicing, particularly piston pumps. They usually work less efficiently with less viscous fluids, where slip or leakage occurs. All positive-displacement pumps can be damaged by pumping against a closed discharge; consequently, some form of pressure relief line is required on the outlet. They are often used for metering purposes. Piston pumps can be of either a single-piston or a multipiston type. A pulsating flow occurs with a single piston and may need some form of damping. As the number of pistons increases the flow rate becomes more constant. Piston pumps require regular maintenance and are not suitable for liquids with suspended solids or abrasive particles. They are very suitable for generating high pressures.

Examples of rotary pumps are gear pumps or worm pumps. They can be used for handling particulate matter under conditions of low shear when operating at low speeds. They deliver a steady flow rate, although they are

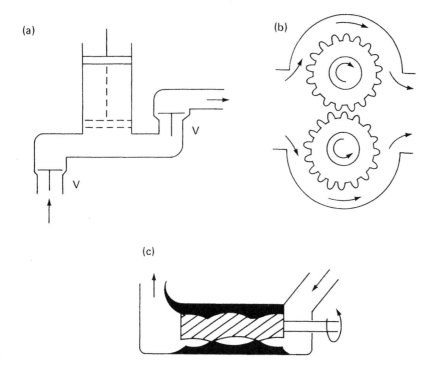

Fig. 3.20 — Some positive-displacement pumps: (a) a piston pump showing valves V; (b) a rotary gear pump; (c) a mono or worm pump.

not always truly positive in nature as the flow rate decreases very slightly as the outlet pressure increases. The characteristics of these pumps are usually represented by plots of volumetric flow rate against speed.

3.14.2 Centrifugal pumps

These are the most widely used type of pump (Fig. 3.21(a)), particularly when high flow rates and fairly low pressures are required. They are suitable for fluids of low viscosity and their efficiency falls when the viscosity becomes too high.

The volumetric flow rate is dependent upon the pressure (head) developed, and the relationship between the two is given by the pump characteristic (Fig. 3.21(b)).

The maximum head developed occurs when there is no flow rate, i.e. when pumping against a closed discharge. This will cause no damage to the pump. As the flow rate increases, the pressure developed decreases. Note that pump characteristics are obtained at constant speed, and most centrifugal pumps are driven by constant-speed motors. Pump characteristics are useful for selecting a pump of appropriate size for an application.

The flow rate generated by a centrifugal pump depends on the pump

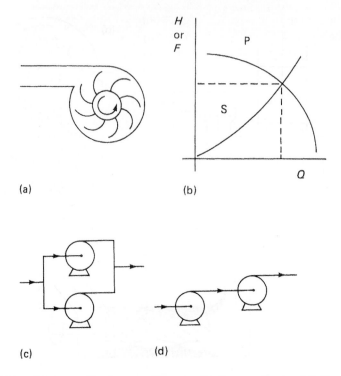

Fig. 3.21 — (a) A centrifugal pump; (b) a centrifugal pump characteristic P and system characteristic S; (c) centrifugal pumps in parallel; (d) centrifugal pumps in series.

characteristics and the system losses and is determined from the point where the two curves intersect (Fig. 3.22(b)). The system characteristics can be altered by a valve in the line; opening or closing the valve will affect the system losses and result in a change in the flow rate. When centrifugal pumps are used in pasteurizers and other continuous heat exchangers, a flow controller is fitted to maintain a constant flow rate (see Fig. 3.23 and section 3.14.4).

The layout of a centrifugal pump is shown in Fig. 3.21. Fluid enters the eye of an impeller which rotates at a fixed speed. The impeller is designed to convert kinetic energy to pressure energy. As the fluid is forced to the periphery of the blades, the cross-sectional area available for flow increases and the velocity falls, resulting in a decrease in kinetic energy. This is converted to pressure energy (see section 3.8). The efficiency of conversion of kinetic energy to pressure energy will depend on the shape and size of the impeller and the number of blades. Centrifugal pumps are cheaper than positive pumps, simpler in construction and very reliable. They can also be connected in series (Fig. 3.21(d)) or parallel (Fig. 3.21(c)).

When two similar pumps are connected in series, they develop a head approximately equal to the sum of the individual heads, but the flow rate would be similar to that from a single pump. When connected in parallel, the flow rate would be double that from a single pump, but the head developed would be the same. Determination of the flow rates and heads for combinations of pumps with different characteristics has been discussed by Holland (1973). By connecting pumps in series, reasonably high pressures can be generated using these pumps.

Although these pumps usually operate at constant speeds, the flow rate, head developed and power requirements are all affected by the speed N:

$$\text{flow rate} \propto N$$
$$\text{head} \propto N^2$$
$$\text{power} \propto N^3$$

3.14.3 Other characteristics of pumps

Most positive pumps are self-priming and will fill with liquid when they are switched on. However, centrifugal pumps are not and need to be filled with fluid prior to operation. For these reasons, it is advisable to install pumps at ground level or below the level of the fluid being pumped (Fig. 3.22). Special provision may be required to prime pumps when forcing liquid from below the level at the pump.

It is also desirable to avoid a phenonenon known as cavitation. Cavitation occurs when bubbles of gas or vapour, which are present in the liquid, collapse in the high-pressure region of the pump. This will cause wearing and mechanical damage to the blades. Cavitation is often recognized by rattling, vibration and excessive noise from the pump.

Bubbles occur for two major reasons. Sometimes the pressure of the liquid may be below atmospheric pressure and air may be sucked in through leaking joints. Also the saturated vapour pressure of the liquid may exceed the local pressure in the pipe and the liquid will boil, thereby forming bubbles of vapour. This is most significant with hot fluid and organic solvents. Correct positioning of pumps and attention to these details should help to reduce the incidence of cavitation. The presence of air with centrifugal pumps will not only reduce the flow rate but also aerate the liquid.

The factors affecting the pressure at the inlet of the pump will now be discussed.

Application of Bernoulli's equation (Fig, 3.22) between the surface 1 of the liquid in the tank and the inlet 2 to the pump gives

$$\frac{p_1}{\rho g} + z_1 + \frac{v_1^2}{2g} + H = \frac{p_2}{\rho g} + z_2 + \frac{v_2^2}{2g} + F$$

It can be assumed that $v_1 = 0$, $z = 0$ and $H = 0$ and therefore

$$\frac{p_2}{\rho g} = \frac{p_1}{\rho g} + z_1 - \frac{v_2^2}{2g} - F$$

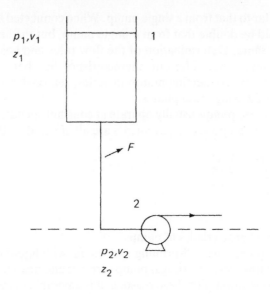

Fig. 3.22 — Factors affecting the pressure at the inlet of a pump.

Care has to be taken to ensure that the local pressure p_2 at the inlet of the pump, or anywhere along the suction side, does not fall below the saturated vapur pressure of the liquid, at the operating temperature.

This may happen if the following occur.

(1) The feed tank is at a reduced pressure.
(2) z_1 is negative (i.e. the pump is positioned above the liquid level).
(3) The flow velocity is very high.
(4) The frictional losses are high.

If cativation occurs, these four points will need to be investigated.

3.14.4 Flow control

With positive-displacement pumps, flow control is relatively simple. With the pump operating at constant speed, and with a fluid of constant viscosity, the flow rate remains virtually constant. The flow rate is changed by changing the speed of the pump.

With centrifugal pumps, some form of mechanical flow controller is required to account for any variations in upstream or downstream pressures. It is normally situated on the outlet side of the centrifugal pump.

One simple form of flow controller is illustrtated in Fig. 3.23. The moving part of the controller is a float mechanism, consisting of a plunger, stem and control button. The liquid flows upwards, through the outer annulus, through the side ports and over the control button. If the flow rate increases, the float rises and the plunger reduces the size of the side

Sec. 3.14] Fluid transportation and pumping 101

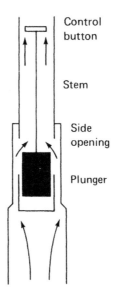

Fig. 3.23 — Flow controller for use with a centrifugal pump, e.g. as used in high-temperature short-time pasteurization.

openings, thereby restricting the flow. The system is self-regulating, and a constant flow rate is maintained. The value of the flow rate can be adjusted by adjusting the control button. This is a useful facility when much higher velocities are required, particularly for cleaning equipment.

3.14.5 Special types of pump

A *homogenizer* is a special type of pump. Many high-pressure homogenizers are of the piston pump type. The liquid is pressurized and passed through one or two very narrow openings, known as the homogenization valves. They are commonly used for emulsions to stabilize the fat globules. Homogenization pressures will have a significant effect on the viscosity of the product. All long-life (UHT) products containing fat will require homogenization. They suffer all the drawbacks of piston pumps and are extremely difficult to clean in place.

Peristaltic pumps work along similar principles to rotary pumps. The liquid is forced along flexible tubing by a squeezing action. They are extremely popular in chromotography systems and for dosage pumps in fermentation and similar systems and are extremely useful for delivering relatively low volumes of liquids in pilot plant equipment.

Extrusion cookers have often been described as inefficiant pumps. Extrusion cookers are used for a wide variety of applications, e.g. mixing, heating and texturizing of solid and semifluid materials. The feed material is transported through a barrel by a specially designed rotor system at high pressures. During transportation, much of the mechanical energy is converted to heat, and further heating or cooling can be induced by jacketed systems. At the end of the barrel the pressure is released as the product is extruded through holes (dies), which causes it to expand and cool as water is flashed off, thereby giving the desired final shape. Products are thus heated at high temperatures for short times, and the final product usually has a low bulk density (Jowitt, 1984).

3.15 VACUUM OPERATIONS

In vacuum evaporation and freeze-drying operations, provision needs to be made for removing large volumes of water vapour and other non-condensible gases. Usually a combination of a vacuum pump and a water-cooled or refrigerated condenser is used. If only a vacuum pump is used, large units would be required to remove the large volumes of vapour removed, particularly at low pressures (see section 11.4).

Mechanical vacuum pumps with rubber seals should not be used with organic solvents, as they ruin the seals. If they have to be used, low-temperature traps, e.g. liquid nitrogen or solid carbon dioxide, will be required to condense these.

An alternative device, often used in food-processing plant, is a steam ejector. Use is made of the energy in high-pressure steam to entrain the vapour from the low-pressure sections.

Vacuum pumps and steam ejectors have been discussed in more detail by Perry and Chilton (1973).

3.16 ASEPTIC OPERATIONS

In the continuous production of commercially sterile products, the product is heated to a high temperature and held for a short time in a holding tube, to achieve commercial sterility (e.g. 145 °C for 4 s).

It is extremely important that the sterilized product picks up no contaminating organisms during cooling and subsequent processing. All the equipment downstream of the holding tube requires sterilization, prior to processing, and any pumps or homogenizers downstream need to be operated under aseptic conditions. Some form of steam seal is generally provided to prevent organisms from entering the product, thereby making it possible to operate these pumps under sterile conditions.

3.17 AVERAGE RESIDENCE TIME

In any processing operations it is important to be able to control the processing time of the fluid. In batch processes this is a simple matter, but in continuous processing operations it is less straight-forward. For example, in the high-temperature short-time continuous pasteurization of milk, the milk requires to be held at a temperature of 72 °C for at least 15 s. It is highly desirable to avoid keeping the milk at this temperature for too long, as this will lead to over-processing and a reduction in the product quality. It is also essential that all the milk is held at 72°C for 15 s, to ensure that the pathogenic and spoilage organisms have been inactivated.

In continuous operations, the required residence times are achieved by passing the heated fluid through a holding tube. It is normal to define an average residence time, which is based on the average velocity and tube length, L.

$$\text{average residence time } t_{av} = \frac{L(\text{m})}{v(\text{ms}^{-1})} \text{ (s)}$$

An alternative way of expressing the average residence time is as follows:

$$t_{av} = \frac{\text{volume of tube}}{\text{volumetric flow rate}}$$

$$= \frac{V(\text{m}^3)}{Q(\text{m}^3 \text{ s}^{-1})} \text{ (s)}$$

If the internal diameter of the tube is equal to D and the average velocity is $4Q/\pi D^2$, the appropriate expression can be derived.

These equations are useful for determining the length or volume of a holding tube required to achieve any desired average residence time for any particular flow rate.

The average residence time will give the time taken for 50% of the material entering the holding tube to pass through. It is effectively the time taken if there was no velocity distribution across the tube; this situation is known as plug or piston flow (Fig. 1.4).

3.18 DISTRIBUTION OF RESIDENCE TIMES

In streamline and turbulent flow there is a velocity profile across the tube; therefore, some elements of the fluid are moving faster than the average velocity and some move more slowly. This gives rise to a distribution of residence times. It is important to know what the distribution of residence time is in the thermal processing of fluids, in order to evaluate sterilization effects or changes in chemical composition.

There are several methods for determining the distribution of residence times, but only one will be mentioned here. This method is illustrated for the following three situations in a pipeline (Fig. 3.24): plug flow; streamline flow; turbulent flow. In all these cases the technique is the same.

The flow rate is adjusted to the desired value, and a pulse of tracer material (dye, salt, etc.) is injected into the tube inlet at time zero. Samples are taken from the outlet at regular time intervals and the concentration of the tracer determined. The residence time distribution function is presented as a graph of tracer concentration against time (Fig. 3.24).

In plug flow, all the tracer emerges from the tube at a time given by the average residence time V/Q. There is no distribution of residence times.

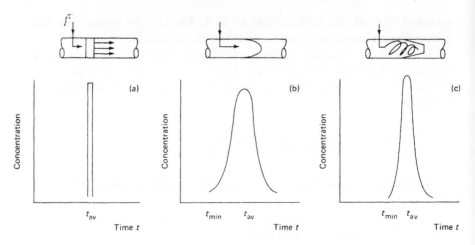

Fig. 3.24 — Distribution of residence times in (a) plug flow, (b) streamline flow and (c) turbulent flow.

Plug flow is an ideal flow situation and is not usually found in practice, except with some highly viscous pseudoplastic fluids (Holland, 1973).

In streamline flow, there is a parabolic velocity distribution, which leads to a fairly wide scatter of residence times. The maximum velocity in the tube is twice the average velocity. Elements of the fluid travelling at the maximum velocity will take the minimum time to pass through the tube.

$$t_{min} = \frac{t_{av}}{2}$$

The minimum residence time is the time corresponding to the emergence of the first amount of traces (Fig. 3.24).

Also the distribution of residence times is quite wide.

In turbulent flow the velocity profile is much flatter, and the maximum velocity and minimum residence time are given by

$$v_{max} = 1.20 v_{av}$$
$$t_{min} = 0.83 t_{av}$$

Compared with streamline flow the distribution of residence times is narrower and the resulting processing would be more uniform.

Once the volume of the holding tube has been ascertained, additional turbulence can be induced by selecting a long tube of small diameter rather than a short tube of large diameter.

In both situations the average residence time can be calculated from these equations. It corresponds to the time required for 50% of the injected material to leave the tube, and in most pipeline situations the average residence time corresponds closely to the time of maximum concentration in the residence time distribution curve.

In practice the residence time distribution curve tends to have a long tail and a very small fraction of the fluid spends considerably long times in the holding tube. However, since it is a very small fraction, it makes very little difference to the quality of the product.

When pasteurising or sterilizing materials in continuous processing operations, it is important that all the fluid is held for at least the minimum required time. Thus, in the pasteurization of milk, all the milk should be treated for at least 15 s, and the stipulated time should correspond to the minimum holding time.

A minimum time of 15 s would require an average residence time of 30 s (streamline flow) or 18 s (turbulent flow). When this is borne in mind, it can be seen that turbulent flow conditions will give a much narrower distribution of residence times and less chance of over-processing.

If the pasteurization process had been designed on an average residence time of 15 s, the fluid would be considerably under-processed. As far as evaluation of sterilization efficiencies is concerned, the average residence time should not be used. For example, heating a sample of milk at 73 °C for 15 s would not have the same effect as dividing the sample in two and heating one portion for 10 s and the other for 20 s and recombining them. In the latter case the number of surviving organisms would be higher, despite the fact that the average heating time is 15 s.

If the residence time distribution function is known, it is possible to evaluate both the sterilization effect and the effect on the nutrients much more accurately. However, there are considerable practical problems when measuring such distribution at the very short holding times found in UHT processes; these times can be as low as 2 s.

3.19 CONTINUOUS STIRRED-TANK REACTOR

A continuous stirred-tank reactor is a tank with an agitator device to ensure good mixing, operating at a constant volume V. Under steady-state conditions the input flow rate will equal the output flow rate. The contents are well mixed if the concentration of any component in the outlet stream equals the concentration in the bulk of the liquid.

The average residence time is given by

$$t_{av} = \frac{V}{Q}$$

If a residence time distribution function is performed (Fig. 3.25), it can be seen that there is an infinite distribution of residence times and that the fall in concentration follows an exponential relationship. A stirred-tank system is completely unsatisfactory for continuous thermal sterilization of liquids but is much used in continuous chemical reactions and fermentation studies.

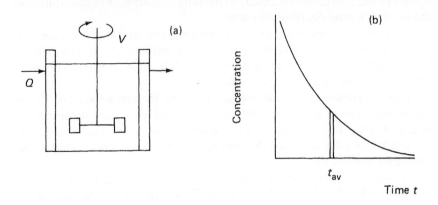

Fig. 3.25 — (a) Flow through a continuous stirred-tank reactor; (b) the distribution of residence times that occur.

This type of analysis can be applied to flow through any piece of processing or analytical equipment. It is interesting to note that plug flow is achieved when transporting solids on conveyor belts or trollys in continuous baking, drying or freezing operations, or in screw conveyors. Descriptions of other techniques for analysing flow situations and more detailed mathematical treatments have been given by Levenspiel (1972) and Loncin and Merson (1979).

3.20 FLOWABILITY OF POWDERS

For a liquid flowing out of a tank, the flow rate is governed by the hydrodynamic considerations. Application of Bernoulli's equation shows that the discharge velocity v will depend upon the square root of the head of fluid.

The flow of powders from bins and hoppers differs from liquids in two respects. Firstly, the flow rate has been found to be almost independent of the height, if the height is 2.5 times greater than the exit diameter. Secondly, powders can resist appreciable shear stresses and, when compacted, can form mechanically stable structures that will completely prevent any flow, despite the existence of a head. Other factors affecting the flow of powders, such as the angles of repose and friction have been discussed by Peleg (1983).

3.21 SYMBOLS

A surface area
C_d drag coefficient
C_D coefficient of discharge

Symbols

d	particle diameter
D	pipe diameter
e	roughness factor
F	frictional loss
F_R	retarding force
h	height or head
H	head developed by pump
j_f	modified bed friction factor
K	friction constant for fittings
K_c	constant (equations (3.11) and (3.15))
L	length
m	mass flow rate
N	rotational speed
p	pressure
Δp	pressure drop
Q	volumetric flow rate
R_W	wall shear stress
S_p	projected surface area
S_0	surface area per unit volume
t	time
v	velocity
v_b	velocity in packed bed
v_e	entrainment velocity
v_F	fluidisation velocity
v_t	terminal velocity
V	volume (packed bed or tube)
z	height

3.21.1 Greek symbols

ε	porosity
μ	viscosity
ρ	density
ρ_m	density of manometer fluid
ρ_S	solid density
ϕ	basic friction factor
ψ	shape factor

3.21.2 Dimensionless groups

Re	Reynolds number
$(Re)_b$	bed Reynolds number
$(Re)_p$	particle Reynolds number

4
Viscosity

4.1 INTRODUCTION

In many food-processing operations it is essential to know the viscosity of the fluid being processed, so that the most suitable equipment can be selected. During some operations the viscosity may change considerably. This is particularly so in processes involving heating, cooling, homogenization and concentration as well as during many industrial fermentations by moulds; these viscosity changes need to be considered when designing these processes.

Measurement of viscosity is often very important for quality control, particularly on products that we expect to be of a particular consistency in relation to appearance or mouth feel, e.g. cream, yoghurt, tomato paste and custards.

Viscosity can be simply defined as the internal friction acting within a fluid, i.e. its resistance to flow. A fluid in a glass, when inverted, is subjected to gravitational forces; some fluids will flow easily out of the glass, some with difficulty and some not at all. Viscosity is also a measure of the rate of flow.

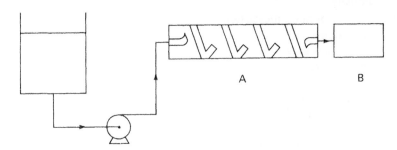

Fig. 4.1— Flow through a heat exchanger A and packaging unit B.

Consider a processing application where milk or fruit juice is being pumped from tank, by pump, through a pasteurizer A, to a bottling plant B or packaging unit (Fig. 4.1). The function of the pump is to supply energy to overcome both the internal resistance within the fluid and the frictional resistance between the fluid and the pipe walls. The factors affecting the energy input will be the pressure (or head) required, the volumetric flow rate and the magnitude of the fluid viscosity and their frictional forces. As the fluid viscosity increases, the frictional forces will increase and more energy would be required. In addition to the frictional forces developed, the viscosity of the fluid may well influence the choice of the pump, heat exchanger and packaging equipment. Before going on to consider viscosity and its measurement in more detail, it is necessary to explain and quantify the concepts of shear stress, shear rate and coefficient of viscosity.

4.2 IDEAL SOLIDS AND LIQUIDS

The concept of an ideal solid and an ideal liquid will now be introduced. Like all ideals, they do not exist in reality; therefore they represent the extremes of behaviour. All substances exert some of the characteristics of these two ideals; the proportion of each shown by an object will depend upon the conditions prevailing and the stresses that the object is subjected to. Under given conditions, the degree of solidity or of fluidity shown would decide whether the substance is classified as a solid or as a liquid.

For example, rock and ice are normally considered to be solids; under large stresses, these substances move, albeit very slowly, and therefore show liquid tendencies.

The ideal solid is represented by the Hookean solid, and the ideal liquid by the Newtonian liquid (Muller, 1973). Both are structureless (there are no atoms), both are isotropic (they have the same properties in all directions) and both follow their respective laws exactly. The laws for solids will be dealt with in section 5.5, whereas the laws for Newtonion liquids will now be considered.

4.3 SHEAR STRESS AND SHEAR RATE

When a shearing force is applied to a fluid, this will cause a deformation; this deformation is termed flow.

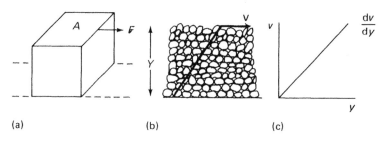

Fig. 4.2 — (a) Shear stress (b) layers (c) velocity gradient established in a fluid.

If a shearing force F is applied over an area A of the surface of a fluid in contact with a flat stationary plane (Fig. 4.2(a)), then the upper layer of the fluid will move with a velocity v, while the layer in contact with the plane will remain stationary. The upper layer in turn drags the layer below, with a slightly reduced velocity; this in turn drags the next layer (Fig. 4.2(b)). Consequently a velocity gradient is set up within the fluid (Fig. 4.2(c)). The nature of the velocity gradient will depend upon the viscosity of the fluid.

The shear stress τ that the fluid is subjected to is given by

$$\tau = \frac{\text{force } F}{\text{area } A} \text{ N m}^{-2}$$

The rate of shear (velocity gradient) is dv/dy (s^{-1}). When fluids are being tested or processed, they can be subjected to a wide range of shear rates, e.g. 10^{-5}–10^5 s^{-1}.

4.4 NEWTONIAN FLUIDS AND DYNAMIC VISCOSITY

For a class of fluids known as Newtonian fluids, there is a linear relationship between the shear stress and the shear rate. The dynamic viscosity (or coefficient of viscosity) is defined as the ratio of the shear stress to shear rate:

$$\text{dynamic viscosity} = \frac{\text{shear stress}}{\text{shear rate}}$$

Data for fluids are often presented in the form of shear stress–shear rate

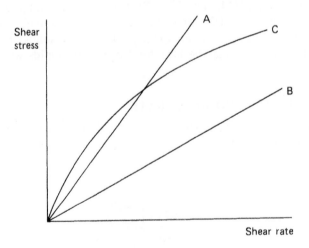

Fig. 4.3 — Rheograms for two different Newtonian fluids (lines A and B) and a non-Newtonian fluid (line C).

diagrams, plotted in either a linear or a log–log form. Such plots are called rheograms; lines A, B and C in Fig. 4.3 are rheograms for three different fluids. The fluids represented by lines A and B are Newtonian, with the dynamic viscosity for line A being greater than that for line B. The fluid represented by line C is non-Newtonian (see Section 4.7) as there is not a linear relationship between the shear stress and shear rate. Occasionally the data are presented as a plot of shear rate against shear stress.

The dimensions of dynamic viscosity are $[ML^{-1}T^{-1}]$. The units in the various systems are given in Table 4.1.

Table 4.1 — Units of viscosity in the various systems

System of units	Shear stress	Shear rate	Dynamic viscosity
SI	N m^{-2}	s^{-1}	N s m^{-2} (or poiseuille (Pl))
cgs	dyn cm^{-2}	s^{-1}	dyn s cm^{-2} (or poise (P))
Imperial	lbf ft^{-2}	s^{-1}	lbf s ft^{-2}

Conversion: 1.488 N s m^{-2} = 1 lbf s ft^{-2}.

In the SI system, a fluid has a viscosity of 1 N s m^{-2} (or 1 poiseuille (Pl)) if a shear stress of 1 N m^{-2} produces a shear rate of 1 s^{-1}. In the literature the poise (P) (or centipoise (cP)) is still commonly used. British units are very rarely encountered:

$$1\ P = 1\ \text{dyn s cm}^{-2} = \frac{10^{-5}\ \text{N s}}{10^{-4}\ \text{m}^2} = 10^{-1}\ \text{Pl}$$

Therefore, 1 cP = 10^{-3} Pl

An alternative way of expressing 10^{-3} P is in terms of megapascals seconds (mPa s); thus,

$$1\ \text{mPa s} \equiv 10^{-3}\ \text{Pl}\ (N\ \text{s m}^{-2}) = 1\ \text{cP}$$

Water has a viscosity of 1 cP at 20.2 °C. Most gases and simple fluids exhibit Newtonian behaviour at the shear rates normally encountered. Table 4.2 gives the dynamic viscosities of some simple fluids.

Table 4.2 — Viscosity values at 20 °C.

Fluid	Viscosity (N s m^{-2})	Fluid	Viscosity (N s m^{-2})
Carbon dioxide	1.48 × 10^{-5}	20% sucrose (g per 100 g of solution)	2 × 10^{-3}
Water	1.002 × 10^{-3}	40% sucrose (g per 100 g of solution)	6.2 × 10^{-3}
Carbon tetrachloride	0.969 × 10^{-3}	60% sucrose (g per 100 g of solution)	58.9 × 10^{-3}
Olive oil	84 × 10^{-3}	Honey (average values after mixing) (25 °C)	6000 × 10^{-3}
Castor oil	986 × 10^{-3}	Milk	2 × 10^{-3}
Glycerol	1490 × 10^{-3}	Ethanol	1.20 × 10^{-3}
		n-hexane	0.326 × 10^{-3}

Gases have the lowest viscosity values. Simple fluids such as water, dilute solutions and organic solvents are considered to be low-viscosity fluids. It should be noted that, as the solids concentration increases, the viscosity increases, so that during certin unit operations, such as evaporation and reverse osmosis, the viscosity increases, the changes being more marked at higher solids contents. In fact, in the concentration of proteins by ultrafiltration, it is the viscosity that limits the extent of the concentration. Proteins have a much more marked effect on viscosity than do salts or minerals in equal concentrations. Milk has a variable viscosity, depending upon its chemical composition, while homogenization increases its viscosity.

Oils are much more viscous than water. Most food-grade oils are Newtonian in behaviour. However, there is considerable variation m viscosity between the different types of oil. Glycerol is also Newtonian and is a medium-viscosity fluid.

4.4.1 Temperature effects

Viscosity is also very temperature dependent; so it is important both to control the experimental temperature during its determination and to state the temperature when quoting viscosity data. All liquids decrease in viscosity as the temperature increases. On average there is about a 2% change in viscosity for each degree celsius change in temperature, but for some substances it changes by greater amounts, e.g. for castor oil is is 8% (20 °C → 21 °C) (Muller 1973), for pitch 30% (20 °C → 21 °C) and for glycerol 0.95 N s m^{-2} (25 °C) or 1.49 N s m^{-2} (20 °C).

Temperatures should be controlled to within ± 0.1 °C during viscosity determinations. The most satisfactory type of empirical relationship, to fit the experimental data, for changes in viscosity with temperature has been found to be of the following form:

$$\log \mu = \frac{B}{T} + C$$

where T is the absolute temperature, and B and C are constants for the fluid.

Again, during a typical processing operation, there may be considerable change in temperature, i.e. in the UHT processing of milk the temperature rapidly rises from 10 °C to 140 °C, followed by rapid cooling to below 20 °C. Most gases increase in viscosity as the temperature increases.

4.5 KINEMATIC VISCOSITY

A further term often encountered is the kinematic viscosity v, where

$$\text{kinematic viscosity} = \frac{\text{dynamic viscosity } \mu}{\text{density } \rho}$$

The dimensions of kinematic viscosity are $[L^2 T^{-1}]$ and the most common unit is the stoke (St).

Table 4.3 gives the conversion factors for converting from dynamic viscosity to kinematic viscosity in cgs and SI units. For example, the

Table 4.3 — Conversion from dynamic to kinematic viscosity.

System of units	Kinematic viscosity	Dynamic viscosity	Density
cgs	St (cm^2 s^{-1})	P (g cm^{-1} s^{-1})	g ml^{-1}
SI	m^2 s^{-1}	Pl (kg m^{-1} s^{-1})	kg m^{-3}

kinematic viscosity in stokes is obtained by dividing the dynamic viscosity in poises, by the density in grams per millilitre. When the commercial U-tube viscometers are used, the kinematic viscosity is measured directly (see section 4.11.2).

The conversion factor from SI to British units is 92.9×10^{-3} m^2 s^{-1} = 1 ft^2 s^{-1}.

4.6 RELATIVE AND SPECIFIC VISCOSITIES

When dealing with emulsions or suspensions, the viscosity is often measured in comparative terms, i.e. the viscosity of the emulsion or suspension is compared with the viscosity of the pure solvent. One viscosity that is commonly used is the *relative viscosity*, where

$$\eta_r = \frac{\eta_s}{\eta_0} = 1 + k\phi \qquad (4.1)$$

η_r is the relative viscosity of the suspension, η_s is the viscosity of the suspension, η_0 is the viscosity of the solvent, ϕ is the volume fraction occupied by the dispersed phase and k is a constant.

This type of relationship was proposed by Einstein in 1906 for dilute solutions of spherical particles. In dilute solutions, the volume fraction is proportional to the solute concentration; so equation (4.1) becomes

$$\eta_r = 1 + kc$$

where c is the concentration.

The *specific viscosity* η_{sp} is defined as the increment due to the addition of the solute, i.e.

$$\eta_{sp} = \eta_r - 1 = kc$$

The ratio of the specific viscosity divided by the concentration is termed the *reduced viscosity* or *viscosity number*. The *intrinsic viscosity* is the limiting value of reduced viscosity as the concentration approaches zero. The units of reduced viscosity and intrinsic viscosity are cubic metre per kilogram (m^3 kg^{-1}) or millilitre per gram (ml g^{-1}).

For spherical particles in dilute solutions the reduced viscosity should equal a constant value. Einstein found this value to be 2.5.

These viscosity values are normally determined using commercial U-tube viscometers. These types of relationship can be used to determine particle shape, solvation ratios and molecular weights for non-linear polymers (Levitt, 1973). Walstra and Jenness (1984) give an equation for estimating the viscosity of milk and related products in terms of the hydrodynamic volumes of the component fractions, i.e. fat, casein, lactose and whey proteins.

4.7 NON-NEWTONIAN BEHAVIOUR

Unfortunately, many fluids are not Newtonian fluids, i.e. there is not a linear relationship between the shear stress and the shear rate. Such fluids are termed non-Newtonian.

Examples of non-Newtonian fluids are concentrated solutions of macromolecules (starches, proteins and gums) and colloidal materials such as emulsions, pastes and suspensions. The viscosity and shearing action are dependent upon a number of factors such as the following.

(a) The nature of the continuous and dispersed phase.
(b) Particle–particle interaction and particle–solvent interaction.
(c) The concentration of particles, their shape, size and chemical composition.

Many natural and formulated foods are colloidal in nature, e.g. milk, cream, mayonnaise and tomato paste. Non-Newtonian fluids are more difficult to deal with and to classify experimentally, as the viscosity will depend upon the experimental conditions selected. The viscosity recorded under those conditions is termed the apparent viscosity μ_a and is equal to the shear stress divided by the shear rate (similar to a Newtonian fluid).

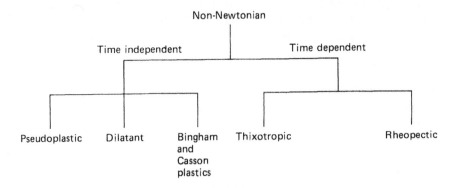

Fig. 4.4 — Types of non-Newtonian behaviour.

They can be further subdivided into two major divisions, namely time-independent and time dependent. Time-independent behaviour is where the apparent viscosity is independent of the previous shearing history (i.e.

how the fluid is being treated), whereas a time-dependent fluid is where the apparent viscosity is dependent upon the previous shearing history; this type of fluid is the most difficult to classify experimentally. The main divisions are shown in Fig. 4.4.

4.8 TIME-INDEPENDENT FLUIDS

The major types of time-independent non-Newtonian behaviour can be

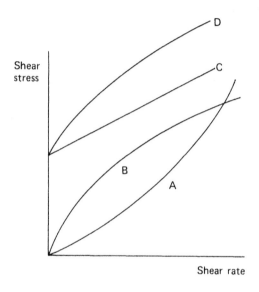

Fig. 4.5 — Rheograms for time-independent fluids: line A, pseudoplastic fluid, line B, dilatant fluid; line C, Bingham plastic fluid, line D, Casson-type plastic-fluid.

represented by lines A, B, C and D on the rheogram, representing four different fluids (Fig. 4.5). Line A is for pseudoplastic fluid. Line B is for a dilatant fluid, lines C and D are for plastic fluids; line C is an example of a Bingham plastic and line D a Casson body.

For pseudoplastic and dilatant fluids, the apparent viscosity μ_a at a fixed shear rate can be defined as the ratio of shear stress to shear rate, i.e. the definition and units are similar to those of Newtonian fluids. Data from the rheogram can be transformed to a graph of apparent viscosity against shear rate. Thus the apparent viscosity at any shear rate is given by the gradient of the straight line joining that point to the origin. Fig. 4.6 shows a typical rheogram for a pseudoplastic fluid. It can be seen that the apparent viscosity is greater at shear rate $(dv/dy)_1$ than it is at $(dv/dy)_2$, as measured by the respective gradients. Therefore, transformation of these data to those shown in Fig. 4.6 shows that a pseudoplastic fluid appears to become less viscous as the rate of shear increases. Pseudoplastic behaviour is sometimes

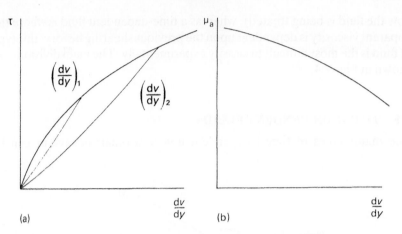

Fig. 4.6 — Rheogram for a pseudoplastic fluid: (a) shear stress against shear rate; (b) apparent viscosity against shear rate.

called shear thinning behaviour and is the most common type of non-Newtonian behaviour encountered. Dilatant fluids exhibit the opposite behaviour, i.e. the apparent viscosity increases as the shear rate increases. Single cream is an interesting example of a fluid that can show pseudoplastic and dilatant characteristics. Dilatant behaviour is much less common than pseudoplastic behaviour.

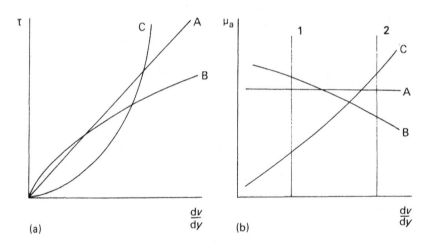

Fig. 4.7 — Comparison of the viscosities of a Newtonian fluid (lines A), a pseudoplastic fluid (lines B) and a dilatant fluid (lines C) at different shear rates: (a) shear stress against shear rate; (b) apparent viscosity against shear rate.

Fig. 4.7 represents the data for three different fluids: lines A are for a Newtonian fluid, lines B for a pseudoplastic fluid and lines C for a dilatant fluid. When the apparent viscosity is determined at different shear rates,

Fig. 4.7 results (the relationships are not necessarily linear). This illustrates some of the problems of determining and comparing the apparent viscosities of non-Newtonian fluids. If a viscometer were used to measure viscosities at a low shear rate denoted by (1), the fluid with the greatest viscosity would be represented by line B, and the fluid with the lowest by line C. However, if another (or the same) viscometer were used at a higher shear rate denoted (2), then the fluid with the highest viscosity would be represented by line C. Therefore, it is necessary to measure the apparent viscosity over a range of shear rate values.

4.8.1 Plastic fluids

The third class of fluid in Fig. 4.5 is a plastic fluid; these are normally defined in terms of a yield stress τ_0 and a plastic viscosity μ_p. The main characteristic of thes types of fluid is that at low shear stresses (below τ_0) they behave like solids and there will be no deformation until a critical (yield) shear stress is reached. Beyond that shear stress, the fluid will flow. A *Bingham plastic* gives a linear relationship between shear stress and shear rate; the plastic viscosity of such a fluid is given by the gradient of the straight line (Fig. 4.5, fluid C). A Casson body gives a parabolic shape (Fig. 4.5, fluid D); this can be transformed to a straight line when the square root of the shear stress is plotted against the square root of the shear rate.

4.9 TIME-DEPENDENT FLUIDS

Further characteristics of time-independent non-Newtonian fluids, shown in

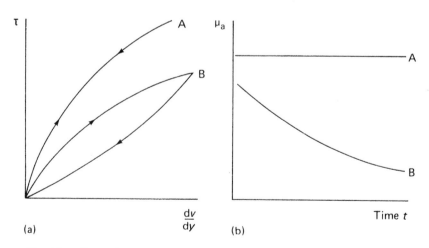

Fig. 4.8 — Comparison of rheograms for a time-independent fluid (lines A) and a time-dependent fluid (lines B): (a) shear stress against shear rate; (b) apparent viscosity against time at a constant shear rate.

Fig. 4.8(a), are illustrated only for a pseudoplastic fluid but are also applicable to dilatant fluids. If shear stress data are obtained as the shear rate increases and also as the shear rate decreases, the two sets of data coincide (illustrated by the two arrows on line A). The shear stress–shear rate relationship is also illustrated for a time-dependent fluid subjected to the same procedure. In this case, the two sets of readings do not coincide; this phenomena is known as hysteresis (shown by line B). Thus, at any fixed shear rate, there are two possible apparent viscosities, depending at what times the readings were taken.

Another way to illustrate this is by operating a viscometer at a constant shear rate over a considerable time period. A time-independent fluid will show a constant apparent viscosity at a fixed shear rate, whereas the viscosity of a time-dependent fluid will appear to change, normally to break down (Fig. 4.8(b)); often an equilibrium value (μ_e) is attained. If the shear is removed, the fluid may recover its structure. However, there are considerable differences between fluids in both the time required to reach equilibrium and the rate at which they recover their structure.

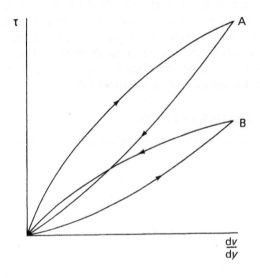

Fig. 4.9 — Rheograms for a thixotropic fluid (line A) and a rheopectic fluid (line B).

The major types of time-dependent behaviour are illustrated in Fig. 4.9. Thixotropic fluids (line A) show a breakdown in structure with continued application of shear whereas rheopectic substances (line B) show a build-up in structure; thixotropic behaviour is more common, e.g. most creams, whereas whipping cream is an example of a rheopectic fluid.

Therefore, when measuring time-dependent fluids, it is necessary to define the conditions under which the sample was measured very precisely, if the results are to be meaningful and reproducible. This should include the shearing conditions and the duration of the shear rate.

Some of the models proposed for predicting the behaviour of some time-independent fluids will now be discussed.

4.10 THE POWER LAW EQUATION

For many fluids, there has been found to be a straight-line relationship when the shear stress is plotted against the shear rate in log–log coordinates.

The equation, relating shear stress and shear rate, under these experimental conditions is known as the *power law equation*; accordingly, it is empirical and, for a fluid exhibiting no plastic behaviour, takes the form

$$\tau = k \left(\frac{dv}{dy} \right)^n \qquad (4.2)$$

where τ (N m^{-2}) is the shear stress, k (N s^n m^{-2}) is the consistency index, dv/dy (s^{-1}) is the shear rate and n is the power law index.

When expanded using the laws of logarithms, this becomes

$$\log \tau = \log k + n \log \left(\frac{dv}{dy} \right)$$

The *power law index* and *consistency index* can be determined by

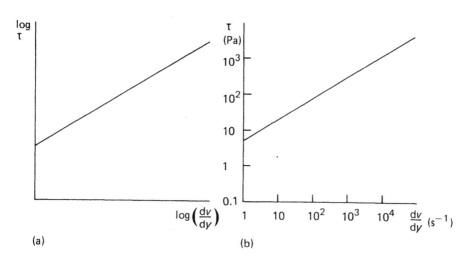

Fig. 4.10 — Rheological data plotted in logarithmic coordinates using (a) ordinary paper and (b) log–log paper.

plotting the experimental data, either in ordinary coordinates (Fig 4.10(a)) or in log–log coordinates (Fig. 4.10(b)). The power law index is the value of the gradient of the straight line, whereas the consistency index is the intercept value of shear stress at a shear rate value of 1 s^{-1} (i.e. log (dv/dy) = 0).

The values of n and k can then be used to characterize the fluid over the experimentally determined range of values, i.e. the apparent viscosity can be determined at any shear rate by substitution into equation (4.2) or by construction of a shear stress–shear rate diagram from the n and k values.

$$\tau = k\left(\frac{dv}{dy}\right)^n$$

and

$$\mu_a = \frac{\tau}{dv/dy}$$

Therefore,

$$\mu_a = k\left(\frac{dv}{dy}\right)^{n-1} \tag{4.3}$$

For Newtonian fluids, $n = 1$; for pseudoplastic fluids, $n<1$; for dilatant fluids, $n>1$. Therefore, it is easy to determine whether a fluid is Newtonian or non-Newtonian. The further the power law index is from one, the greater is the deviation from Newtonian behaviour.

Table 4.4 — Power law and consistency index values.

Product	Temperature (°C)	k (N s^n m^{-2})	n	Reference
Custard	80	7.24	0.36	Milson and Kirk (1980)
Gravy	80	2.88	0.39	Milson and Kirk (1980)
Tomato juice (12.8% total solids)	32	2.0	0.43	Toledo (1980)
Tomato juice (25.0% total solids)	32	12.9	0.40	Toledo (1980)
Tomato juice (30.0% total solids)	32	18.7	0.40	Toledo (1980)
Sunflower meal	25	—	0.41	Launay and Lisch (1979)
Faba bean flour	25	—	0.32	Launay and Lisch (1979)

Some examples of k and n values are given in Table 4.4. k and n values have been widely used to characterize time-independent non-Newtonian fluids. The power law relationship often holds over two or more logarithmic cycles of shear rate; the range of shear rates that the fluid is subjected to should always be recorded, to prevent erroneous results from extrapolation outside this range.

For fluids showing some plastic behaviour, the power law equation becomes

$$\tau = k\left(\frac{dv}{dy}\right)^n + \tau_0$$

where τ_0 (N m^{-2}) is the yield stress. The power law equation is one of the simplest equations that has been found to fit the experimental data for a variety of fluids. Not all fluids obey the power law, and many other more complicated models have been proposed to fit the experimental data more closely, for both time-independent behaviour. Some of these have been discussed by Wilkinson (1960, Coulson and Richardson (1977), Launay and Lisch (1979) and Bourne (1982).

4.11 METHODS FOR DETERMINING VISCOSITY

4.11.1 Streamline flow methods

For a fluid flowing along a horizontal tube, under streamline flow conditions, Poiseuille has shown that there is a relationship between the pressure drop and the volumetric flow rate of the form (see section 3.10)

$$Q = \frac{\Delta p \pi D^4}{128 L \mu}$$

where Q (m^{3}s^{-1}) is the volumetric flow rate, Δp ($= p_1 - p_2$) (N m^{-2}) is the pressure drop, D (m) is the tube diameter, μ (N s m^{-2}) is the dynamic viscosity and L (m) is the tube length. Therefore, if the pressure drop and the flow rate can be measured, the dynamic viscosity can easily be determined. However, it is important to check that the flow is streamline and that the fluid is Newtonian in behaviour.

This can easily be done by taking the pressure drop at a number of flow rates and plotting pressure drop against the flow rate in either ordinary coordinates (Fig. 4.11(a)) or in log–log coordinates (Fig. 4.11(b)). If a straight-line relationship results from Fig. 4.11(b) and the gradient of the log–log graph is 1.0, then the flow is streamline, and the fluid is most probably Newtonian. If, however, the relationship is non-linear, then this indicates either non-Newtonian behaviour or the onset of turbulent flow. Further work would be necessary to distinguish between these two possibilities. Reducing the diameter of the capillary tube is one way towards making the flow streamline. As a first approximation, if the fluid has a low viscosity, it is probably Newtonian and the flow is turbulent; if it has a high viscosity, then the fluid is probably non-Newtonian.

One simple practical arrangement is to set up a fine-bore capillary tube with a constant-head device. The position of the constant-head device can be

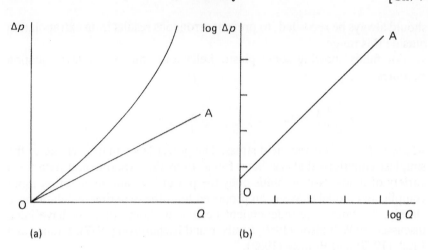

Fig. 4.11 — The relationship between pressure drop and flow rate (a) on ordinary graph paper and (b) on log–log paper.

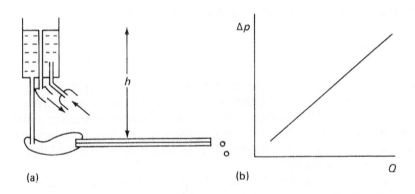

Fig. 4.12 — (a) Apparatus for measuring the coefficient of viscosity using a length of capillary tube; (b) calculation of the coefficient of viscosity from the pressure drop–flow rate diagram.

changed to vary the pressure drop over the tube (see Fig. 4.12(a)). For long tubes of fine bore, the pressure drop over the tube approximates to *pgh* (this takes no account of the kinetic energy of the issuing fluid). The volumetric flow rate can be measured directly and readings are taken for several pressure drops.

For Newtonian fluids, under streamline flow conditions, there will be a linear relationship between the flow rate and the pressure drop (Fig. 4.12(b)) and the dynamic viscosity can be calculated from

$$\mu = \frac{\Delta p}{Q} \frac{\pi D^4}{128L}$$

where $\Delta p/Q$ is the gradient of the straight line.

The largest source of experimental error is in the measurement of the capillary diameter, as this raised to the power 4 in Poiseuille's equation.

An alternative way of presenting the data is in the form of a rheogram, where

$$\text{wall shear stress } \tau_w = \frac{\Delta p \; D}{4L}$$

and

$$\text{wall shear rate } \frac{dv}{dy} = \frac{32Q}{\pi D^3} \text{ or } \frac{8v}{D}$$

The flow rate-pressure drop data can be expressed in these terms, and the shear stress plotted against shear rate. On log–log paper a gradient of 1.0 will indicate Newtonian behaviour; a gradient different from 1.0 will indicate non-Newtonian behaviour.

For a non-Newtonian fluid obeying the power law equation and flowing under streamline flow conditions, the following relationship exists between the pressure drop and flow rate:

$$Q = \frac{n\pi D^3}{8(3n+1)} \cdot \left(\frac{D \; \Delta p}{4Lk} \right)^{1/n}$$

If the fluid is non-Newtonian, capillary flow viscometers can be used, but the procedure is more complicated; a good account has been given by Toledo (1980). More rapid results would probably result by using rotational viscometers (see section 4.12).

4.11.2 Capillary flow viscometers

Capillary flow viscometers are normally in the form of a U-tube. Some typical designs of capillary flow viscometers are shown in Fig. 4.13. The simplest design is the Ostwald viscometer (Fig. 4.13(a)) (sometimes called the U-tube viscometer). The viscometer is accurately filled with the test fluid to the level of an inscribed mark A; fluid is sucked up the other limb, through the capillary tube, until it is above mark B. The fluid then flows through the capillary tube, under the influence of the induced pressure head. The time is recorded for the fluid to fall from mark B to C, as it flows through the capillary tube. This time is then multiplied by a constant for the instrument to determine the kinematic viscosity of the fluid:

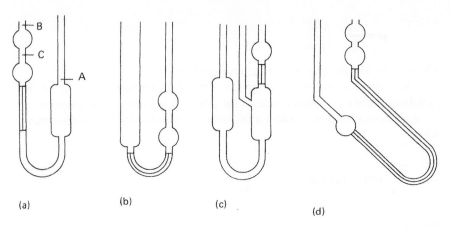

Fig. 4.13 — Types of capillary flow viscometer: (a) an Ostwald viscometer; (b) a reverse-flow viscometer; (c) a suspended-level viscometer; (d) a Cannon–Fenska viscometer.

kinematic viscosity (cSt or m s^{-1}) = instrument constant (cSt s^{-1} or m^2 s^{-2}) × time (s)

The time taken for the fluid to flow will be inversely proportional to the density of the fluid and directly proportional to the dynamic viscosity, provided that the flow is streamline:

$$\text{time} \propto \frac{\mu}{\rho}$$

Because the time is dependent upon both the density and the viscosity of a fluid, capillary flow viscometers give a direct reading of the kinematic viscosity.

U-tube viscometers may be provided with or without the calibrated instrument constant; if it is not provided, it can easily be determined using a fluid of known kinematic viscosity.

Ostwald viscometers can be obtained in a wide variety of capillary sizes and the correct capillary size should be selected for the particular application. If the capillary diameter is too large, then the fluid will flow very quickly and it is likely that the flow will be turbulent. The reproducibility of the results will be poor. However, if the capillary diameter is too small, it will take a long time to obtain the determinations, but the accuracy will be much improved. Times between 100 s and 500 s offer a reasonable compromise between speed and accuracy. The capillary flow viscometer can deal with reasonably small samples, it is cheap and it is one of the most suitable viscometers for accurately determining the viscosity of low-viscosity fluids, such as water, organic solvents, very dilute solutions and milk, and for monitoring changes brought about in these fluids by, for example, heat or homogenization. They are not suitable for telling whether a fluid is Newtonian or non-Newtonian, and for non-Newtonian fluids they will give an apparent viscosity at the shear rate that the fluid is subjected to.

Care should be taken to ensure that the correct volume of fluid is introduced into the Ostwald viscometer and that the instrument does not deviate from the vertical by more than 1°. There have been several modifications to the basic design to overcome these problems. The suspended-level viscometer (Fig. 4.13(c)) avoids the complication of needing to introduce the correct volume and thereby lessens the variation in hydrostatic head during the measurement. This may be particularly useful when taking a series of measurements at different temperatures, because it avoids the need to readjust the working volume between determinations. The Cannon–Fenske modification (Fig. 4.13(d)) reduces the problems that arise because the viscometer is not vertical; here the two bulbs lie in the same vertical axis, the capillary being disposed at an angle to the vertical to make this possible. A final viscometer of interest is the reverse-flow viscometer. Most of the other U-tube designs necessitate forcing the liquid through the capillary tube before the measurement is made; this could very well alter the viscosity of the fluid (time dependent). The reverse-flow viscometer allows measurement of the fluid viscosity as it is being subjected to a shearing force for the first time (Fig. 4.13(b)).

More sophisticated capillary viscometers work by applying an external pressure to the fluid or by forcing the fluid at a predetermined flow rate and measuring the pressure drop. For highly precise measurement, the reader should refer to articles by Dinsdale and Moore (1962) and Johnson *et al.* (1975) for the precautions to be taken and the corrections required.

In all capillary flow techniques, the fluid is subjected to a variable shear stress (i.e. the maximum stress is exerted at the tube wall and a zero stress is exerted at the centre of the tube).

4.11.3 Falling-sphere viscometers

When an object falls through a fluid, it is subjected to a number of forces. In a downward direction, there is the force of gravity and, in an upward direction, a viscous drag and an upthrust (equal to the weight of fluid displaced by the object). When equilibrium is attained, the upward forces and downward forces are balanced and the object moves at a constant velocity (the terminal velocity). If the flow is streamline for a spherical particle of diameter D, these forces can be represented as

$$\text{weight} = \text{upthrust} + \text{viscous drag}$$

$$\frac{\pi D^3 \rho_2 g}{6} = \frac{\pi D^3 \rho_1 g}{6} + \frac{6\pi D \mu v}{2}$$

This simplifies to

$$v = \frac{D^2(\rho_2 - \rho_1)g}{18\mu}$$

where v (m s^{-1}) is the terminal velocity, D (m) is the particle diameter, ρ_2 (kg m^{-3}) is the particle density, ρ_1 (kg m^{-3}) is the fluid density and μ (N s m^{-2}) is the dynamic viscosity.

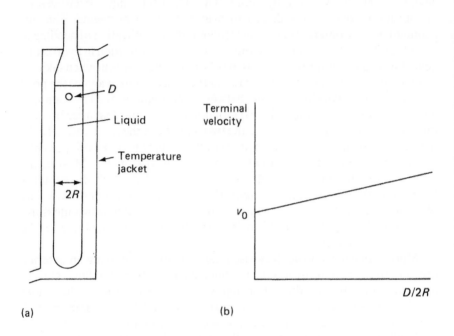

Fig. 4.14 — (a) A falling-sphere viscometer; (b) wall effect correction evaluation.

This equation is known as Stokes' law (see section 3.11). If the terminal velocity of the particle can be calculated, it is possible to determine the dynamic viscosity of the fluid. It is more suitable for viscous fluids, where the terminal velocity is low but can be adapted for low-viscosity fluids by use of specially constructed spheres. Again, it is not easily possible to determine whether a fluid is Newtonian or non-Newtonian using this technique.

Normally the fluid being measured is contained in a long tube, with the tube itself being enclosed in a constant temperature vessel. The time is taken for the particle to fall between two marks a known distance apart, one near the centre of the tube, and another towards the bottom. Care should be taken to ensure that the particle has reached its terminal velocity by the centre of the tube. The terminal velocity can easily be calculated.

If the particle size approaches that of the tube diameter, then a correction term needs to be applied to account for wall effects. The correction for the wall effect is of the form

$$v\left(1 + k\frac{D}{2R}\right) = v_0 \tag{4.4}$$

where v is the observed terminal velocity, k is a constant, R is the radius of the tube and v_0 true terminal velocity (no wall effects).

The terminal velocity is determined for the same particle in tubes of different diameters. The terminal velocity is plotted against the value $D/2R$; when the straight line is extrapolated back to $D/2R = 0$, the true terminal velocity, when no wall effects are present, can be obtained (see Fig. 4.14). This is then substituted in equation (4.4) to determine the dynamic viscosity. Other corrections have been discussed by Dinsdale and Moore (1962).

The falling-sphere principle has been incorporated into the Hagberg falling-number apparatus, for determining the α-amylase activity in wheat varieties; the more active the enzyme is, the more quickly will they convert a viscous starch suspension into a less viscous sugar solution.

4.12 ROTATIONAL VISCOMETERS

It has been seen that to characterize a non-Newtonian fluid it is necessary to determine the viscosity at a number of different shear rates. The two common situations in which a fluid is subjected to shear is in the flow of a fluid along a tube and in the mixing or agitation of a fluid. In a flow situation the shear rate is a function of v/D where v is the fluid velocity and D is the pipe diameter; in an agitation situation the shear rate is proportional to the rotational speed. Therefore, in both these situations, if it is possible to measure the shear stress as the shear rate is changed, it should be possible to characterize the fluid. This is the principle behind the rotational viscometer.

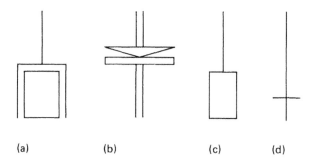

Fig. 4.15 — Types of measuring system used with rotational viscometers: (a) concentric cylinder; (b) cone and plate; (c) single spindle; (d) T-piece spindles (for high-viscosity fluids and gels).

Rotational viscometers can take several forms; they are normally described in terms of their measuring systems. The major types (Fig. 4.15) are the concentric-cylinder type, the cone-and-plate type and the single-spindle type. The principle is the same for all types.

4.12.1 Concentric-cylinder viscometers

A concentric-cylinder viscometer (Fig. 4.15(a)) consists of two cylinders with a narrow gap between them. One of the cylinders (the inner or the outer) is attached to a variable-speed motor. The fluid to be measured is situated in the gap. When this cylinder rotates, a velocity gradient is set up and the fluid transmits a torque to the second cylinder and drags it around. Some means is provided for measuring the force or torque exerted on the second cylinder. The fluid can be characterized in terms of the torque (related to the shear stress) and the angular velocity (related to the shear rate). For this particular geometry, the shear stress is given by

$$\tau = \frac{3\Gamma}{2\pi R^2 h}$$

Where Γ is the torque, R the radius of inner cylinder and h the height of cylinder;

$$\frac{dv}{dy} = \frac{2\pi RN}{x}$$

where x is the gap width and N the rotational speed.

In many cases the torque is measured by the angular deflection θ of a calibrated spring (of spring constant k_s):

$$\Gamma = k_s \theta$$

By measuring the angular deflection, and knowing the spring constant, the rotational speed and the geometry of the system, the shear rate and shear stress can be calculated.

The equation for the shear rate is almost exact if the gap width is small. Further, if the gap width is small compared with the radius of the cylinder, the shear stress is almost constant throughout the material being sheared (compare this with fluid flow in a capillary tube).

The shear rate can be predetermined by changing either the speed or the geometry of the system. Most rotational viscometers are suitable for viscous liquids and non-viscous liquids at high rates of shear. Many of the measuring units are not sensitive enough for measuring low-viscosity fluids at low rates of shear.

The expression for the shear stress takes into account only the torque exerted on the curved surface of the inner cylinders; it ignores the forces on the ends of the cylinder. These end effects are normally present and it is customary to regard them as increasing the effective height of the cylinder. Measuring systems should be designed to eliminate, as far as possible, these end effects; for accurate absolute measurements and for studying non-

Newtonian materials, it is necessary to eliminate, calculate or determine experimentally the end corrections. These alternatives have been discussed by Dinsdale and Moore (1962).

4.12.2 Cone-and-plate viscometers

The cone-and-plate viscometer (Fig. 4.15(b)) operates on a similar principle to the concentric-cylinder viscometer. The cone rotates at a selected speed and the fluid is in the gap between the cone and plate and transmits a torque to the plate, which is measured:

$$\tau = \frac{3\Gamma}{2\pi R^3}$$

and

$$\frac{dv}{dy} = \frac{2\pi RN}{x}$$

where R is the radius of the cone and x the gap width.

In this case the equation for the shear rate is exact, and all elements of the fluid are subjected to exactly the same shear stress. These measuring heads tend to be more expensive than the concentric-cylinder design because of the precision involved in the machining of the cone. The cone angle is usually less than 1°. At high shear rates, there is a danger that low-viscosity dilitant fluids will be thrown out of the gap. The viscometers are simple to operate, easy to fill and clean and suitable for small samples.

4.12.3 Single-spindle viscometers

In this case (Fig. 4.15)c)) a bob or spindle is rotated in the test fluid and the torque necessary to overcome the viscous resistance is measured. These viscometers are sometimes called infinite-fluid viscometers (similar to the concentric-cylinder viscometers but with a wide gap width). It is not possible to state the exact shear rate that the fluid is subjected to; so it is not possible to present shear stress–shear rate data. One such commercial viscometer is the Brookfield viscometer; this is widely used in the food industry. Normally, it operates at eight different speeds, and it is a matter of trial and error to select a suitable spindle and a rotational speed for a particular fluid. The steady-state deflection is noted and a conversion chart is provided to estimate the apparent viscosity under the test conditions. It is possible to determine the apparent viscosity at different speeds (shear rates) but, because it is not possible to predict the exact shear rate, the results are normally presented in the form of apparent viscosity against rotational speed.

When viscosity data are obtained using a Brookfield viscometer, the model number, the spindle size, the rotational speed and the temperature should always be quoted; otherwise, it would be difficult for such data to be reproduced. It is also possible to determine whether a fluid is time-dependent or time independent. If the viscometer is operated at a constant speed and the angular deflection changes, then the fluid is time dependent; in such cases the shearing history should also be mentioned when results are reported.

The Brookfield viscometer is the cheapest of the rotational viscometers available. It is easy to use, robust and reasonably accurate. There are additional attachments available for handling small volumes, for measuring low-viscosity fluids, for measuring pastes and suspensions and for moving the spindle slowly either up or down in the fluid, so that it is always rotating in fresh fluid; (this is particularly useful with the T-piece spindles (Fig. 4.15(d)). A range of calibration fluids is available for checking the accuracy of the equipment; this should be done at regular intervals.

4.13 VISCOMETER SELECTION

Simple viscometers can be constructed very cheaply, e.g. using a length of capillary tube, a few small metal spheres and a clear tube (even from laboratory filter funnels); there is no reason why such viscometers should not be reasonably accurate (certainly for quality control purposes), but they might not necessarily be quick to use.

At slightly more expense, but still relatively cheap, are the U-tube capillary flow viscometers. Each viscometer has a narrow measuring range, but within that range they are very accurate. Therefore, several U-tube viscometers would be required to cover the viscosities of typical food materials.

Rotational viscometers are much more expensive but, at the cheaper end of the range, the Brookfield viscometer is extremely useful and popular. As the viscometer became more sophisticated, it became possible to program them, and to take a fluid through a shear stress–shear rate cycle, to increase or decrease the shear rate at fixed rates or to hold at a constant shear rate for a fixed time. The results can be plotted on an x–y recorder (shear stress against shear rate) or an x–t recorder (apparent viscosity or shear stress against time) at a constant shear rate.

4.14 VISCOSITY DATA

The measurement of viscosity is not necessarily a simple matter, and many factors have to be taken into consideration. The results are widely spread throughout the literature. Obviously, it is not possible to compile all the viscosity data in a book of this nature and so I would like to make a few general comments on the viscosity characteristics of some of the major classes of fluid foodstuffs (namely dairy products, oils and fats and sugar solutions), and materials that are used in food formulations (such as hydrocolloids and proteins).

4.14.1 Milk and milk products

Milk is an interesting colloidal system; it consists of an aqueous phase containing lactose, minerals, soluble proteins (whey proteins), water-soluble vitamins and trace elements. Dispersed in the aqueous phase as very small dropets or globules is a fat phase. The characteristic milky appearance is due to a colloidal dispersion of milk protein (casein) and calcium in the solution.

There are considerable differences between the composition of milk from different species (Porter, 1975). Furthermore, milk from a particular species can show a considerable fluctuation in composition from animal to animal and from season to season. Ordinary cows' milk contains about 3.8% fat and 3.3% protein (2.6% casein). As such, the dynamic viscosity of full-cream milk is of the order of 2×10^{-3} N s m^{-2} (2 cP) at 20 °C and under moderate conditions of shear it acts as a Newtonian fluid.

Milk can be processed by a variety of techniques to prolong its shelf-life and to convert it to milk-based products. Most processing techniques may alter the condition of one or both of the dispersed and aqueous phases, and therefore the dynamic viscosity. Table 4.5 gives an indication of the range and viscosities of some milk-derived products.

Table 4.5 — Viscosities of some milk and dairy products.

Product	Temperature (°C)	Viscosity
Whey (5% lactose)	20	1.30×10^{-3} N s m^{-2}
Whey (5% lactose)	72	0.35×10^{-3} N s m^{-2}
Skim-milk (5% lactose)	20	1.70×10^{-3} N s m^{-2}
Skim-milk (5% lactose)	72	0.60×10^{-3} N s m^{-2}
Whole milk (5% lactose)	20	2.10×10^{-3} N s m^{-2}
Whole milk (5% lactose)	72	0.75×10^{-3} N s m^{-2}
Skim-milk (20% total solids)	25	3.8×10^{-3} N s m^{-2}
Skim-milk (33% total solids)	25	13×10^{-3} N s m^{-2}
Concentrated whey (65% total solids)	10	5000×10^{-3} N s m^{-2}
Concentrated milk (48% total solids)	20	1000×10^{-3} N s m^{-2}
Cream (20% fat)	20	6.1×10^{-6} m^2 s^{-1}
Cream (35% fat)	20	14.0×10^{-6} m^2 s^{-1}
Cream (45% fat)	20	35.0×10^{-6} m^2 s^{-1}
Cultured milk (12% total solids)	20	18×10^{-6} m^2 s^{-1}
Butter oil	30	60×10^{-6} m^2 s^{-1}

Adapted from the data of Kessler (1981).

Heat treatment of milk normally gives a slight increase in the viscosity, probably because of the denaturation of whey proteins. Homogenization will increase the viscosity of full-cream milk by up to 15%. Milk that is to be sterilized or UHT treated will require homogenization to prevent fat separation during storage. Some pasteurized milk is also homogenized. Homogenization appears to give the milk a creamier mouth feel.

Cream products also require homogenization. Single cream (18% fat) undergoes considerable homogenization (up to 200 bar) to improve the consistency and to give the cream some body. Whipping cream (35% fat) requires very little, or no, homogenization and appears to have a runny consistency before it is whipped; homogenization would impair the whipping behaviour of the cream. Double cream (48% fat) requires a low homogenization pressure (around 30 bar); if the homogenization pressure is too high, the cream may well solidify in the pack. The rheology of cream products is extremely complicated and the final viscosity of the cream will depend upon such factors as the separation temperature, fat content, heat treatment, rate of cooling and storage conditions.

Skim-milk, full-cream milk and cheese whey are evaporated to as high a solids content as possible, prior to spray drying. The final concentration may be limited by either the viscosity of the feed or the solubility limits of lactose. Such fluids can also be concentrated by membrane techniques such as reverse osmosis and ultrafiltration. Again the extent of the concentration is governed by the viscosity characteristics of the concentrate.

Any process that results in the souring of milk normally increases the viscosity of the product.

For further details on the rheological properties of dairy products the reader should consult Prentice (1979), Webb et al. (1974), Tschubik and Maslow (1973), Sherman (1979) and Lewis (1986b).

4.14.2 Oils and fats

Oils and fats are essentially esters of glycerol and fatty acids, derived from plant and animal sources. Different oils will have different fatty acid compositions and hence different viscosities. Oils are normally liquid at ambient temperature; fats are normally solid.

Oils are generally more viscous than aqueous solutions. They are usually Newtonian in behaviour, although they might show pseudoplastic behaviour at high shear rates. In general terms, the viscosity will increase as the amount of long-chain fatty acids increase and as the degree of saturation increases. Thus, hydrogenation will increase the viscosity.

Viscosity is not one of the physical tests that is used to characterize an oil, but the viscosity will be important when processing oils and fats. The viscosity values are recorded for some common vegetable oils in Table 4.6.

Further viscosity data can be obtained from Swern (1964) and Tschubik and Marlow (1973). Swern (1964) also gives the viscosities of the major saturated fatty acids and their methyl and ethyl esters, and some of the more common triglycerides.

For oils and sugar-based products, viscosity results are occasionally presented in seconds (Saybolt Universal) (s (Saybolt Universal)). The relationship between data in s (Saybolt Universal) and data in cP is as follows:

$$\text{s (Saybolt Universal)} = \frac{\text{cP} \times 4.55}{\text{specific gravity}} \text{ or } \frac{\text{N s m}^{-2} \times 4550}{\text{specific gravity}}$$

Table 4.6 — Kinematic viscosities of oils.

Oil	Temperature (°F)	Kinematic viscosity (cSt)
Almond	100	43.2
Olive	100	46.7
Rape seed	100	50.6
Cotton seed	100	35.9
Soya bean	100	28.5
Linseed	100	29.6
Sunflower	100	33.3
Castor	100	293.4
Coconut	100	29.8
Palm kernel	100	30.9

Adapted from the data of Swern (1964).

4.14.3 Sugar solutions

There are a whole range of sugar solutions available to the food processor. The viscosity characteristics of single sugars depend upon the temperature and the concentration. As the temperature decreases and the solids content increases, the viscosity will increase. Most single sugar solutions are Newtonian in behaviour. For corn syrups, high-fructose corn syrups and invert sugar solutions, an extra factor affecting the viscosity is the degree of conversion or inversion. Some data have been presented by Pancoast and Junk (1980), together with limited data on some sucrose-corn syrup blends.

Toledo (1980) and Bourne (1982) give some power law data (k and n values) for a wide variety of concentrated fruit juice and purées. Most of these fluids were pseudoplastic at high concentrations. There is now considerable interest in glucose–galactose syrups, obtained from hydrolysis of lactose. Further viscosity data can be obtained from Tschubik and Maslow (1973).

4.14.4 Hydrocolloids

Hydrocolloids are polymeric materials that are soluble or dispersible in water. Some examples are shown in Table 4.7. They are usually added to food formulations to increase their viscosity or to obtain a gelled consistency. In very dilute solutions, most of the listed materials are Newtonian in behaviour. As the concentration increases, the viscosity rapidly increases and there is often a transition from Newtonian to non-Newtonian behaviour; many of them form gels at relatively low concentrations. They are derived from a wide variety of plant or animal sources, or by the process of fermentation.

Many of these hydrocolloids can be modified chemically or enzymatically to control their thickening action and are available in a wide variety of grades. For example, the viscosity of ten guar grades, made up to a 1%

Table 4.7 — Viscosity data for some hydrocolloids.

Hydrocolloid	Viscosity of solutions (cP or mPa s)		Temperature (°C)	Transition concentration from Newtonian to non-Newtonian behaviour (%)	Quantity of hydrocolloid to obtain a viscosity of 100 c (mg per 100 ml of water)
	1% solution	5% solution			
Gum arabic	1.2	2.7	30	40	Very high
Gum ghatti	2.0	288	25	—	—
Gum karaya	800	2×10^4	25	0.5	—
Gum tragacanth	500–3000	—	—	0.5	—
Locust bean-gum	150	—	25	—	455
Guar gum	5000	—	—	0.5	500
Carrageenan	120 (gelation)	—	40	—	Gelation
Xanthan	1000	—	25	—	—
Cellulose gums	5–600	60–>30000	20	—	780–4600

Compiled from the data of Whistler (1975) and Adrian (1976).

solution and measured at a shear rate of 100 s^{-1} ranged between 5 cP and 525 cP (Ellis, 1982).

The hydration time can be quite significant for some materials; for example, a 1% guar gum solution will take over 24 h to reach its maximum or near-maximum viscosity. The viscosity of many hydrocolloids may be significantly affected by the pH of the medium and the presence of substances such as salts, sugars and proteins.

Considerable viscosity data are available for the hydrocolloids and some are listed in Table 4.7. (Whistler, 1975). This table shows, where available, the viscosity of 1% and 5% dispersions, the experimental temperature and the transition concentration from Newtonian to non-Newtonian behaviour. Also included is the quantity of hydrocolloid required to obtain a viscosity of 100 cP (Adrian, 1976). Whistler also presents further viscosity data on psyllium and seed quince gums, starches and their derivatives and scleroglucan, as well as general information on these and other hydrocolloids other useful sources of information are the work by Glickman (1969, 1982) and by Blanshard and Mitchell (1979). The viscosity and consistency of starches have been discussed in more detail by Radley (1976).

The gelling behaviour of compounds such as agar, pectins, alginates and starches have also been discussed in these texts. The Brabender (amylograph) viscometer is often used for such materials. This is a rotational viscometer operating at a constant speed which takes the material through a heating and cooling cycle, continuously recording the resistance to flow throughout this period (Section 5.6).

Proteins form a special class of polymeric material. The flow behaviour of dilute and concentrated proteins in solution depends upon pH, ionic strength and temperature (as is the case for many of the hydrocolloids). Some of these factors have been reviewed by Lee and Rha (1979); this is particularly relevant with the advent of membrane and ion exchange techniques for concentrating proteins and extracting them from solutions in an undenatured form. Extrusion and texturization of protein are now popular techniques; again the rheological properties have to be considered for evaluating these processes. Some of these aspects have been discussed by Hermansson (1979).

4.15 SOME SENSORY ASPECTS

At some particular point it becomes necessary to distinguish between solid and liquid food. For liquids and semiliquids we describe mouth-feel sensation in terms of viscosity or consistency, whereas for solids we use the term texture. It has been suggested that a convenient division is the force due to gravity. If an object flows under the influence of gravity, it is a liquid; if not, it is a solid. This does raise a difficulty with a plastic material, for which it will be necessary to classify it both below and above its yield shear stress.

Many forms of viscosity measurement are used as a quality control checks for different products. However, if the consumer cannot distinguish between the mouth feel of two batches of a product that appear to give different instrumental readings, then the instrumental measurement loses some of its validity as a quality assurance measurement. Sherman (1975) has performed some interesting work on psychological aspects connected with fluids; he has taken a wide variety of fluids with known shear stresses and shear rate characteristics and used these with sensory assessors to see whether they are capable of distinguishing between the fluids by testing, pouring and stirring. From the data, he has estimated the shearing conditions that the fluid is subjected to during these particular activities. This should provide useful guidelines as to whether people can distinguish between different fluids, by a sensory measurement, if the flow characteristics are known.

The vocabulary of descriptive terms for mouth feel is also of considerable interest, particularly to those interested in the organoleptic assessment of food products. Some of the problems are discussed for beverages by Szczesniak (1979). Texture measurement is considered in more detail in Chapter 5.

4.16 SYMBOLS

A	surface area
B	constant
c	concentration
C	constants
D	diameter
F	shear force
h	height
k	consistency index
L	length
n	power law index
N	rotational speed
$\triangle p$	pressure drop

Q	volumetric flow rate
R	radius
T	temperature
v	velocity
dv/dy	shear rate
x	gap width

4.16.1 Greek symbols

Γ	torque
η_r	relative viscosity of suspension
η_s	viscosity of suspension or emulsion
η_{sp}	specific viscosity of suspension or emulsion
η_0	viscosity of solvent
μ	coefficient of viscosity, or dynamic viscosity
μ_a	apparent viscosity
μ_e	equilibrium viscosity
μ_p	plastic viscosity
ν	kinematic viscosity
ρ	density
τ	shear stress
τ_0	yield shear stress
ϕ	volume fraction occupied by the dispersed phase

5

Solid rheology and texture

5.1 INTRODUCTION

Rheology has been defined as the deformation of objects under the influence of applied forces. Solids, like liquids, can be subjected to a large number of different types of force. In this chapter the forces that solids are subjected to and the resulting deformation will be covered in more detail.

Foods are subjected to forces during processing, particularly during size reduction operations and expression (pressing) processes; in addition, packaging material such as cans and flexible pouches will undergo stress, particularly during heating, as the contents expand. From the mechanical point of view, equipment should be designed to withstand high pressures. Also of extreme importance is the mechanical properties of food and their relationship to texture.

5.2 THE PERCEPTION OF TEXTURE

As consumers, we are all acutely aware of texture when we eat or drink solids or liquids, and there can be no doubt that texture is an important determinant of food quality. In the mastication process the forces that a food is subjected to are complex. Chewing breaks the food down and makes it more digestible. During this process, information is transmitted from various sensory receptors in the mouth to specific parts of the brain, where it is integrated with other incoming information as well as information stored in the memory to give an overall impression of texture. If this does not conform to what we would expect from that particular food, we may well be disappointed with its quality. For example, we expect our steak to be tender and juicy, and not chewy or tough; liver should be smooth, and not hard or fibrous; apples should be crisp, firm and juicy, and not starchy, soft or dry; ice-cream should be smooth and creamy, and not icy or gritty. We prefer our biscuits to be crisp or crunchy rather than soft, and we like our bread to be soft or doughy or perhaps hard and crusty. We like our margarine to spread easily, preferably straight from the refrigerator.

We have recently witnessed the introduction of texturized vegetable protein which is physically or chemically manipulated so as to change the impression of texture that is ultimately perceived, such that it is similar to the texture of more familiar products such as meat. Meat replacers derived from fungal protein possess an important property in comparison with most other vegetable proteins in that they already have a similar fibrous structure to meat products.

Texture of foods is related to physical and chemical properties, perceived via eyes prior to consumption, the sense of touch in handling the food, various sensory receptors in the mouth during consumption and the sense of hearing. Thus the consumer is aware of a whole host of textural characteristics which derive from various physicochemical properties of the food such as overall size and shape, particle size, fat content, structure and mechanical properties. There have been many attempts to define texture concisely. One generally accepted definition is that texture describes the attribute of a food material resulting from a combination of physical and chemical properties, perceived largely by the senses of touch, sight and hearing. For a more detailed explanation of the sensory characteristics of foods, the reader is referred to an article by Thomson (1984).

5.3 TEXTURE ASSESSMENT BY SENSORY METHODS

The most important thing to bear in mind is that texture is more of a psychological or sensory characteristic than a physical property. Therefore, strictly speaking, texture can only be measured by sensory methods involving the use of both trained and untrained sensory assessors. However, sensory methods are very time consuming and the reliability of the results depends very much on correct design and skilful implementation of the experiments, and the aptitude and cooperation of the sensory assessors.

One method which has been developed over the last 20 years and is now extensively used is sensory texture profiling. Bourne (1982) makes a strong case that well-conducted texture profile techniques are objective tests, i.e. they are free from hedonic bias, and that results from different panels are reproducible to a high degree.

Texture profiling involves the use of a panel of trained sensory assessors to develop a list of 'texture words' to describe the textural characteristics perceived in a range of samples that are typical of a particular food or product. Testural characteristics are often associated with fairly distinct stages of the chewing process, i.e. the first bite impression, impression during mastication and residual impression. Once a complete list of appropriate descriptors has been decided by the panel, the magnitude of each textural characteristic can then be quantified on rating, category or graphic scales (see Pigott, 1984). Thorough interpretation of texture profile data usually requires fairly sophisticated statistical techniques, such as analysis of variance and principal component analysis (Piggott, 1984). A generalized texture form is given in Table 5.1.

It should be noted that textural characteristics are divided into three

Table 5.1 — Basic texture profile ballot.

PRODUCT   DATE

I INITIAL (perceived on first bite)
   (a) Mechanical
      (i) Hardness (1–9 scale)
      (ii) Fracturability (1–7 scale)
      (iii) Viscosity (1–8 scale)
   (b) Geometrical
   (c) Other characteristics (Moistness, oiliness)

II MASTICATORY (perceived during chewing)
   (a) Mechanical
      (i) Gumminess (1–5 scale)
      (ii) Chewiness (1–7 scale)
      (iii) Adhesiveness (1–5 scale)
   (b) Geometrical
   (c) Other characteristics (moistness, oiliness)

III RESIDUAL (changes induced during mastication and swallowing)
   Rate of breakdown
   Type of breakdown
   Moisture absorption
   Mouth coating

---

with permission from Dr. A. Szezniak.

classes, according to their likely origins, namely from mechanical geometrical and other properties of the food. These are listed in Table 5.2

Before using taste panels, consideration should be paid to the following points; selection of panel members, training of the panel, an accurate description of the textual characteristic and establishing standardized scales for various descriptive textural terms. Bourne (1982) gives examples of reference materials for establishing scales for hardness (Table 5.3). Materials may vary from place to place, depending upon availablity. Examples of reference materials, used in the USA and Bogota, to standardize a nine-point harness scale, are shown in this table.

Other sensory-testing methods, such as threshold tests, difference tests and scaling, have been reviewed by Brennan (1984). Further details on sensory-testing methods have been given by Piggott (1984) and Jellinek (1985).

5.4 TEXTURE EVALUATION BY INSTRUMENTAL METHODS

An attribute of texture is a manifestation of (or results from) a combination of physical and chemical properties; these include the size, shape, number, nature and arrangement of the constituent structural elements. They are often a reflection of the structure of a material, and so a structure of a material can often lead to a better understanding of its physical properties

Table 5.2 — Classification of textural characteristics according to their likely origins.

Mechanical characteristics	Geometric characteristics		Other characteristics
Hardness	Powder		Moistness
Cohesiveness	Chalky		Oiliness
Viscosity	Grainy	relating to particle size and shape	
Sponginess (elasticity)	Gritty		
Adhesiveness	Coarse		
Fracturability	Lumpy		
Chewiness	Beady		
Gumminess	Flaky		
	Fibrous	relating to shape and orientation	
	Pulpy		
	Cellular		
	Aerated		
	Puffy		
	Crystalline		

Table 5.3 — Hardness and viscosity scales in New York and Bogota.

Hardness scale value	Product (New York)	Product (Bogota)
1	Philadelphia cheese (Kraft)	Philadelphia cheese (Alpina)
2	Cooked egg white	Cooked egg white
3	Frankfurters (Mogen David)	Cream cheese (Ubate)
4	Processed cheese (Kraft)	Frankfurters (Suiza)
5	Pickled olives (Cresca)	Mozzarella cheese (LaPerfecta)
6	Peanuts (Planters)	Peanuts (La Rosa)
7	Carrot (raw)	Carrot (raw)
8	Peanut brittle (Kraft)	Candied peanuts (Colombina)
9	Rock candy	Milk candy (Columbina)

Adapted from the data of Bourne (1982).

and ultimately its textural characteristics. This concept is the basis for most instrumental methods for texture evaluation. Such instruments may measure a single physical property, but more often a composite of a number of physical properties of the material under test is determined. Thus, they are indirect methods of texture measurement and the results are only meaningful if they can be shown to relate conceptually and statistically to those obtained by sensory methods, which must be regarded as definitive.

Despite the problems involved, a large number of instruments have been developed, for texture evaluation, that rely upon the deformation of a sample. These are usually classified under three headings, namely fundamental methods, imitative methods and empirical methods. Each of these will now be discussed in turn.

5.4.1 Fundamental methods

This group consists of methods designed to measure one (or more) well-defined physical property of a sample under test and to relate this property to textural characteristics assessed sensorily. Such physical properties are as follows.

(1) Stress–strain relationships.
(2) Viscoelastic behaviour.
(3) Plastic–viscoplastic materials (see section 4.8.1).
(4) Liquid foods: viscous behaviour (see Chapter 4).

In all these cases the rheological behaviour of the food can be described mathematically and related to sensory characteristics.

Bourne (1982) concludes that there is generally a poor correlation between these and sensory methods. However, many of these tests will also give useful information about the behaviour of materials in such operations as crushing, grinding and pumping.

5.4.2 Imitative methods

This group of instruments which attempt to simulate, to some degree, the forces and deformations that the food is subjected to whilst the food is being consumed. Brennan (1984) states that texture is mainly assessed from the sensations caused when the food comes into contact with the hard and soft parts of the mouth. In addition to the teeth, the tongue, cheeks, palate and all other oral structures play some part in the masticatory process and in texture measurement. However, despite the extensive research, comparatively little is known about masticatory processes that can be applied to texture measurement, and the influence of this work on the design of texture-measuring instruments has been small. Recently, Bourne (1982) has produced a comprehensive review of mastication processes, with some interesting quantitative data on chewing efficiency, chewing rates, saliva production and composition, maximum forces exerted between teeth (including dentures) and average forces exerted when chewing different foods. The general foods texturometer, described in section 5.9.3, has been designed to simulate the biting process.

5.4.3 Empirical methods

These methods measure properties of materials that are often not well defined and cannot easily be expressed in fundamental terms. However, for certain types of food the results have been found to relate to one (or more) textural attribute and so they can be used as an indirect measurement of that attribute. The force can be applied in a wide variety of ways, such as penetration, shear, compression, extrusion, cutting, flow and mixing. Some of these mechanisms are indicated in Fig. 5.1 and are discussed in more detail in section 5.9.

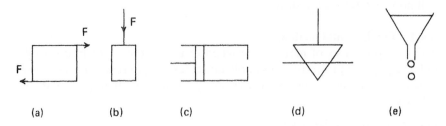

Fig. 5.1 — Illustration of some of the different methods available for subjecting foods to an applied force: (a) shear; (b) compression; (c) extrusion; (d) penetration; (e) flow.

An important aspect of all these tests is that the pattern of forces established within the test sample is difficult to establish, and so it is not possible to express the results in fundamental units. The results obtained are usually characteristic for that particular material, under the prescribed experimental conditions, and it is difficult to predict the effect of altering the probe size or speed without extended experimentation. Therefore, to ensure that any experimental work is reproducible, it is necessary to report the probe size, sample temperature, probe speed and any other pertinent information together with the actual experimental results.

5.4.4 Chemical and microscopic methods

In addition to rheological methods, some chemical methods are useful for evaluating texture, particularly for fruit and vegetables. Biochemical changes occur whilst the fruit and vegetables are growing and during subsequent storage, many of which have a direct effect on the texture. Analyses are performed to determine such factors as soluble pectin, total alcohol-insoluble solids, acidity, sugar, acid-to-sugar ratio and starch-to-sugar ratio. Further details of the relationship between these have been given by Arthey (1975) and Bourne (1983).

Examination of materials under the microscope may also produce useful information about the structure of the materials. In recent years the use of the microscope for textural studies has received much more attention. An excellent review on electron microscopy has been given by Kalab (1983). The use of the microscope for products such as emulsions, dairy products, baked products, meat and fish has been described by Peleg and Bagley (1983).

The rheological properties of foods obviously play an important role in textural evaluation as well as in the selection of the correct equipment in

Sec. 5.5] Fundamental properties 143

operations involving the deformation of foods. The more fundamental rheological properties will now be described.

5.5 FUNDAMENTAL PROPERTIES

When an object is subjected to a force in one direction, the object is said to

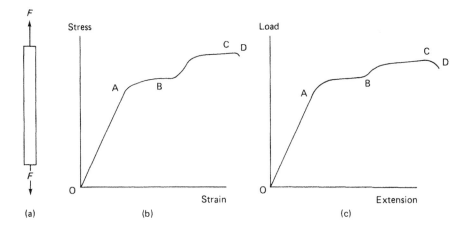

Fig. 5.2 — (a) Strain produced in a thin wire; (b) stress-strain relationship for stretching a material; (c) load–extension realtionship.

be stressed or under traction. This will produce a deformation or strain. Obviously, in the example shown in Fig. 5.2(a), the application of a force F results in the extension ΔL of the wire:

$$\text{stress} = \frac{\text{force}}{\text{area}}$$

$$\text{strain} = \frac{\text{extension}}{\text{original length}}$$

In the simplest situation the stresses and strains set up are in one direction. However, it should be apparent that, if the length increases, the cross-sectional area will decrease. Therefore, stresses and strains will be set up in a direction at right angles to the main force. If the material is prepared in the form of a thin wire, then the main stresses take place along the wire whereas, if the material is in the form of a cube, a considerable three-dimensional stress will be set up. One-dimensional stress situations are much easier to handle than are three-dimensional situations. Thus, by careful choice of experiemental conditions, the complex stress situation can be simplified, enabling only one stress component to be considered.

Further simplifications can be made by considering an ideal solid; the ideal solid is structureless (i.e. it has no atoms) and isotropic (i.e. it has the same properties in all directions and follows the respective laws relating shear stress to strain). Unfortunately, most foods do not fall into this category; they are anisotropic. Such ideal solids are often described as elastic. When an elastic material is subjected to a varying shear force (load) and the extension (or strain) is measured, the following pattern emerges (Figs. 5.2(b) and 5.2(c)).

Over the range OA, there is a linear relationship between the stress and the strain. Point A represents the elastic limit for the material. Within this range, the material immediately returns to its original length when the load is removed. If the material is taken above its elastic limit, very little additional force produces a large extension. At this point the material becomes plastic, and this transition is marked by the yield point. If the stress is further increased, the extension increases rapidly along the curve BC, until the wire snaps. The breaking stress is the *corresponding* force per unit area at point C. Materials taken above their elastic limit will be permanently deformed, once the stress is removed.

Substances that elongate considerably after the elastic limit and undergo plastic deformation are termed ductile (e.g. lead and copper). Substances that break just after the elastic limit has been exceeded are termed brittle (e.g. glass and high-carbon steel). Some materials appear to have no yield point; they increase in length beyond the elastic limit as the load is increased, without the appearance of a plastic stage.

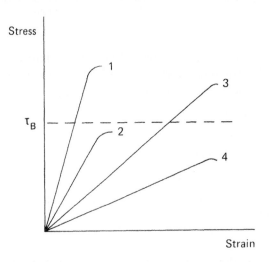

Fig. 5.3 — The relationship between stress and strain for a hard material (lines 1 and 2) and a soft material (lines 3 and 4): line 1, hard strong material; line 2, hard weak material; line 3, soft strong material; line 4, soft weak material.

Often the terms hard and strong are used to describe materials (together with the respective opposites soft and weak). A hard substance displays substantial resistance to deformation or breaking, i.e. a large shear stress produces a relatively small strain, whereas for soft materials a large shear stress will produce a large strain. Fig. 5.3 shows the stress–strain relationship for a hard material (lines 1 and 2) and a soft material (lines 3 and 4).

A strong material requires a large force to break it, whereas a weak material will break at a low shear stress. Materials which break at shear stresses below τ_B (Fig. 5.3) would be considered soft, and those which break above τ_B hard. (Note that τ_B is arbitrary to simplify the explanation.)

Strength and hardness are very important properties governing size reduction operations. The Mohs scale is used for minerals (Loncin and Merson, 1979), whereas reference materials used for foods have been listed by Bourne (1982), with examples shown in Table 5.3. In this case, hardness is defined as the force required to compress a substance between the molar teeth (for solids) or between the tongue and palate (for semisolids).

A material breaks or cracks along defects in the structure. For a large piece, which has many defects, a small stress may cause breakage but, as the size is reduced, fewer defects remain and the breaking strength increases; this makes grinding operations more difficult as the particle size decreases. Energy requirements during grinding are considered in more detail in section 5.10.

Over a reasonable range of shear forces, many materials are elastic. A further property of elastic materials is that, as soon as the shear stress is removed, the strain is instantaneously removed, i.e. there are no time-dependent effects (see section 5.6).

5.5.1 Young's modulus E
An ideal solid is one in which the deformation produced is proportional to the force producing it.

The stress or traction in a uniform cylinder of cross-sectional area A pulled by a force F is given by

$$\text{stress} = \frac{\text{force } F}{\text{area } A}$$

This produces a deformation, known as strain. The strain produced is given by e/L, where e is the extension and L is the original length. The modulus E of elasticity (also known as Young's modulus) for a material is given by

$$E = \frac{\text{stress}}{\text{strain}}$$

The modulus of elasticity can be likened to viscosity; as the modulus value increases, the material deforms less on application of a fixed force.

The simple model for an ideal solid is given by a spring (Fig. 5.4(a)). The equation describing such a model is

$$\text{stress} = E \times \text{strain}$$

146 Solid rheology and texture [Ch. 5

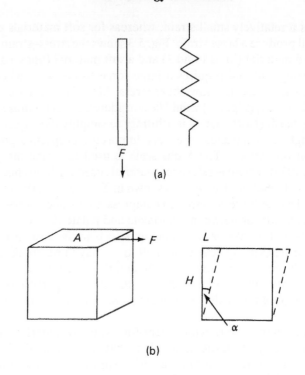

Fig. 5.4 — (a) Simple model for an ideal solid, i.e. a spring; (b) shear modulus.

Thus, it appears that the value of Young's modulus is a measure of the hardness of a material.

Muller (1973) describes a simple experiment in which Young's modulus can be measured for spaghetti, apple or potato tissue by stretching or bending. Some values of Young's modulus are given in Table 5.4. Finney

Table 5.4 — Values of Young's modulus

Material	Young's modulus E (N m^{-2})	Material	Young's modulus E (N m^{-2})
Soft foam rubber[a]	10^2	Carrots[b]	$(2–4) \times 10^7$
Rubber[a]	8×10^5	Pears[b]	$(1.2–3) \times 10^7$
Dry Spaghetti[a]	0.3×10^{10}	Potatoes[b]	$(0.6–1.4) \times 10^7$
Lead[a]	1.0×10^{10}	Apples raw[b]	$(0.6–1.4) \times 10^7$
Concrete[a]	1.7×10^{10}	Gelatin (gel)[b]	0.02×10^7
Glass[a]	7×10^{10}	Peach[b]	$(0.2–2) \times 10^7$
Iron[a]	8×10^{10}	Banana[b]	$(0.08–0.3) \times 10^7$
Steel[a]	25×10^{10}		

[a] From the data of Muller (1973).
[b] From the data of Finney (1973).

(1973) discusses how the modulus of elasticity can be used to assess the texture of bread (staling), chicken tenderization and the moisture and turgor in fresh vegetables.

5.5.2 Shear modulus

A further situation in which the three-dimensional stress situation is considerably simplified is shear. Imagine a cube of material subjected to a shearing force F over a surface area A (Fig. 5.4(b)). Then the shear stress τ is given by

$$\tau = \frac{F}{A}$$

The amount of shear is given by L/H or $\tan \alpha$. For shear, the modulus is called the shear or rigidity modulus G and is given by

$$G = \frac{\tau}{\tan \alpha}$$

For small angles, than $\alpha = \tan \alpha$, where α is the angle in radians, and so

$$G = \frac{\tau}{\alpha}$$

If the material is isotropic, the values of the shear modulus and Young's modulus are all that are required to describe its behaviour. However, for convenience, two further constants are often measured. These are Poisson's ratio and the bulk modulus.

5.5.3 Poisson's ratio

When a body is stretched or compressed, its width will almost always change. On compression of a cylinder, its diameter will increase and, on stretching, its diameter will decrease.

Poisson's ratio μ_P is the ratio of the lateral contraction (as a fraction of the diameter) to the longitudinal strain:

$$\mu_P = \frac{d/D}{e/L}$$

$$= \frac{dL}{De}$$

For a material where no volume change takes place when it is stretched or compressed, Poisson's ratio is 0.5. For materials which can be compressed, with no change in diameter (usually because there is much air within the structure, Poisson's ratio is zero. Some values of Poisson's ratio are given for common materials in Table 5.5. For fruit and vegetable tissue, Poisson's ratio decreases as the amount of air within the tissue increases, i.e. as the particle density decreases (see section 2.2).

Table 5.5 — Values of Poisson's ratio for some common materials.

Material	Poisson's ratio μ_P	Material	Poisson's ratio μ_P
Cheddar cheese	0.50	Copper	0.33
Potato tissue	0.49	Steel	0.30
Rubber	0.49	Glass	0.24
Apple tissue	0.37	Bread-crumbs	0.00
Apple	0.21–0.34		

5.5.4 Bulk modulus

When a body sinks in water, the pressure exerted on the body due to the increasing depth of water compresses the body. Conversely, when bubbles of air rise, they increase in size. The deformation caused by the hydrostatic pressure from all sides is called the volumetric strain. The ratio of the hydrostatic pressure to the volumetric strain is termed the bulk modulus K:

$$K = \frac{\text{hydrostatic pressure (N m}^{-2})}{\text{volumetric strain } \Delta V/V}$$

where ΔV is the change in volume and V is the original volume. The bulk modulus is a measure of compressibility of a substance. Some bulk modulus values are recorded in Table 5.6.

Table 5.6 — Values of Bulk modulus

Material	Bulk modulus k	Material	Bulk modulus k
Dough	1.4×10^6 N m^{-2}	Silver	10^{11} N m^{-2}
Rubber	1.9×10^7 N m^{-2}	Steel	1.6×10^{11} N m^{-2}
Liquid	1×10^9 n m^{-2}	Glass	5.8×10^6 lb in^{-2}
Granite	3×10^{10} N m^{-2}	Steel	26.0×10^6 lb in^{-2}

Taken from the data of Mohsenin (1970) and Muller (1973).

5.5.5 Relationship between these properties

If the material is isotropic, the four elastic constants that we have considered are interconvertible and, if two of them are known, then any of the others can be calculated. If the properties of solids differ in different directions (anisotropic) more constants are required. The relationship between these properties are summarized as follows:

$$\text{shear modulus } G = \frac{9EK}{9K - E}$$

$$\text{Young's modulus } E = \frac{9GK}{3K + G}$$

$$= 2G(1 + \mu_P)$$
$$= 3K(1 - 2\mu_P)$$

$$\text{bulk modulus } K = \frac{E}{3(1 - 2\mu_P)}$$

$$= \frac{EG}{9G - 3E}$$

$$= \frac{2G(1 + \mu_P)}{3(1 - 2\mu_P)}$$

$$\text{Poisson's ratio } \mu_P = \frac{E - 2G}{2G}$$

$$= \frac{(1 - E)/3K}{2}$$

These properties can be determined for materials by a series of relatively simple experiments some of which have been described by Muller (1973). It might be expected that such properties would have an effect on food texture.

5.6 VISCOELASTIC BEHAVIOUR

A viscoelastic material is one that exhibits both elastic and viscous properties; it differs from a plastic fluid, however, in the sense that both properties are exhibited at the same time (contrast the plastic fluid, where above a yield shear stress the substance behaves like a viscous fluid, whereas below the yield stress it behaves like a solid (see section 4.8.1)). A further property of a viscoelastic material is that, when the shear stress is removed, the strain in the material is not immediately reduced to zero (Fig. 5.5)

Many food materials have been described as being viscoelastic; some examples are dough, cheese and most gelled products.

The behaviour of many of these materials is extremely complex, but there are five major techniques that have been used to analyze them.

(1) Examination of deformation against time at a constant shear stress during the application of a load and after removal of the load.
(2) Stress relaxation at a constant shear.
(3) Use of the Weissenberg effect.
(4) Oscillation techniques.
(5) Empirical methods.

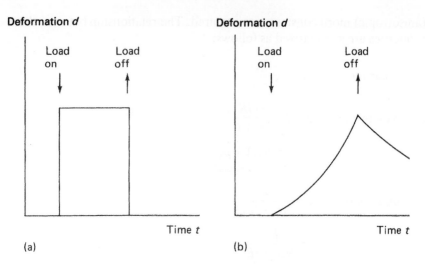

Fig. 5.5 — Curves of deformation d for (a) an elastic material and (b) a viscoelastic material.

5.6.1 Deformation–time relationships

Fig. 5.6 shows how the strain or extension changes with time when a viscoelastic material is subjected to a constant shear stress, followed by removal of the shear stress.

When the load is initially applied, there is a fairly rapid elastic deformation followed by a more retarded deformation BC. When the load is removed, there is an almost instantaneous elastic recovery CD, followed by a retarded recovery DE. The material never fully reaches its original length, and a permanent deformation E is recorded.

Creep function is the term used to describe the change in strain at a constant stress, and creep compliance is the ratio of the stress divided by the strain.

When such experiments are performed at various shear stress values, a series of curves results (Fig. 5.7). A sample is said to exhibit linear viscoelasticity if there is a straight-line relationship between the strain and the applied stress, when the samples subjected to the different stresses are measured after the same time interval.

At high shear stress values the sample may rupture. As the shear stress is reduced, the nature of the response changes and the permanent deformation decreases. Again, to ensure that results are capable of being reproduced, all experimental variables should be controlled as far as possible and described.

5.6.2 Stress relaxation experiments

In this type of experiment the material is subject to a shear stress and the stress is then allowed to relax at a constant strain, i.e. the length is kept constant. Fig. 5.8 shows the decay in stress for a viscoelastic material at a constant strain.

The relaxation time is the time required for the stress at constant strain to

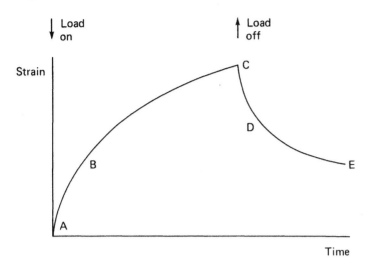

Fig. 5.6 — The relationship between deformation and time for a viscoelastic material: AC, application of load; CE, after removal of load.

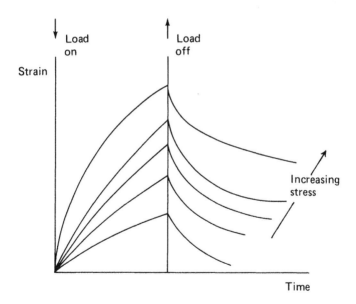

Fig. 5.7 — Deformation curve for different shear stress values.

fall to 1/e (where the exponential function e equals 2.718) of its original value (i.e. 36.8% of its original value). If this time is excessive, some other arbitrary relaxation time may be considered, e.g. the time for the stress to fall to 70% or 50% of its original value.

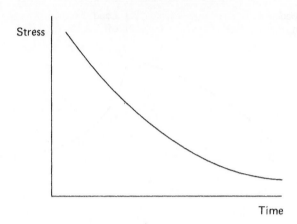

Fig. 5.8 — The relationship between stress and time during the application of a constant strain.

Muller (1973) has stated that the relaxation time is equal to the ratio μ/G where μ is the viscosity coefficient and G is the shear modulus. High relaxation times are associated with materials in which the viscous forces predominate and low relaxation times where the elastic forces predominate.

5.6.3 Weissenberg effect
When a Newtonian fluid is agitated in a beaker, the circular motion causes a

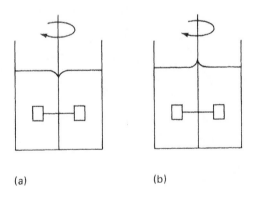

Fig. 5.9 — (a) The behaviour of a normal fluid in a stirred tank; (b) the behaviour of a viscoelastic fluid, exhibiting the Weissenberg effect.

vortex (Fig. 5.9(a)). If the procedure is repeated for a viscoelastic fluid, the fluid may climb up the rod (Fig. 5.9(b)). This phenomena, called the Weissenberg effect, has been noticed in aged condensed milk, cake batter, eucalyptus honey, wheat flour dough and malt extract. This is due to the production of a normal force acting at right angles to the rotational forces, which in turn acts in a horizontal plane. The normal force has the effect of pushing the fluid up the shaft. Viscoelastic fluids can be characterized by a modified cone-and-plate viscometer known as a rheogoniometer. The modifications are made so that both the normal rotational forces (the torsion forces) and the normal forces can be measured.

5.6.4 Oscillatory methods
Oscillatory methods involve subjecting the sample to a harmonic shear

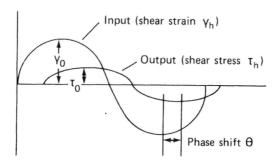

Fig. 5.10 — The relationship between harmonic shear strain γ and harmonic shear stress τ for a viscoelastic material.

strain and measuring the corresponding shear stress set up in the sample (Fig. 5.10). Cone-and-plate or concentric-cylinder viscometers are useful for this purpose with the test fluid being placed in the gap between the measuring unit.

Consider a harmonic shear strain γ_h of amplitude γ_0 gives rise to a harmonic shear stress τ_h of amplitude τ_0 which is out of phase with the shear strain by an angle θ. The harmonic shear strain is set up in one of the elements and the corresponding shear stress is detected by the other element. If the phase shift is θ, then $\theta = 90°$ for an ideal viscous liquid and $\theta = 0°$ for an ideal elastic material, where $\tan \theta$ is known as the loss factor.

A storage modulus and loss modulus are defined as follows:

$$\text{storage modulus} = \frac{\tau_0 \cos \theta}{\gamma_0}$$

which is high for elastic materials;

$$\text{loss modulus} = \tau_0 \frac{\sin \theta}{\gamma_0}$$

which is high for viscous materials.

Rheological and viscoelastic properties are useful because not only do they describe the flow characteristics of materials but also they can be used to monitor certain processes, e.g. curd formation in cheese and yoghurt, as well as in the quality control and assessment of products. Further information has been provided by Muller (1973) and Sherman (1979).

5.6.5 Empirical methods
Empirical methods play a very important role, particularly in bakery applications, because they give results that allow the prediction of the handling and baking quality of different flours.

Two systems which are commonly used for viscoelastic materials are the Simon extensometer and the Brabender system. Other empirical systems are mentioned later in this chapter.

5.6.5.1 The Brabender system
The Brabender system is basically a mixing unit which is commonly used for flours; the flour is placed in a mixing bowl and water is run in from a burette. As the dough is being mixed at a constant temperature (30 °C), the torque required is automatically recorded. When a consistency of 500 empirical units is reached, the mixer is stopped and the amount of water needed is recorded; this is known as the water absorption of the flour. The experiment

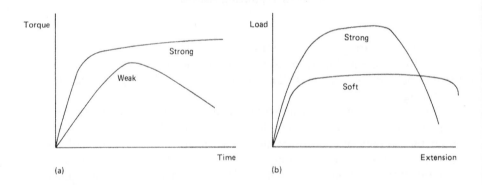

Fig. 5.11 — The contrast (a) between a strong flour and a weak flour using the Brabender system and (b) between a strong and a soft flour using an extensometer.

is repeated by adding the determined amount of water to the flour at once, followed by mixing of the material The corresponding trace of the torque against time for a strong (12% protein) and weak (8% protein) flour is shown in Fig. 5.11(a).

It may be interesting to know that one Brabender unit (1 BU) is equivalent to a torque of one meter gram-force (1 m gf), i.e.

$$1 \text{ BU} = 1 \text{ gf m} = 9.8 \times 10^{-3} \text{ N m}$$

The Brabender amylograph is also widely used for determining the viscosity of starch pastes and starch-based cereal gruels. In this case the starch suspension or gruel is heated at a controlled rate (usually 1.5 °C min^{-1}) whilst it is being mixed, from ambient temperature to 95 °C, and the torque is continuously monitored. It is then held at 95 °C for approximately 20 min, followed by cooling at a controlled rate (Fig. 5.12). Collison (1968), shows some traces for some different starch products. The interpretation of these curves has been described by Radley (1976), as follows.

(1) 'The highest viscosity that the user may encounter during the preparation of a usable paste is indicated irrespective of the temperature (peak viscosity).
(2) The viscosity of the paste, when it reaches the temperature of 95 °C in relation to the peak viscosity, reflects the ease of cooking starch.
(3) After cooking for 1 h at 95 °C, the viscosity curve indicates the stability or breakdown of the paste.
(4) The viscosity of the cooked paste after cooling to 50 °C is a measure of the thickening produced by cooling.
(5) The final viscosity after stirring for 1 h at 50 °C indicates the stability of the cooked paste to mechanical treatment.'

Other components such as proteins, sugars and fats affect the viscosity of starch suspensions. These interactions can be checked using the Brabender system. The Brabender system has been criticized on the grounds of poor reproducibility. Using starch pastes, Kempf and Kalender (1972) have compared the Brabender system with the Haake rotational viscometer and concluded that the results from the Haake viscometer were much more reproducible.

5.6.5.2 The extensometer or extensograph
This instrument forms a cylinder of the dough containing the flour, and a standard amount of water, determined using the farinograph. After a standard resting time at a constant temperature, the dough is stretched until it tears. Suring the stretching, the load is plotted against the extension. Typical plots for a strong and soft flour are shown in Fig. 5.11(b).

5.7 GELATION
Gelation has been described by Glicksman (1982) as the association or cross-linking of long polymer chains to form a three-dimensional continuous network which traps and immobilizes the liquid within it to form a rigid structure that is resistant to flow, under pressure. The main constituents responsible for gelation are hydrocolloids and proteins. All hydrocolloids have thickening properties, but relatively few have gelling properties. Glicksman (1982) gives a useful table showing the gelling properties and characteristics of hydrocolloids; tabulated are the solubilities, effects of electrolytes and heat, gelling mechanisms, additives, appearance, type of

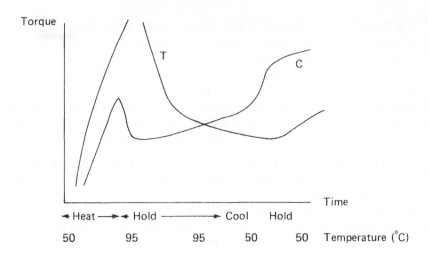

Fig. 5.12 — Typical results from a Brabender system for two different types of starch: T, tapioca; C, corn.

texture, and food applications. For example, pectin is a thermal setting gel which requires sugar and acid to gel but which is not affected by the presence of electrolytes; the gel is clear and spreadable and is used in jams and jellies. Similar information is given for κ-carrageenan, locust bean gum, furcellaran agar, sodium alginate, gum arabic and starches.

The mechanisms of gelation remain poorly understood. Most carbohydrates require heat for gelation; this is believed to provide energy to alter the structure of the molecule, making it more amenable to cross-linking. Often specific ions are required to complete the gelling process; furthermore the gel network may then shrink during storage, liberating free water, this process being known as weeping or syneresis. Factors involved in gelation are the formation of random chains and cross-linking of the chains, either by hydrogen bonds or by salt bridges. Some gels are clear and thermally reversible, e.g. gelatin; others are cloudy or opaque, e.g. egg white.

Alginates and low methoxyl pectin will produce instantaneous gels by addition to calcium ions; this is known as chemical gelation and it forms the basis of the production of reconstituted fruit pre-fillings from a non-uniform staring material. The fruit pieces are macerated with alginate and extruded into a batch of calcium salt. Mechanisms have been discussed by Glicksman (1982).

The most well known of the protein gelling agents is gelatin; gelatin forms a reversible gel, i.e. one which sets at low temperatures and melts at higher temperatures over the full pH range without the addition of sugar or metallic cations. No syneresis occurs. This is in contrast with egg white protein which produces a white irreversible gel. Kinsella (1976) has reviewed the gelation properties of proteins extracted from milk, whey vegetable proteins such as soya, cotton seed and leaf protein. More recently, Howell and Lawrie (1984) have studied the gelation of fractionated samples of blood plasma. The gelling ability of most proteins is affected by the pH, the ionic environment and the presence of other substances. The mechanism involved in gel formation for proteins are complex and may involve salt bonds between amino and carboxyl groups in the side chain, hydrogen bonding between peptides and disulphide cross-linkage.

Gelation can be measured in two ways. The first involves measuring the time for a solution of the material to form a gel when it is heated at a constant temperature. For proteins, the concentration used is between 8% and 12% and temperatures between 80 °C and 140 °C. The second method involves measuring some rheological property of the gel such as strength, hardness or compressibility.

Enzymes such as those present in rennet will also cause the casein in milk to gel; this is very important in cheese making; gelation is affected by the previous heat treatment and the concentration of soluble calcium. Gelation mechanisms have been reviewed by Webb et al. (1974) and Fox (1982).

Gelation may be unwanted in certain circumstances. UHT milk may gel during storage if the bacterial proteases in the raw milk are not completely inactivated by the heating process (Mehta, 1980). This is most likely to occur if the milk has been stored for long periods under refrigerated conditions, where psychrotrophic bacteria are most active; they excrete proteases into the milk and, if these are present at high concentrations, they may not be completely inactivated, even at temperatures exceeding 140 °C.

5.8 MODEL SYSTEMS

The rheological behaviour of quite complex materials can be explained in terms of the basic elastic property (i.e. E) and the basic viscous propery (i.e. μ) by means of model systems.

The Hookean solid is represented by a spring. The deformation produced by a shear stress τ is proportional to the force applied and takes place instantaneously.

The Newtonian liquid is represented by a dash pot where a plunger is pulled through a liquid. The rate of deformation is proportional to the shear stress applied. When the stress is removed, there is no change in the deformation.

An additional element known as the St Venant slider is included; this introduces the concept of a limiting frictional force. Below the limiting force, no deformation is produced; above the limiting it slides freely, offering no resistance.

These three elements are shown in Fig. 5.13, together with their respective deformation–time responses (the St Venant slider is shown below the limiting force). Models for complex materials are made up of combinations of these elements arranged in series or parallel. Figs. 5.13(d)–5.13(f) shows such examples for a Bingham model, a Kelvin–Voigt model and a

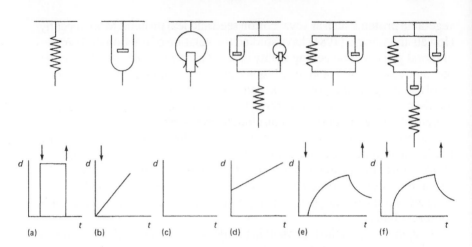

Fig. 5.13 — Model systems showing deformation d–time t responses: (a) Hookean solid; (b) Newtonian liquid; (c) St Venant slider; (d) Bingham plastic; (e) Kelvin–Voigt model; (f) Burgess model.

Burgess model. In each case the deformation–time response is given in the diagram, on application of a force and after removal of that force.

More information about the use of models has been given by Muller (1973) and Mohsenin (1970).

5.9 OBJECTIVE TEXTURE MEASUREMENT: EMPIRICAL TESTING

The two major types are force- and distance-measuring equipment. Force-measuring equipment measures the force required to puncture, crush, deform or extrude a food. The probe is pushed into the food, causing irreversible crushing or flow of the food; the depth of penetration is kept constant and the force exerted by the food is indicated or recorded. In distance-measuring equipment, the material is subjected to a constant force and the deformation is measured. One of the most widely used instruments in this category is the cone penetrometer. Empirical testing equipment varies widely in cost and complexity. Many of the more sophisticated instruments have the capacity to utilize both systems.

5.9.1 Penetrometer

The penetrometer is a reasonably simple device for measuring the distance that a cone or rod penetrates into a food, in a fixed time. In its simplest form the cone C is positioned on the surface S of the food and is then released for a fixed time (Fig. 5.14). At the end of the time the probe is clampled Cl and the penetration depth is measured using a dial gauge D. The penetration depth will depend upon the weight of the cone and the angle, the type of the material, its temperature and the penetration time. Experimental con-

Sec. 5.9] Objective texture measurement: empirical testing

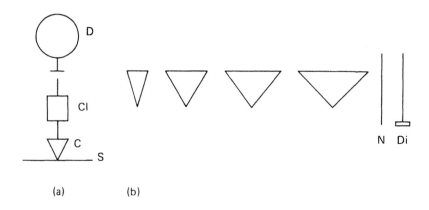

Fig. 5.14 — (a) The cone penetrometer showing the cone C, the clamp Cl, the dial D and the surface S of the food; (b) variation in cone size and the use of other shaped attachments such as needles or rods N and disks Di.

ditions should be uniform throughout. For materials such as butter and margarine, several readings should be taken at different positions on the surface of the sample to minimize deviations which arise because the samples are not homogeneous.

The penetrometer has been found to be extremely useful for assessing the yield stress of plastic materials, such as butter, margarine and similar spreads. The International Dairy Federation (1981) gives the following relationship between penetration depth and apparent yield stress:

$$\tau_0 = \frac{gm}{P^2 \tan^2(\tfrac{1}{2}\alpha_C)}$$

where P (m) is the penetration depth after 5 s, m (kg) is the mass of the cone, α_C (deg) is the cone angle and τ_0 (Pa) is the apparent yield stress. The apparent yield stress was found to correlate well with spreadability, assessed by sensory methods. Another type of instrument widely used on fruit and vegetables is the puncture tester. These instruments have been described in more detail by Bourne (1982).

5.9.2 Extrusion and extruders

In extrusion, the force required to extrude a material through a small orifice or an annulus is determined. One piece of equipment developed by Prentice (1954), designed specifically for extrusion, is the FIRA–NIRD (Food Industry Research Association/National Institute for Research in Dairying) extruder. A sample is removed from the bulk material in a cylindrical borer about $1\tfrac{1}{2}$ in long and with a diameter of $\tfrac{1}{2}$ in. This method enables a representative sample to be drawn without damage. The sample is then forced from this cylinder through an orifice of diameter $\tfrac{1}{8}$ in by a plunger moving at a uniform speed. The thrust required to extrude the material is recorded on a moving chart throughout the test, and this chart serves as a record of the firmness and various other rheological properties of the material The force required to extrude the material is measured by a series of interchangeable leaf springs.

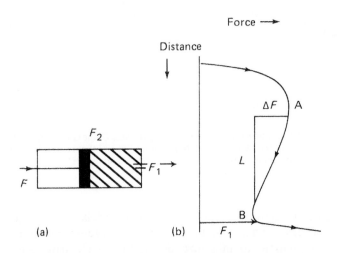

Fig. 5.15 — (a) The forces set up during extrusion, namely the extruder thrust F_1 and the frictional force F_2, giving a total force of F_1 and F_2; (b) the relationship between force and time during the extrusion process.

When the instrument is switched on, the carriage starts to move and causes a deflection of the spring via the thrust rod (Fig. 5.15(a)). This produces a gradual linear increase in the thrust on the piston, which is shown as a sloping straight line on the chart (Fig. 5.15(b)). As soon as this thrust is sufficient, the material begins to be extruded from the orifice and the trace begins to curve over (point A). the recorder thrust is then a sum of the forces required to overcome the frictional resistance between the plug of material and the wall of the sample cylinder and the pressure required to force the material through the orifice. As more and more of the material passes out of the cylinder, the frictional resistance decreases and the recorded thrust falls to a minimum value at the point where a projection on the thrust rod hits a stop on the travelling carrieage (point B). The trace then follows another sloping straight line parallel to the first as the pressure rises rapidly until the machine is automaticaly switched off. The extruder thrust is F_1 and the extruder friction (slope) is $\Delta F/L$. The values of extruder friction and extruder thrust are recorded for each sample, together with the sample temperature.

For edible fats the extruder thrust has been found to correlate very closely with subjective assessments by a skilled taste panel of the 'spreadability' of the fat and to correlate inversely with the subject assessment of firmness. Extruder friction can be regarded as a measure of 'stickiness' and

Sec. 5.9]   Objective texture measurement: empirical testing

will be a useful measure for some materials. Many of the more sophisticated instruments, such as the Instron machine also have extrusion units as one of the test principles (section 5.9.4).

5.9.3 General Foods texturometer
The instrument is designed to simulate the mastication process. The food is compressed twice, simulating the first two bites, by pushing a plunger a standard distance, usually 35 mm into the food and measuring the force exerted. There are two mastication speeds, and bite-size objects are used. The equipment has been described in more detail by Bourne (1982); he has also given a wide range of applications. The major parameters measured are hardness, cohesiveness, adhesiveness and crispness for which definitions relevant to this instrument have been given by Bourne (1978).

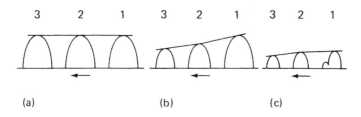

Fig. 5.16 — The response for the General Foods texturometer towards different types of food: (a) perfect elasticity; (b) retarded elasticity, irrecoverable; (c) brittle and fracturable. (Adapted from the data of Bourne (1983).)

Typical response curves for a perfect elastic food, a retarded elasticity and a brittle and fracturable food are shown in Fig. 5.16.

5.9.4 Instron universal testing machine
The Instron machine is used for a wide variety of materials for studying stress–strain relationships by either compressing or stretching. It can also be used for more sophisticated tests, such as hysteresis stress relaxation, stress recovery, strain rate sensitivity and rupture, and to assess the energy required for deformation. With food materials it is usually used in the compressive mode.

It consists of a drive unit, which drives a horizontal cross-unit in a vertical direction, and a force-sensing and recording system consisting of a series of interchangeable load cells, the output from which is fed to a chart recorder. A typical unit will allow full-scale deflection over the force range from 2 N to 5 kN. A detailed description for calibrating the load cell has been given by Bourne (1982). The load cell can be attached either to the moving cross-head or to the fixed base. A wide variety of test principles are available, e.g. puncture, snap, extrusion and deformation. When the load cell is attached to the moving cross-head, the moving part of the test cell is also attached to the cross-head and the fixed part of the test cell to the base.

Bourne (1982) has summarized the use of the Instron machine for a wide variety of foods, together with the most appropriate testing conditions, in a very comprehensive table. For example, canned green beans were evaluated by means of back extrusion using a cylinder with an internal diamter of 10.15 cm with an annulus of 4 mm, a cross-head speed of 300 cm min^{-1}, a chart speed of 10 cm min^{-1} and a full-scale force of 1000 N. The most important parameter used for characterization was the maximum plateau force. Bananas were measured using deformation, apple slices using puncture and chocolate bars using snap. All these conditions chould be recorded, together with environmental conditions such as temperature and relative himidity (for hygroscopic materials).

5.9.5 Other instrumental methods

Texture profile analysis involves quantifying a number of different parameters from the stress–strain curves obtained in one test. Thus the general foods texturometer has been used to evaluate hardness fracturability, cohesiveness (see section 5.9.4) and the Instron machine to assess fracturability, hardness, stringiness and springiness. A generalized texture profile curve has been given for the Instron machine by Bourne (1978). These descriptive terms have also been described by Arthey (1975) and in the *British Standard Glossary of Rheological Terms* (British Standards Institution, 1975).

A wide variety of empirical equipment is available, which makes use of the basic test principles of penetration, shear, compressing extrusion, flow and mixing (see Fig. 5.1). One instrument worthy of note is the tenderometer. This measures the force required to crush peas and similar agricultural produce and is widely used to determin the maturity of the crop. The tenderometer reading affects whether the peas are used for freezing, canning and dehydration and also the price paid to the producer by the food processor. Further details of the construction and calibration of the instrument have been given by Arthey (1975). Bourne (1982) has given a comprehensive account of empirical testing equipment, together with addresses of suppliers and approximate prices. Some of the instruments described are listed in Table 5.7.

There is a vast amount of literature in which textural measurements on different food products are described. Some representative examples are given in Table 5.8.

5.10 SIZE REDUCTION AND GRINDING

Size reduction is a very important operation in food processing, its main effect being to reduce the size and to increase the surface-area-to-volume ratio. This will help to increase the rate of diffusion processes which occur in dehydration and solvent extraction. It is commonly used as a pre-treatment operation for raw materials prior to canning, freezing and dehydration or for

Table 5.7 — Some empirical testing systems

Ottowa measuring system	TUC cream corn meter
Bloom gelometer	Succulometer
Steven's texture analyser	Maturometer
Bostwick consistometer	Christel texture meter
Ridgel meter	Armour tenderometer
Haugh meter	Marine colloids gel tester
Adams consistometer	

Table 5.8 — Texture evaluation for a range of foods

Commodity	Examples	Methods	References
Fruit and vegetables	Peas, apples, strawberries	Puncture testing, extrusion deformation. chemical methods, microscopy	Arthey (1975) Bourne (1975) Bourne (1983)
Dairy products	Yoghurt, cheese, butter	Extrusion, flow, penetration	Taneya et al. 1979) Prentice (1979) Deman et al. (1979) International Dairy Federation (1981 Prentice (1984)
Animal produce	Meat, fish	Shear tensile, microscopy	Stanley (1976) Howgate (1977) Stanley (1983)
Cereals	Flour, gruels, dough	Viscoelasticity, gelation flow properties	Baird (1983) Zangger (1979)
Oils and fats	Vegetable oils, butter, margarine, low fat spreads	Flow, extrusion	International Dairy Federation (1981) Deman et al. (1979) Deman (1976)
Texturized and processed goods	Bread, cake, snack foods, dehydrated products		Schweingruber et al. (1979) Flink (1983) Taranto (1983)
Confectionery	Chocolate, fondant, caramel		Tscheuschner and Wunsche (1979) Kleinert (1976)

the production of finer powders which are subsequently more easily handled or transported.

The effectiveness of a size reduction operation can be assessed by particle size analysis, using sieves, microscopy or more recently a Coulter counter which detects a change in impedance due to the small particles. Results are normally presented as a distribution of particle sizes. It may be necessary to define criteria for the grinding process. One may want to produce particles within a narrow particle size range or a wide range, or to ensure that all particles are below a certain size. A combination of grinding and some separation technique may be necessary to achieve the final objective. Air classification is a technique which is currently being investigated for separating powders, based on their particle size and density. Sieving methods are satisfactory in most cases.

The type of equipment used will depend upon the nature of the material, particularly its rheological properties, and the size reduction required. The cost of the grinding operation will depend very much on the final particle size. Some of the major mechanisms for effecting size reduction have been discussed by Loncin and Merson (1979). As described, foods vary considerably in their rheological properties, ranging from meats which may be soft and fibrous through to other materials which are extremely hard.

The energy required for grinding processes in theory is governed by the surface energy σ requirements, i.e. the force required to produce a new surface σA; only sufficient energy is required to exceed the crushing strength of the material.

However, in practice, it has been estimated that this only accounts for 0.06–1% of the total requirement. Much energy is lost in friction and other inefficiencies within the process. The resulting high temperature may well affect the properties of food.

Most of the equations used for estimating energy requirement for crushing are based on the following equation:

$$\frac{dE_r}{dD} = -\frac{K_m}{D^n}$$

where dE_r is the energy required to produce a small dange dD in diameter and K_m is a characteristic of the material. The three main equation result from the use of three different values of n: for $n = 2$,

$$E_r = K_m \left(\frac{1}{D_2} - \frac{1}{D_1} \right) \quad \textit{Rittinger's Law}$$

for $n = 1$,

$$E_r = K_m \ln \left(\frac{D_1}{D_2} \right) \quad \textit{Kick's law}$$

and for $n = \frac{3}{2}$,

$$E_r = 2K_m\left(\frac{1}{\sqrt{D_2}} - \frac{1}{\sqrt{D_1}}\right) \quad \text{Bond's law}$$

where D_2 is the final diameter and D_1 in the initial diameter. Equipment for size reduction has been described in more detail by Brennan *et al.* (1976) and Coulson and Richardson (1978).

5.11 EXPRESSION

Expression is the act of expelling a liquid from a solid, either by squeezing or by compaction. It is used for a variety of purposes such as recovering fruit and vegetable juices, recovering oil from seeds and consolidating cheese by removal of the whey. Further examples together with the pressure ranges used are given in Table 5.9.

Table 5.9 — Force required for a range of commodities.

Commodity	Pressure (lbf in^{-2})
Oil seeds	16000
Sugar cane extraction	600
Sugar beet extraction	500–1000
Spent coffee grounds	1000–2000
Fruit juice extraction	200–450
Hard cheese pressing	6–12

Expression processes can be either batch or continuous, and their efficiency is monitored by the yield and solids content of the liquid obtained. Expression can usually be divided into an induction period, during which the air is expelled from the press cake pores and the pores gradually fill with exuded liquid and an outflow period. Some of the factors which affect the rate of expression, such as the properties of the cake, e.g. its particle size distribution and degree of compaction, and the pressure drop over the cake which depends upon the applied pressure are discussed by Schwartzberg (1983).

5.12 SYMBOLS

A	surface area
d	deformation
D	diameter

e	extension
E	Young's modulus
E_r	energy requirement
F	force
F	extruder thrust
g	acceleration due to gravity
G	shear or rigidity modulus
H	height
K	bulk modulus
K_m	material characteristic (size reduction)
L	length
ΔL	change in length
m	mass
P	penetration depth
V	original volume
ΔV	change in volume

5.12.1 Greek symbols

α	angle of deformation
α_c	cone angle
γ_h	harmonic shear stress of amplitude γ_0
θ	phase shift
μ_P	Poisson's ratio
σ	surface energy
τ	shear stress
τ_B	breaking shear stress
τ_h	harmonic shear stress of amplitude τ_0
τ_0	apparent yield stress

6
Surface properties

6.1 INTRODUCTION

Many food systems are either two-phase systems or mixtures of immiscible liquids. Therefore there is a boundary between the two phases; this boundary is called either a surface for gas–liquid systems or, an interface for liquid–liquid systems.

Many food systems are also colloid in nature; colloid science is concerned with systems where one or more of the components has dimensions within the range from 1 nm to 1 µm, i.e. systems containing large molecules or small particles.

The phases are distinguished by the terms dispersed phase (for the phase forming the particles) and continuous phase or dispersion medium (for the medium in which the particles are distributed). Some colloidal systems are listed in Table 6.1.

Table 6.1 — Examples of food colloidal systems

Dispersed phase	Continuous phase	Name	Example
Liquid	Gas	Fog, mist, aerosol	Spray in spray drying
Solid	Gas	Smoke, aerosol	Smoke
Gas	Liquid	Foam, bubble	Whipped cream
Liquid	Liquid	Emulsion	Mayonnaise
Solid	Liquid	Sol, colloidal solution, gel suspension	Cloudy fruit juice, chocolate-flavoured drinks
Gas	Solid	Solid foam	Meringue, bread

Milk is a complex colloid in which the fat is dispersed as fine globules in an aqueous phase containing lactose, minerals and proteins in true solution, together with a colloidal dispersion of the protein casein in micellar form.

The casein micelles are stabilized by colloidal calcium phosphate. Milk is potentially unstable, the fat will easily separate and the casein can be precipitated by lowering the pH, adjusting the ionic environment or freezing the milk. Removal of the calcium by ion exchange will cause the casein micelles to dissociate and the casein becomes soluble. The colloidal system in milk has been discussed in much more detail by Webb *et al.* (1974), Dickinson and Stainsby (1982), Graf and Bauer (1976) and Walstra and Jenness (1984).

One important aspect of all colloidal systems is their high surface area-to-volume ratios and the high contact area between the dispersed and continuous phase. When a food dispersion is produced, e.g. a mayonnaise is made or milk is homogenized, there is a drastic increase in the surface area of the dispersed phase.

Table 6.2 shows some of the changes that take place when a spherical

Table 6.2 — Changes in the properties of a sphere on size reduction

Original radius (cm)	Number of spheres	Volume per spheres (cm^3)	Total area (cm^2)	Number of water molecules per sphere	Fraction of total water molecules at the surface
1	1	4.19	1.26×10^{-1}		
1.25×10^{-1}	512	8.18×10^{-3}	1.01×10^2	2.19×10^{24}	8.37×10^{-8}
3.13×10^{-2}	3.28×10^4	1.28×10^{-4}	4.02×10^2	3.42×10^{19}	3.35×10^{-7}
10^{-4}	10^{12}	4.2×10^{-12}	1.26×10^5	1.40×10^{11}	2.09×10^{-4}

Adapted from the data of Heimenz (1977).

water droplet of radius 1.0 cm is dispersed into a number of smaller spheres of equal size. As the number of spheres increases, there are drastic increases in the total surface area, and also the fraction of the total number of molecules at the surface.

We shall see later that we need to put in energy to extend the surface; furthermore, there is a large increase in the free surface energy and, because of this, such systems are thermodynamically unstable; they may need to be artificially stabilized to maintain their new structure. Any operation in which new surfaces are formed will involve overcoming the forces predominant on those surfaces.

If we take two immiscible liquids (e.g. oil and water) and mix them, there is the possibility of forming an oil-in-water emulsion or a water-in-oil emulsion. The type of emulsion will depend upon the relative volumes of the two phases, the order of addition and the type of emulsifying agent used. During some processing operations, phase inversion may occur. One example is in the churning of cream (a 40% fat-in-water emulsion) to form butter (water-in-oil emulsion, containing about 84% fat). This releases 'butter-milk', which is the remainder of the aqueous phase of the original cream.

Another important aspect of the surface reactions is concerned with the wetting of solid surfaces by liquids and the use of substances which will dissolve water-soluble and fat-soluble components; this is particularly important in the cleaning and sanitizing of both internal and external surfaces of food processing equipment by cleaning-in-place techniques.

6.2 SURFACE TENSION

We shall now examine the forces acting on a gas–liquid interface.

Within the bulk of a fluid, short-range van der Waals forces of attraction exist between molecules. A molecule in the bulk of the solution will be subjected to forces of attraction from all directions (Fig. 6.1). These

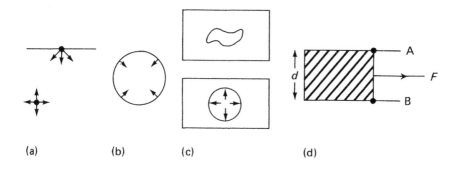

Fig. 6.1 — (a) Forces acting on molecules in the bulk liquid and at the surface;(b) forces acting on a spherical droplet; (c) forces acting on a piece of string placed on the surface of a fluid before and after the addition of a surface-active agent to the fluid enclosed by the string; (d) the force required to extend a surface.

effectively cancel each other out. However, a molecule on the surface of the liquid will be subjected to a net attraction by the molecules in the bulk of the solution. Hence the surface of the liquid is in a state of tension. Work would be required to extend the surface or for a molecule to escape from the surface, e.g. by evaporation. Therefore, there are forces acting on the surface which are attempting to minimize the surface area. Most droplets are spherical in shape because the sphere is the shape that offers the minimum surface area for a given volume (Fig 6.1(b)). The surface has a characteristic of a membrane that holds the liquid in place.

These forces acting on the surface and attempting to minimize the surface of the fluid are known as surface tension forces. Water has a very high surface tension value. Liquids which have high surface tension values also have high latent heat values.

Water will form spherical droplets on a waxy leaf, or a clean glass surface, or when it emerges from a tap, orifice or spray nozzle. However, if the droplets are too large, the surface tension forces will be counteracted by gravitational forces and the droplet will break up or, if the surface forces are reduced, then again the droplets may break up or the liquid will spread over the surface. Compounds that reduce the surface tension of a liquid are said to be *surface active*; in contrast, substances that increase the surface tension of a liquid are said to be *surface inactive*.

The first definition of surface tension is the force per unit length acting in a particular surface, acting at right angles to one side of a line drawn on the surface.

Therefore the units of surface tension are newtons per metre (N m^{-1}) or dynes per centimetre (dyn cm^{-1}):

$$\frac{1N}{1m} = \frac{10^5 \text{ dyn}}{10^2 \text{ cm}} = 10^3 \text{ dyn cm}^{-1}$$

SI units are often expressed in terms of millinewtons per metre (mN m^{-1}), where 1 mN m^{-1} = 1 dyn cm^{-1}.

If a piece of string with the ends tied is laid out on a water surface in a random fashion (Fig. 6.1(c)) and a surface-active substance, e.g. detergent, is added to the surface of the fluid enclosed by the string, then the string is subjected to outward forces due to the differences in surface tension, and it will assume a circular shape. Forces are acting at right angles to the surface for this to occur.

Another way of viewing surface tension is in terms of the work required to extend a surface under isothermal conditions. Consider a box with two smooth wires A and B a distance d apart (Fig. 6.1(d)), with a slider capable of moving along the wires to enclose a film within the framework of wires. A force of F needs to be applied to counteract the surface tension forces (i.e. a Langmuir balance (Levitt, 1973).

At equilibrium,

$$F = 2\gamma d$$

(Note that there are two points of contact between the wire and film.) If the slider is advanced to the right by a distance L m, the work done is FL. The surface area increases by $2Ld$. Therefore,

$$\frac{\text{work done}}{\text{increase in area}} = \frac{FL}{2Ld} = \frac{F}{2d} = \gamma$$

Thus the surface tension can be regarded as the work done per unit area in increasing the surface area under isothermal conditions. In this form it is often referred to as the free surface energy. As in all thermodynamic situations, reactions are favoured when free energy changes are negative. It is expressed in terms of either joules per metre (J m^{-1}) or newtons per metre (N m^{-1}).

As would be expected, the surface tension will depend upon the nature of the liquid and the forces of attraction operating within that liquid. The surface tension values for some common liquids are shown in Table 6.3. The surface tension values of some aqueous solutions have been given by Weast

Table 6.3 — Surface tension values of some liquids at 20 °C

Liquid	Surface tension (mN m^{-1})	Liquid	Surface tension (mN m^{-1})
Water[a]	72.75	Milk[b]	42.3–52.1
Ethyl alcohol[a]	22.75	Skim-milk (0.04% fat)[b]	51.0
Methyl alcohol[a]	22.65	Whole milk (2.4% fat)[b]	46.7
Chloroform[a]	27.14	Cream (34% fat)[b]	44.8
Carbon tetrachloride[a]	26.95	Cotton seed oil[c]	35.4
Glycerol[a]	63.4	Coconut oil[c]	33.4
Mercury[a]	435.5	Olive oil[c]	33.0
		Oleic acid[c]	32.5

[a]From the data of Weast (1982).
[b]From the data of Jenness et al. (1974).
[c]From the data of Powrie and Tung (1976).

(1982). Surface tension data have not appeared to receive as much attention as many of the other physical properties.

Surface tension values and interfacial values of fatty acids have been given by Swern (1964).

Liquid metals have the highest surface tension values. Water also has a high surface tension value compared with other liquids. In contrast, oils and organic solvents, for which bonding is predominantly convalent, have low surface tension values. Milk has a surface tension of between 42.3 and 52.1 mN m^{-1}. It is considerably lower than the value for water because of the presence of fat and natural surface-active agents.

Water is not a good surface tension standard for calibrating equipment because it is easily contaminated by surface-active agents.

6.3 SURFACE ACTIVITY

Many materials such as short-chain fatty acids and alcohols are soluble in both water and liquids that are immiscible with water (such as oils and organic solvent). Normally such materials have a polar or hydrophilic moiety, e.g. COOH and OH, and non-polar or lipophilic groups. These substances will orientate themselves at the interface between the two phases with the polar group dissolved in the aqueous phase, and the non-polar group dissolved in the oil phase, because this is the most stable configuration form from an energy standpoint. The strong adsorption of these materials at surfaces or interfaces, often as a monomolecular layer, is termed surface activity. Surface-active components or surfactants will decrease the surface tension of water considerably at very low concentrations, and the concentration of surface-active components will always be higher at the interface or surface than in the bulk of the solution. many naturally occurring food constituents exhibit surface activity, e.g. alcohols, fatty acids, phospholipids, proteins and tannins.

Surfactants can be classified as anionic, cationic or non-ionic, according to the charge carried by the surface-active part of the molecule. An example of an anionic surfactant is sodium stearate $CH_3(CH_2)_{16}COO^-Na^+$; an example of a cationic surfactant is laurylamine hydrochloride $CH_3(CH_2)_{11}NH_3^+Cl^-$; an example of a non-ionic surfactant is polyethylene oxide $CH_3(CH_2)_7C_6H_4(O \cdot CH_2 \cdot CH_2)_8OH$.

Surface-active agents are widely used as emulsifying agents and detergents.

As small amounts of a surface-active component are added to a solution, the surface tension decreases until a point is reached where further addition results in no more decrease. The concentration, corresponding to this levelling-off, is known as the critical micelle concentration. At concentrations above the critical micelle concentration, the molecules are clustered together in micelles, which act as a pool or reservoir of molecules. Molecules are free to dissociate from the micelle and move to the surface of interest.

The critical micelle concentration for sodium dodecyl sulphate is 0.0081 mol l^{-1}. For most surfactants, critical micelle concentration values fall between 0.004 and 0.15 mol l^{-1}. This decreases in the presence of salt. The critical micelle concentration has been considered in more detail by Rosen (1978).

The formation of an adsorbed layer is not an instantaneous process but is governed by the rate of diffusion of the surfactant through the solution to the interface.

In certain cases, the addition of some components to water increases the surface tension. This is known as negative adsorption. Here the solute–surface forces of attraction are greater than the solvent–solvent attractive forces, and the solute molecules tend to migrate away from the surface into the bulk of the liquid. Such compounds are referred to as being surface inactive. Sodium chloride and sucrose are examples. The surface tensions of some common salts, acids and sugars have been given at different concentrations by Weast (1982). For dilute solutions, there is a relationship between the surface tension γ, the bulk concentration c_b and the surface excess concentration τ_b:

$$\tau_b = \frac{c_b}{RT}\frac{d\gamma}{dc_b}$$

The excess surface concentration can be regarded as the mass of solute absorbed per unit area of the surface. The tendency of surface-active components to pack into an interface favours expansion of the interface; at equilibrium, this must be balanced against the normal surface tension forces. If Π is the expanding or surface pressure of an adsorbed layer, then the surface tension will be lowered to a value γ_0, where

$$\gamma_0 = \gamma - \Pi$$

$$\Pi = \gamma - \gamma_0$$

Thus the surface pressure of a monolayer is the reduction in surface tension due to the monolayer. Surface pressures and the properties of monomolecular films can be studied using the Langmuir–Adam surface balance. Further details have been given by Shaw (1970) and Heimenz (1977).

6.4 TEMPERATURE EFFECTS

The surface tensions of most liquids decrease as the temperature increases and, at temperatures not too near the critical temperature, the relationship is almost linear. In the region of the critical temperature, the surface tension value becomes very low, as the intermolecular cohesive forces approach zero.

Some surface tension values for ethyl alcohol and water over the temperature range 0–30 °C are given in Table 6.4.

Table 6.4 — Surface tension values of ethyl alcohol and water in the temperature range 0–30 °C

Liquid	Surface tension (mN m^{-1}) at the following temperatures			
	0 °C	10 °C	20 °C	30 °C
Ethyl alcohol	24.05	23.61	22.75	21.89
Water	75.60	74.22	72.75	71.18

Swern (1964) has observed an almost linear relationship between surface tension and temperature for most oils and fatty acids, with surface tension decreasing as temperature increases.

One relationship, known as the Ramsay–Shields equation is as follows (Powrie and Tung, 1976):

$$\gamma \frac{M}{\rho} = k\,(T_c - T - 6)$$

where γ is the surface tension; M is the molecular weight; T is the experimental temperature; ρ is the density; k is the Eötvös constant and T_c is the critical temperature.

6.5 METHODS FOR MEASURING SURFACE TENSION

There are many methods for determining surface tension. Levitt (1973) has classified these methods into six groups, with the names of some of the methods that fall in the respective groups. These groups are as follows.

(1) Direct measurement of capillary pull (Willhelmy's, du Nouy's or the Cambridge method).
(2) Capillary rise, in single tubes, differential tubes, parallel or inclined plates.
(3) Bubble pressure (Jaeger's, Sugden's or Ferguson's methods).
(4) Size of drops (volume or weight).
(5) Shape of drops or bubbles.
(6) Dynamic methods (ripples, etc).

The important practical points have been concisely summarized by Levitt (1973): 'In almost every case the elementary theory of the methods needs correction if accurate results are required. The choice of method will depend on whether the primary consideration is extreme absolute accuracy, practical convenience, speed of working, small size of sample or study of time effects. The capillary rise method is the most accurate absolute method, but results accurate to better than 0.5% can be obtained much more conveniently by the drop-weight, ring detachment or bubble pressure methods. As elsewhere in physical chemistry it is rarely necessary to use laborious absolute methods, since most of the difficulties can be avoided by calibrating the apparatus with a liquid of known surface tension.' For this purpose, water should be avoided.

'Scrupulous cleanliness is essential in surface tension work; aqueous solutions in particular are highly susceptible to contamination by minute traces of grease or detergent with large reduction in their surface tension. Consequently from this point of view the methods in which fresh drops or bubbles are repeatedly formed offer an advantage over static methods such as capillary rise. For the same reason a form of apparatus that can be readily cleaned and protected from contamination is desirable and, in addition, it should be suited to immersion in a thermostat to control the temperature within 0.1 degC.'

Some of the methods for measuring surface tension will now be examined.

6.5.1 Capillary rise

When a capillary tube is immersed in water vertically, with one end touching the water, the water rises in the tube to a height above the surface (Fig. 6.2(a)); the narrower the tube, the greater is the height h to which the water rises. This phenomenon is known as capillarity; it is commonly observed in most porous materials, e.g. ink soaking into blotting paper, or water soaking into porous solids as in the reconstitution of freeze-dried foods.

The angle between the tangent to the surface of a liquid at the point of contact with a surface, and the surface itself, is known as the angle of contact (Fig. 6.2(b)). For water in contact with clean glass the angle of contact is 0°; for dirty glass it may be as high as 8°. The angle of contact between clean glass and most other aqueous fluids and alcohol is also 0°.

When equilibrium is achieved, the upward forces of surface tension

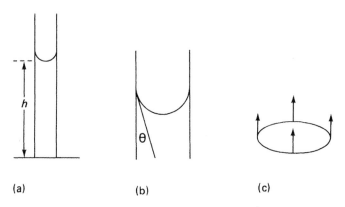

Fig. 6.2 — (a) Rise of a fluid in a capillary tube; (b) the angle of contact; (c) forces acting on the surface of a fluid.

acting round the circumference of the fluid balance the downward forces due to the height of the column of liquid. Thus,

$$2\pi r\gamma = \pi r^2 h \rho g$$

Therefore,
$$\gamma = \frac{rh\rho g}{2}$$

where γ (N m^{-1}) is the surface tension; r (m) is the tube radius; h (m) is the height to the bottom of the meniscus, ρ (kg m^{-3}) is the fluid density and g (m s^{-2}) is the acceleration due to gravity. This equation does not take into account the weight of liquid above the bottom of the meniscus. For more accurate work, particularly as the capillary diameter increases, this should be taken into consideration. The correction factor, obtained by considering a spherical surface, is

$$\gamma = \frac{g\rho h r}{2}\left(h + \frac{r}{3}\right) \tag{6.1}$$

These formulae only apply when the angle of contact is 0°. For other fluids which have an angle θ of contact with glass, equation (6.1) is modified to

$$\gamma = \frac{g\rho h r}{2\cos\theta}$$

The angle of contact can be fairly easily estimated experimentally (Tyler, 1972). The capillary method itself is therefore very simple. The height of the meniscus and the tube radius are normally measured using a travelling microscope. Care should be taken to ensure that there is no sticking of the liquid in the tube, by slowly moving the tube upwards or downwards in the liquid and observing that the liquid runs smoothly; if not, the tube is probably dirty and results are likely to be in error.

Note that mercury exhibits capillary depression and in this case the capillary rise is a negative quantity.

6.5.2 Bubble methods

If we consider the forces acting on air bubbles in a liquid, it can be seen that contraction (Fig. 6.3.(a)) or enlargement (Fig., 6.3.(b)) is prevented by the

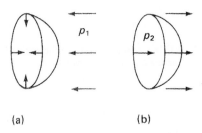

Fig. 6.3 — Forces acting at a spherical surface: (a) causing contraction; (b) causing enlargement.

forces acting on the bubble, which are due to the liquid pressure p_1 plus the surface tension forces acting to reduce the size of the bubble, whereas the force due to the internal pressure p_2 acts to increase the size of the bubble. At equilibrium,

$$p_1 \pi r^2 + 2\pi r \gamma = p_2 \pi r^2$$

$$p_2 - p_1 = \frac{2\gamma}{r}$$

Therefore, $$\gamma = \frac{r}{2}(p_2 - p_1) \qquad (6.2)$$

Thus, there is an excess pressure within the air bubble. The magnitude of the excess pressure depends directly upon the surface tension and inversely on the size of the bubble. Such forces are important in foaming and aeration of liquids. If the bubble is in the form of a film e.g. a soap bubble, there are two surfaces to consider and the excess pressure is given by

$$p_2 - p_1 = \frac{4\gamma}{r}$$

For the simple air bubble, equation (6.2) can be further modified, when the angle of contact is not 0°, to give

$$\gamma = \frac{r}{2\cos\theta}(p_2 - p_1)$$

If the excess pressure inside a bubble can be measured, and the radius is known, then the surface tension of the liquid can be determined. One simple practical way of setting this up is to use Jaeger's method (Fig. 6.4).

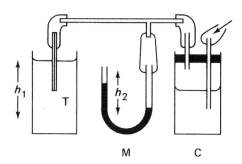

Fig. 6.4 — Experimental layout for Jaeger's method: T, capillary tube; M, manometer; C, container.

Water is supplied to a large container C, thereby slowly displacing air from that container. This air is directed to a piece of glass capillary tubing T, that is immersed below the surface of the test liquid. A manometer M is incorporated into the air line (normally rubber tubing) to measure the pressure of the air (i.e. the pressure in the bubble during its formation and bursting). The water rate is regulated to give a slow but steady stream of air bubbles. The maximum pressure is recorded just as the bubble breaks away from the tubing (it should be noted that the manometer reading will fluctuate; however, it should not be too difficult to read the maximum pressure recorded). This is the pressure p_2 in the air bubble. If a manometer is used, with one end open to the atmosphere, and the recorded height difference is h_2, then the bubble internal pressure is $p + h_2\rho_2 g$, where p is the atmospheric pressure. The pressure of the surrounding fluid is given by the depth h_1 of immersion of the tubing. The liquid pressure is $p + h_1\rho_1 g$. Therefore the excess pressure in the bubble is $g(h_2\rho_2 - h_1\rho_1)$.

The surface tension is determined from equation (6.2);

$$\gamma = \frac{gr}{2}(h_2\rho_2 - h_1\rho_1)$$

It is assumed that the radius r of the bubble is the same as the radius of the capillary tubing, from where it issues. Again this is measured with a travelling microscope. (ρ_1 and ρ_2 are the densities of the test liquid and manometer fluid, respectively.)

For dilute aqueous solutions, when water is used as the measuring fluid (as is often the case), the $\rho_2 = \rho_1$ and

$$\gamma = \frac{gr\rho_1}{2}(h_2 - h_1)$$

With a capillary tube of fixed radius the value $h_2 - h_1$ should be a constant for different depths h_1 of immersion. This provides a useful check as the experiments are being performed.

Strictly speaking, the method will not give an accurate absolute value since the theory applies to a static situation rather than to a dynamic situation, as occurs during bubble formation. Nevertheless the method is simple and yields reliable results. The accuracy of the final results will depend upon the accuracy of the individual measurements. Probably the largest source of error will be in the measurement of the internal bubble pressure. The use of an inclined-tube manometer or pressure transducer may help to improve this. It is an extremely useful technique for comparing the surface tension of liquids or examining the effect of temperature on the surface tension.

6.5.3 Drop weight techniques

The simplest drop-weight technique relies upon measuring the volume or mass of a droplet as it forms at the end of a capillary tube. The mass of the drop is directly related to the surface tension of the liquid. The drops should be allowed to form very slowly and normally five, ten or twenty drops are collected and their masses determined, on the assumption that the drop has a cylindrical form, as shown in Fig 6.5(a), i.e. no spreading takes place when

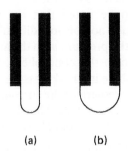

(a)   (b)

Fig. 6.5 — Drop-weight methods: (a) formation of a droplet when no spreading occurs; (b) formation when spreading occurs.

it is about to break away.

Tyler (1972) has shown that, by balancing the forces acting upon the droplet as it breaks, the surface tension is given by

$$\gamma = \frac{mg}{\pi r}$$

where m is the mass of one droplet and r is its radius. This expression is deduced on the assumption that the drops break away under ideal statical conditions. The problem, however, is complicated by dynamic considerations; taking these into account, Rayleigh has shown that a closer approximation is given by the formula

$$\gamma = \frac{mg}{3.8r}$$

As is normally observed, the drop will spread over the tube (Fig. 6.5(b)) and there may be advantages in measuring the external radius, particularly if the tube wall is thick. Levitt (1973) considered that no simple theory was applicable; he used a correction factor F_c and expressed the surface tension as follows;

$$\gamma = \frac{F_c mg}{r}$$

The correction factor is dependent upon the ratio V/r^3, where V is the volume of one drop and r is the external radius of the tube. A table relating V/r^3 values to F_c is given for V/r^3 values ranging from 0.865 to 17.7. F_c was within the range 0.24–0.265.

Problems of absolute determinations can also be overcome by using a fluid of known surface tension. If m_1 and m_2 are the drop masses of two fluids of surface tension γ_1 and γ_2, then the ratio

$$\frac{\gamma_1}{\gamma_2} = \frac{m_1}{m_2}$$

will apply.

Fluids of known surface tension are readily available (Table 6.3). The method can also be easily modified to measure interfacial tension (see section 6.6).

All these three techniques involve the use of very simple equipment but, provided that sufficient care is taken, accurate results can be obtained for the surface tension of liquids.

6.5.4 Direct measurement of capillary pull

For many purposes, the surface tension of a liquid can be determined quickly and with sufficient accuracy by measuring the force F required to detach a horizontal platinum wire ring or a microscope slide from the surface of a liquid.

A simple torsion wire device using a microscope slide is described in Tyler (1972). In this case the force F required to tear the slide from the surface is determined by using masses (Fig. 6.6(a)). If the length of the slide is l and its width is b and the force required is mg, then $mg=2\gamma(l+b)$.

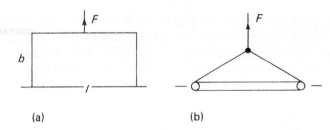

Direct pull method for measuring surface tension, using (a) a microscope slide with $F=2\gamma(l+b)$ and (b) a platinum ring with $F=4\pi R\gamma$.

For tearing a ring from the surface of a liquid, the force should be equal to twice the perimeter of the ring multiplied by the surface tension (Fig. 6.6(b)). More precisely, however, a correction factor F_d needs to be applied ranging from 0.7 to 1.0. The magnitude of correction factor depends upon the ratios R/r and R^3/V, where R is the radius of the ring and r is the radius of the wire and V is the volume of the liquid raised (Becher, 1965; Shaw, 1970). Values of F_d are given for an R/r value of 42 and different R^3/V values. (Note that the volume of liquid raised is $F/\rho g$). An R/r value of 42 corresponds to the du Nouy ring technique which is often used in practice.

One commercial piece of equipment operating on these principles is the du Nouy surface tension balance. The method is appropriate when the surface tension of a large number of small samples is to be measured within $\pm 0.1 \times 10^{-3}$ N m^{-1}. It is not easy to control the temperature; so samples should be measured at about ambient temperature or incubated at the desired temperature and then very quickly determined. Further details have been given by Levitt (1973)

Other methods for measuring the surface tension, in particular from the shape of drops are discussed by Heimenz (1977) and Sherman (1968).

6.6 INTERFACIAL TENSION

At the boundary of two immiscible liquids, there will be an imbalance of intermolecular forces; this gives rise to the phenomenon of interfacial tension. Interfacial tensions usually lie between the individual surface tensions of the liquids in question.

The drop-weight method can easily be modified to measure interfacial tension. A microsyringe or microburette is used to dispense the more dense fluid into the less dense fluid, through a capillary tube of radius r. The interfacial tension γ is calculated from the following equation

$$\gamma = \frac{V(\rho_1 - \rho_2)g}{r} F_c$$

where V is the volume of one drop, ρ_1 and ρ_2 are the densities of the heavier and lighter fluids, respectively, r is the capillary radius and F_c is the correction factor (see section 6.5.3).

Further practical details have been given by Levitt (1973) and Tyler (1972). Some interfacial tension values between water and a range of organic solvents and oils are shown in Table 6.5.

Table 6.5 — Interfacial tension values between water and the following components.

Component	Temperature (°C)	Interfacial tension (mN m^{-2})
n-hexane[a]	20	51.1
Carbon tetrachloride[a]	20	45.0
Benzene[a]	20	35.0
Trioleum[b]	25	14.6
Mercury[c]	20	375.0
Peanut oil[b]	25	18.1
Cotton seed oil[b]	25	14.9
Olive oil[b]	25	17.6
Coconut oil[b]	25	12.8
n-octanol[c]	20	8.5

[a]From the data of Kaye and Laby (1973).
[b]From the data of Powrie and Tung (1976).
[c]From the data of Shaw (1970).

Generally speaking, the higher the interfacial tension, the lower is the solubility of the solvents in each other. To facilitate emulsification, the interfacial tension between water and oil should be reduced to below 10 mN m^{-1}. Many of the available emulsifying agents and detergents work in this way by lowering the interfacial tension. Table 6.6 shows how the different

Table 6.6 — Interfacial tension between water and butter oil at 40 °C with a selection of milk proteins.

Protein	Interfacial tension (mN m^{-1})
Water–butter oil	19.2
0.2% euglobulin	18.0
0.2% β lactoglobulin	14.0
0.2% α-lactalbumin	11.0
0.2% interfacial protein	11.0

Adapted from the data of Powrie and Tung (1976).

milk protein fractions lower the interfacial tension in a water–butter oil mixture at 40 °C.

Berger (1976) has reported values for the interfacial tension of palm kernel oil against water containing various emulsifier–stabilizer systems used in ice-cream, at 70 °C. Some examples are given in Table 6.7.

These stabilizers considerably reduce the interfacial tension; a further slight reduction is brought about by the addition of a small amount of protein.

Before some practical applications of surface phenomena are described, the principles of adhesion, cohesion and spreading will be discussed.

6.7 WORK OF ADHESION AND COHESION

The work W_{AB} of adhesion between two immiscible liquids corresponds to the work required to separate unit area of the liquid–liquid interface and to form two separate liquid–air interfaces (Fig. 6.7(a))

It can be considered to be equal to the increase in total energy of the system, i.e. the final energy minus the initial energy. Therefore,

$$W_{AB} = \underset{\text{final}}{\gamma_A + \gamma_B} - \underset{\text{initial}}{\gamma_{AB}}$$

It can be seen that the work of adhesion will increase as the interfacial energy decreases; when the interfacial energy becomes zero, the two liquids are completely miscible. The work of adhesion for some organic substances and water are given in Table 6.8.

The work of cohesion for a single liquid is equal to the work required to pull apart a column of liquid of unit cross-sectional area (Fig. 6.7(b)):

$$W_{AA} = 2\gamma_A$$

When a drop of oil is added to water, it may form a lens or spread over the surface, depending upon which forces predominate (Fig. 6.8).

If the lens is thin, spreading will occur if

$$\gamma_w > \gamma_o + \gamma_{o-w}$$

where γ_w is the surface tension of water, γ_o is the surface tension of oil and γ_{o-w} is the interfacial tension of oil in water.

A spreading coefficient (S) for oil on water has been defined as the difference between the work of adhesion and the work of cohesion for the oil:

$$S = W_{o-w} - W_{o-o} = \gamma_w - (\gamma_o + \gamma_{o-w})$$

where W_{o-w} is the work of adhesion between oil and water and W_{o-o} is the work of cohesion in oil

Thus, if the spreading coefficient is positive, spreading will occur spontaneously, i.e. the oil adheres more strongly to the water than it coheres to itself. When spreading occurs, there is an increase in area of the oil–water

Table 6.7 — Interfacial tension between PKO and water containing various emulsifier–stabilizer systems used in ice-cream.

Emulsifier-stabilizer system	Interfacial tension (mN m^{-1})
Water–palm kernel oil	30
Water+6.0% commercial glycerol monosterate–palm kernel oil	4.2
Water+6.0% commercial glycerol monosterate+0.1% casein–palm kernel oil	1.8
Water+1.0% vegetable lecithin–palm kernel oil	4.8

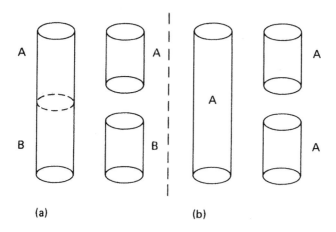

Fig. 6.7 — (a) work of adhesion; (b) work of cohesion.

Table 6.8 — Work of adhesion for some organic compounds and water.

Organic compound	Work of adhesion (mN m^{-1})
Paraffins	36–48
Aromatic hydrocarbons	63–67
Esters	73–78
Ketones	85–90
Primary alcohols	92–97
Fatty acids	90–100

Adapted from the data of Taylor (1961).

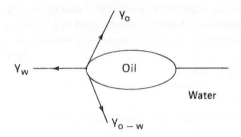

Fig. 6.8 — Forces acting during the spreading of a liquid.

and oil–air interfaces and a corresponding decrease in area of the water–air interface. The corresponding work done in increasing the surface area is equal to the Gibbs free energy change ΔG for the process; for spreading to occur, G must be negative. (Note that $\Delta G = \gamma_o + \gamma_{o-w} - \gamma_w$.)

If the oil is contaminated and the surface tension γ_o is reduced, spreading is more likely to occur. However, if the aqueous phase is not clean, γ_w will be reduced significantly less than γ_{o-w} and spreading is less likely to occur.

When oil spreads on water, it forms either a thin uniform film known as a duplex film, which is thick enough for the two interfaces to be independent, or a monolayer, leaving excess oil as lenses in equilibrium.

6.8 EMULSIONS

An emulsion is a two-phase system consisting of two immiscible liquids with one being distributed as finite globules within the other. Most food emulsions are a mixture of water and oil. In an oil–in–water emulsion droplets of oil are dispersed in an aqueous phase. In a water–in–oil emulsion, droplets of water are present in the oil phase

Some examples are given in Table 6.9.

Table 6.9 — Some examples of oil–in–water and water-–in–oil emulsions.

Oil–in–water emulsions	Water–in–oil–emulsions
Milk	Margarine
Cream	Low-fat spreads, e.g. Gold, Outline
Mayonnaise	Cake batters
Ice-cream	

The formation and stabilization of emulsions are favoured by the addition of emulsifying agents. These work by reducing the interfacial tension and therefore the amount of work required to produce the emulsion. This is further stabilized by preventing the dispersed phase from coalescing and by ensuring that the droplets are small and of uniform size; a high-viscosity continuous phase, a similar uniform charge and the production of strong mechanical films will increase the stability.

There are many different emulsifying agents available; some of these are listed in Table 6.10, together with some of their hydrophile–lipophile

Table 6.10 — Some emulsifying agents and their HLB values.

Ionic	HLB value	Non-ionic	HLB value
General emulsifiers			
Salts of oleic acid		Glycerol esters	2.8
e.g. potassium	20.0	Polyglycerol esters	
e.g. sodium	18.0	Propylene glycol fatty acid esters	3.4
Sodium stearoyl-2-lactylate		Sorbitol fatty acid esters	4.7
Phospholipids		Polyoxyethylene sorbitol fatty acids	14.9, 15.9
e.g. lecithin			
Protein			
e.g. gelatin			
e.g. egg albumin			
Hydrocolloids			
Pectin		Guar	
Alginates		Locust bean	
Xanthan		Carboxymethyl cellulose	
Tragacanth	11.9	Hydroxypropyl cellulose	
Agar		Methyl cellulose	10.5
Carrageenan			
arabic			

balance (HLB) values (i.e. the rate of the weight percentage of hydrophilic groups to the weight percentage of hydrophobic groups in the emulsifier molecule). They can be subdivided into ionic and non-ionic emulsifiers, Hydrocolloids usually function by increasing the viscosity of the continuous phase.

Ionic emulsifiers suffer the disadvantage of reacting with other oppositely charged particles, particularly hydrogen and metallic ions, to form complexes which may have both a reduced solubility and a reduced emulsification capacity. For this reason, non-ionic emulsifiers are more widely used in food systems.

For selection of the best emulsifier for a particular application, an approach known as the HLB system has been developed. HLB values can be determined experimentally.

Emulsifiers with HLB values below 9 are lipophilic, those with HLB values between 8 and 11 are intermediate, and those with HLB values between 11 and 20 are hydrophilic.

HLB values in the range 3–6 will promote water-in-oil emulsions, whereas oil-in-water emulsions are formed with emulsifiers having HLB values between 8 and 18.

For example, a vegetable oil-in-water emulsion requires an HLB value of between 7 and 17, whereas a water-in-oil emulsion requires a value of about 5. HLB values for some common emulsifiers are given in Table 6.10. The application and dispersibilities of emulsifiers are related to the HLB value and these are given in Tables 6.11 and 6.12, respectively. Generally a

Table 6.11 — Application of emulsifiers.

HLB value	Applications
3–6	Water-in-oil emulsions
7–9	Wetting agents
8–15	Oil-in-water emulsions
13–15	Detergents
15–18	Solubilizers

Taken from the data of Shaw (1970, with permission.

Table 6.12 — Dispersibilities of emulsifiers.

HLB value	Dispersibility in water
1–4	Nil
3–6	Poor
6–8	Unstable milky dispersion
8–10	Stable milky dispersion
10–13	Translucent dispersion or solution
>13	Clear solution

Taken from the data of Shaw (1970), with permission

combination of emulsifiers is necessary to achieve a stable emulsion, thereby giving a blend of hydrophobic and hydrophilic groups. For mixed emulsifiers, approximate proportionality holds.

Thus the HLB value of a mixed emulsifier system, denoted $(HLB)_M$, containing emulsifiers A and B with respective values of $(HLB)_A$ and $(HLB)_B$ will be as follows:

$$(HLB)_M = m_A(HLB)_A + m_B(HLB)_B$$

where m_A and m_B are the mass fractions of A and B, respectively.

At a given HLB, emulsifying stability can vary, depending upon the choice of emulsifiers used.

These guidelines will provide initial information for choosing appropriate emulsifiers for preliminary experimental work, with a view to optimizing the system. Additional information on the determination of HLB values has been given by Sherman (1968) and Krog and Lauridsen (1976).

Factors affecting the stability of concentrated emulsions are discussed by Sherman (1980).

Oil–in–water and water–in–oil emulsions can be distinguished by the simple tests given in Table 6.13.

Table 6.13 — Tests for distinguishing between oil–in–water and water–in–oil emulsions.

Oil–in–water emulsion	Water–in–oil emulsion
Can be diluted by the additon of water	Can be diluted by the additon of oil
Has a reasonably high electrical conductivity	Has a low electrical conductivity
Will take up water-soluble dyes	Will take up oil-soluble dyes

Water–in–oil emulsions also tend to have a much oilier consistency. Compounds are tested for their suitability as emulsions by determining their emulsification capacity and the stability of the emulsion. For an oil–in–water emulsion, the emulsification capacity is usually expressed as the amount of oil emulsified by unit mass of the emulsifier. The stability is estimated by determining the oil content in the top of an emulsion which has been allowed to stand for a specified time period. These measurements are empirical in nature, but such an approach is useful for screening potential emulsifiers. Further details on emulsification and other functional properties of proteins have been reviewed by Kinsella (1976).

Sometimes, it may be necessary to destabilize an emulsion. Some of the possible ways of doing this are as follows.

(1) Increasing or decreasing the temperature, including freezing.

(2) Mechanical means, e.g. centrifugal separation.
(3) Addition of chemicals or enzymes to reduce surface effects.
(4) Antagonistic operations, e.g. addition of emulsifiers which would tend not to promote stability of the particular emulsifier.

Further information on factors affecting the stability of emulsions has been given by Friberg (1976).

6.9 YOUNG'S EQUATION (SOLID/LIQUID EQUILIBRIUM)

Consider a liquid droplet on a solid surface (Fig. 6.9). The forces acting are

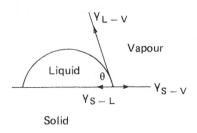

Fig. 6.9 — The forces involved in Young's equation.

shown. At equilibrium the forces are balanced

$$\gamma_{S-V} = \gamma_{S-L} + \gamma_{L-V} \cos \theta$$

where the subscripts S, V and L denote solid, vapour and liquid respectively.

The implications and limitations of this equation have been discussed by Heimenz (1977).

6.10 DETERGENCY

The function of a detergent is to remove 'soil' from the surface of food-processing equipment, floors and walls. The presence of 'soil' may affect the performance of equipment and will look unsightly; additionally, it will act as a substrate for micro-organisms or may become dislodged and contaminate the food.

Soil can be extremely variable in its nature; the composition and amount will depend upon the product being processed and the processing conditions. For example the extent of deposit formation when heating milk will depend upon the temperature and the heat stability of the proteins.

Soil includes residues of one or more of the food ingredients or their reaction products together with deposited minerals and is frequently impregnated with food particles. The types of soil and changes induced on heating are summarized in Table 6.14.

In addition the soil may be microbial in nature, consisting of mould, filamentous bacteria of algae.

Precautions can be taken to reduce the amount of soil prior to treatment, e.g. hard water can be softened and water used in heat exchangers can be filtered or sanitized.

After use, food-processing equipment must be cleaned, either manually or by cleaning–in–place techniques. Provision for easy and efficient cleaning must be made at the design stage and all surfaces in contact with the food should be readily accessible. With cleaning–in–place techniques, correct design procedures will ensure that all the pipework is adequately flushed and that there are no dead-spaces, such as instrument pockets or T-pieces which cannot be properly cleaned. Hygienic design of food processing equipment has been considered in more detail by Timperley and Lawson (1980) and Troller (1983). High fluid velocities during the cleaning regime create high shear stresses at the wall and accelerate the cleaning process. The kinetics of cleaning have been discussed by Loncin and Merson (1979). The selection of detergents will depend upon the nature of the soil.

Detergents are surface-active agents which work by lowering the surface tension of the liquid, thus promoting spreading (see section 6.9). They act both as a wetting agent to allow better spreading over the surface and as an emulsifying agent to dissolve grease and fat. The detergent molecules consist of a hydrophobic region and a hydrophilic region. The hydrophobic region normally consists of a long-chain fatty acid with between 10 and twenty carbon atoms (branching of this chain will reduce the tendency to foam). The hydrophilic region can either be sodium salt of a carboxylic acid (soapy detergent) or based on the sodium salt of an alkyl or aryl sulphonate (anionic detergents); these are the most common in use. If the hydrophilic part of the molecule is an alcohol or ester, the detergent is non-ionic. The composition and major properties of the detergents are summarized in Table 6.15.

Cartionic detergents have a good bacteriocidal action. Their use should be avoided with equipment handling fermented products, e.g. cheese and yoghurt, as they can be difficult to remove and may well inhibit the bacterial cultures used for these products.

When a detergent is added to water, there is a reduction in the surface tension until the critical micelle concentration is reached. At concentrations below the critical micelle concentration the detergent molecules in the bulk of the solution behave as individual molecules and there is no detergent action. Above the critical micelle concentration, the molecules are grouped together as micelles which act as a reservoir of molecules. Any solid containing dirt, suspended in this solution, would be readily wetted and dissolved by these molecules; fat globules would be emulsified and dispersed in the aqueous phase. Proper detergent action depends upon keeping the concentration above the critical micelle concentration. For alkyl and aryl sulphonates the critical micelle concentration is about 0.7 mol l^{-1}; it is affected by temperature and often reduced in the presence of salts.

Commercial detergents used for cleaning processing plant are either

Table 6.14 — Types of soil and changes induced on heat treatment.

Type of soil	Solubility	Change induced on heating
Sugar	Water soluble	Easily removed; more difficult if caramelized
Fats	Alakali soluble, water insoluble	Removed with difficulty; polymerization will make removal more difficult
Protein	Water solubility variable, alkali soluble, slightly acid soluble	Difficult to remove; increased on denaturation
Mineral salts	Water solubility variable, most are acid soluble	Interactions may make them more difficult

From the data of Tamplin (1980) and Cheow and Jackson (1982).

Table 6.15 — Properties of the major types of detergents.

Typical formula	Name	Properties
∿∿∿∿∿–$COO^- Na^+$	Soap	Forms a scum in hard water
∿∿–⟨O⟩$SO_3^- Na^+$	Anionic	Not affected by hard water
∿∿∿∿∿–$NH_4^+ Cl^-$	Cationic	Bacteriocidal action; not commonly used
∿∿∿∿∿–alcohols or esters or ethers	Non-ionic	Good detergent power, low foaming, easily rinsed

based on acids (e.g. nitric or phosphoric acid) or alkali (e.g. caustic soda) and should be handled with care. Caustic soda and nitric acid alone do not result in a decrease in surface tension and have poor wetting properties. A typical cleaning regime after UHT processing would involve flushing the product from the plant with water, circulating an alkali detergent (1–2%) for 20 min at 70–80 °C, rinsing with water, circulating an acid detergent for 20 min, rinsing with water and draining. Recommended conditions, e.g. times, temperatures and concentrations, are usually supplied with proprietory detergents. The plant should be dismantled and inspected at regular intervals to check the efficiency of the cleaning programme. The equipment can be sanitized using hot water at 90 °C after cleaning and/or before processing. The International Dairy Federation (1980) has given a hygienic code of practice for the dairy industry.

Often proteinaceous material is difficult to remove using ordinary detergents. Proteolytic enzymes have been incorporated into detergent formulations to assist removal of the protein by hydrolysis. Such enzymes are produced by bacterial fermentation processes, and their optimum pH activity is generally on the alkaline side of neutrality whilst the optimum temperature is at 50–60 °C. Equipment is soaked with these detergents or cleaned in place. Enzymes are incorporated into detergent formulations between 1% and 10%.

They are particularly useful for removing difficult proteins such as blood, egg, milk and vegetable proteins from surfaces which may have had prolonged contact with these substances. One particular application is the removal of deposits from reverse osmosis and ultrafiltration membranes, particularly when the membrane material has a poor tolerance to extremes of pH and high temperatures, e.g. cellulose acetate. One problem is that enzymes are inactivated almost immediately by free chlorine. Further technical and safety aspects of enzyme detergents have been discussed by Barfoed (1983).

The formulation for a typical heavy-duty (general-purpose) detergent is as follows.

Soap.
Non-ionics.
Sodium tripolyphosphate.
Sodium perborate.
Sodium silicate.
Sodium carboxymethyl cellulose.
Optical brightners, perfume.
Enzyme formulation.
Sodium sulphate, water, etc.

The function of other substances which may be incorporated in a detergent formulation are listed in Table 6.16.

The manufacture of detergents has been discussed by Moore (1967) and the cleaning of tanks by the use of spray nozzles has been described by Loncin and Merson (1979).

6.11 FOAMING

A foam is defined as a two-phase system consisting of a mass of gas bubbles dispersed in a liquid or a solid (Glicksman, 1982) with the gas bubbles being separated from each other by thin films of liquid or solid. However, in contrast with emulsions, the dispersed phase is not colloidal, and it has consequently been described as a coarse dispersion where the bubbles may be quite large. However, the film separating the bubbles can only be a few nanometers thick, and the surface of the liquid has been extended enormously in opposition to surface tension forces. Hence, such a system is

Table 6.16 — Function of additives in detergent formulations.

Additive	Function
Polyphosphates	Soften water by removing Ca^{2+} and Mg^{2+} by formation of soluble complexes; keep dirt in suspension
Sodium carboxymethyl cellulose	Helps to suspend dirt
Sodium sulphate Sodium silicate	Makes powders free flowing
Sodium perborate	Bleaching agent

potentially very unstable and requires the presence of an emulsifying agent for stability (substances which are good emulsifying agents are often effective for stabilizing foams).

Shaw (1970) has described two types of foam: dilute foams consisting of almost spherical bubbles separated by rather thick films of somewhat viscous liquid; concentrated foams comprised of polyhedral gas cells, separated by thin films.

Most of the foam tends to arrange itself in the form of stacked regular dodecahedra; these are twelve-sided figures each side of which is a regular pentagon and which is shared with an adjacent bubble. The foam stability will depend upon a number of factors.

Consider the liquid film depicted in Fig. 6.10.. The pressure in the liquid

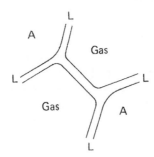

Fig. 6.10 — The gas–liquid interface in a foam.

film L at the junctions A is lower than at other points in the film because of the curved surface. Consequently, there is a tendency for liquid to flow to these points, resulting in thinning of the film and an increased tendency to rupture. In addition the van der Waals attractive forces favour film thinning,

Sec. 6.11] Foaming 193

whereas the overlapping of similarly charged electrical double layers opposes film thinning. If the foam is unstable, the film will rupture. If it is stable, the forces will counteract each other to produce an equilibrium film thickness.

One additional effect is the Marangoni effect. As the film stretches, there will be a corresponding decrease in the excess concentration of the foaming agent and then a local increase in the surface tension; this may persist long enough to allow the film to recover its initial thickness.

In addition the existence of a surface tension gradient has been used to suggest a mechanism by which surfactant spreads along the surface, thereby dragging a significant amount of the underlying solution.

The three major elements required for a liquid for producing a stable foam system have been summarized by Glicksman (1982).

(1) A low vapour pressure to prevent evaporation; this will help to retain the gaseous phase within the liquid and to minimize their tendency to break out or rupture the surrounding membrane.
(2) A low surface tension to prevent contraction and to hold more air in each of the air cells.
(3 Gelation or insolubilization of the coating phase to give some rigidity to the foam and to minimize the escape of entrapped gases.

Many foams are stabilized by the production of solid crystals in the film such as in the freezing of ice-cream (ice plus fat), cream fillings (fat and sugar) and marshmallows (setting of gelatin). The quantity of air incorporated into many of these types of product is important as it has a significant effect in the mouth feel and texture of the product. Some foams are stabilized by heat, as in the production of meringues, cakes and bread.

In whipped desserts and toppings, gum stabilizers can be employed to provide the desired stiffness and to provide stability to synthesis during processing. One important development is in the production of frozen desserts, where freeze–thaw stability is important.

Proteins are extremely useful foaming agents, e.g. egg white. In this respect, they play an important role in foam stabilization by undergoing surface denaturation during the foaming process. Graham and Phillips (1976) have discussed the role of proteins at the interface.

Proteins are now being extracted from a wide variety of food-processing byproducts, e.g. soya beans, cotton seed, cheese whey and blood, by techniques such as isoelectric precipitation, ultrafiltration and ion exchange to produce concentrates and isolates containing up to 90% protein (dry-mass basis). When operating conditions are mild, the proteins are undenatured and possess useful foaming properties. The foaming properties of proteins are strongly influenced by the pH, the temperature and the presence of other components.

The foaming capacity of a solution has been measured in terms of the total volume of foam produced, under fixed aeration or agitation conditions, or the stability of the foam by measuring the amount drained in a given time. As with emulsification, such tests are empirical in nature but are useful for assessing new materials or making comparisons between different foaming agents. In all cases the ultimate test is how the material performs when incorporated into the food itself.

Other measurements that can be taken are the surface tension of the liquid–gas interface or the stiffness or elasticity of foam, using different types of viscometer (Bourne, 1982).

The principle of foaming can be used to separate surface-active ingredients. The liquid is aerated and the foam that collects is skimmed from the top of the vessel. This is useful for removing toxic or bitter components from foods. A foam fractionation application has been described by Johnson and Peterson (1974).

Foaming can be a problem in certain circumstances, e.g. in aerobic fermentation systems and automatic washing machines. In fermenters the foam is detected and controlled by mechanical foam breaking devices for knocking it down; power consumption can be considerable and these devices are only usually used on smaller fermenters, or by the addition of antifoaming agents. These act by reducing the surface tension. Examples of foam inhibitors are polymethyl siloxane, tributyl phosphate, polyethylene glycol and lard oil. Synthetic inhibitors are effective at concentrations of 0.1% whereas lard oil is used at about 0.5% (Wiseman, 1982). The additon of an antifoaming agent may affect the oxygen transfer characteristics. Foaming may also cause problems in the vapour–liquid separator of an evaporator; hence the foam may carry over with water vapour, causing a loss of solids in the final condensate.

In front-loading automatic washing machines, foaming is not necessary and can lead to a poor washing action and leakage from the machine. Detergents for these have foam control agents incorporated in the formulation.

6.12 WETTABILITY AND SOLUBILITY

Two very important properties are wettability and solubility. When a powder is sprinkled on the surface of water, it will either float on the surface or sink fairly rapidly. Factors that affect this behaviour are the size and density of the particles and the surface properties. If there is a high interfacial tension between the particle and the liquid, it will not wet easily; usually the rate of wetting is the rate-controlling step in the dissolution of powders.

Skim-milk powder taken straight from the drier is extremely fine; the lactose is in an amorphous or glassy form. Consequently the powder has poor wetting properties and takes a long time to dissolve. A process known as 'instantizing' has been developed to make the powder more wettable. The powder is rewetted using steam or water to 15% moisture; this causes clumping of the small particles which are then dried to below 5% moisture in a fluidized-bed drier. This also encourages crystallization of the lactose. The wetting properties and dispersibility of the powder are considerably improved by a larger particle size and the adsorption of water into the structure by capillary action. However, instantizing does not necessarily mean that the solubility is increased. Powders containing fat present an additional problem, in that the fat globules present on the surface may repel water. A wetting agent can be incorporated into the milk before spray drying to reduce the interfacial tension; antioxidants can also be added to reduce oxidation during storage. Wetting can also be facilitated by keeping fat in the liquid form during rewetting. For reconstitution of full-cream milk, water temperatures between 45 °C and 60 °C have been found to provide good wetting conditions and to minimize lump formation (Mettler, 1980). Further details on the wetting properties of dairy powders have been discussed by Graf and Bauer (1976). Agglomeration and instantizing have been considered in more mathematical detail in Loncin and Merson (1979).

The solubility of a milk powder will depend upon its composition, the processing conditions during dehydration and the conditions used during reconstitution, including the temperature and hardness of the water and the method of mixing. These principles will also apply to most other powders.

Solubility is considered to be a very important functional property of protein concentrates and isolates (Kinsella, 1976). Nitrogen solubility at the isoelectric point (e.g. pH 4.6 for milk proteins) is usually taken as an indicator of the extent of denaturation of the protein in the powder.

6.13 STABILIZATION (DISPERSION AND COLOUR)

A suspension is defined as a stable dispersion of finely divided material in a liquid medium. If the dispersed material is of a colloidal size, settling will not occur. However, in practical situations, particles tend to be larger than the colloidal size, and the function of the stabilizer in this case is to control and delay the natural settling process. Therefore the production of a stable dispersion requires the smallest possible particle size and a liquid medium having the maximum suitable viscosity. Hydrocolloids are extremely useful for this purpose. Other factors which will affect the stability of such a dispersion are the interfacial characteristics, the concentration of the particles and the density difference between the phases. Most food sols are hydrophilic in nature.

When apples are pressed, the resulting juice is cloudy. Much of the cell debris is held in suspension and stabilized by the presence of pectin, although the larger particles may separate out. Heating the juice will tend to produce more flocculation, and the sedimented material can be removed by high-speed centrifugation or filtration. The resulting juice is still cloudy because of the colloidal dispersion and remains stabilized by the pectin. Subsequent treatment will depend upon whether a clear or cloudy fruit juice is the desired product.

If a clear juice is required, it is necessary to remove the pectin. This is most easily accomplished by treatment with a pectolytic enzyme. The juice is incubated with the enzyme for 24–28 h at 40–45 °C and, as the pectin is broken down, the cell debris will form a sediment, leaving a clear solution. Additional hazes can be formed because of the presence of starch and protein. Starch is removed by addition of amyloglucosidase enzyme at the same time as the pectolase. The levels of starch and pectin are measured in the pressed juice to ascertain the amount of enzyme treatment required. Protein hazes are less common in apple juice but, if they occur, they can be removed by tannin or tannic acid.

If the level of polyphenols is too high, gelatin can be added to precipitate them, thereby preventing the formation of brown insoluble compounds and the production of off-flavours during storage. Bentonite ($Al_2O_3.4SiO_2.H_2O$) has also been found useful for clearing beverages. Browning can be controlled by pasteurizing the juice immediately after pressing and by the addition of ascorbic acid and/or sodium metabisulphite. The acidity, tannin content and sugar content of the juice will depend upon the variety of apple chosen; a reasonable balance should be obtained for production of a high-quality juice. Following sedimentation of the colloidal material, the juice is filtered, pasteurized and packaged.

Stabilizers are added to a wide range of foods. For example, an ice-cream, they prevent the crystallization of water and other solids, and in chocolate drinks they improve the texture by keeping insoluble particles in suspension. The factors affecting the stability of ice-cream have been discussed by Berger (1976).

6.14 OTHER UNIT OPERATIONS

Some unit operations rely on distributing a liquid as fine droplets into a gas stream or, conversely, distributing a gas as small bubbles into a liquid. Some examples are given in Table 6.17.

Table 16.17 — Some unit operations involving dispersion of one phase into another.

Operation	Type of dispersion	Function
Spray drying	Liquid, paste or suspension–air	Water removal
Spray mixing	Steam–fat–air powder	Instantizing or fat addition
Aeration	air–fermentation medium	Oxygen transfer
Hydrogenation	Hydrogen–oil	Hardening of oils

The objective is to promote mass transfer by increasing the interfacial area. In the fractional volume occupied by the disperse phase is V, the surface-area-to-volume ratio S is given by

$$S = \frac{6V}{D_p}$$

where D_p is the mean spherical diameter of one bubble of the dispersion.

Similar principles apply in the contact of a liquid with a second immiscible liquid.

The forces operating will be examined when a sparingly soluble gas such as air contacts an aqueous fermentation medium, with a view to maximizing the rate of dissolution. The air is distributed through a fine orifice, a series of orifices or a perforated plate. When the gas is sparingly soluble, the major resistance to oxygen transfer is in the liquid film surrounding the bubble. The oxygen transfer rate OTR is given by the equation

$$\text{OTR} = k_L a (c_i - c_b)$$

where k_L (m s^{-1}) is the mass transfer coefficient (see section 13.7), a (m^2 m^{-3}) is the interfacial area per unit volume, c_i (kmol m^{-3}) is the concentration of gas at the interface and c_b (kmol m^{-3}) is the concentration of gas in the bulk liquid.

The gas enters as a sheet, but surface tension forces cause the sheet to collapse into a dispersion of bubbles.

The forces acting upon the bubble can be summarized as the surface tension forces, which resist dispersion and distortion, and the dynamic forces, which act unevenly over the surface and cause the bubble to disperse. The ratio of these forces will help to determine the maximum stable bubble size possible. The dynamic forces can be regarded as a shear stress τ. The dimensionless Weber number has been defined as

$$\frac{\tau D_{max}}{\gamma}$$

where D_{max} is the maximum stable bubble size, τ is the shear stress and γ is the surface tension of the liquid.

In many situations it has been found that there is a characteristic Weber number at which equilibrium is achieved between the surface tension and the shear forces. Under these conditions the maximum stable bubble size will increase as the surface tension increases or as the shear stress is reduced. The effect of this is to decrease the interfacial area per unit volume and to decrease the oxygen transfer rate.

The situation in a fermenter is further complicated by bubble coalescence, bubble–cell contact and in a stirred-tank fermenter by the additional shearing action of the impeller. The dispersion of gases into mechanically agitated mixing vessels has been discussed in detail by Calderbank (1967). In this case the Weber number We is given by

$$\text{We} = \frac{N^2 D^3 \rho}{\gamma}$$

where N is the rotational speed and D is the impeller diameter.

The break-up of liquids falling through an immiscible liquid or a gas is as follows:

$$D_{max} \propto \left(\frac{\gamma}{\Delta\rho}\right)^{0.5}$$

where $\Delta\rho$ is the density difference.

Reduction in the surface tension will lead to a reduction in the diameter of the drop. Correlations for mean particle diameters have been presented for pressure nozzles, two fluid nozzles and atomizers by Masters (1972).

The use of the Weber number in connection with high-pressure homogenizers has been discussed by Loncin and Merson (1979).

6.15 SYMBOLS

a	interfacial area per unit volume
b	slide width
c_b	bulk concentration
c_i	concentration of gas at the interface
d	distance
D	impeller diameter
D_p	mean spherical diameter
D_{max}	maximum stable bubble diameter
F	force
F_c	correction factor (drop-weight method)
F_d	correction factor (direct pull method)
g	acceleration due to gravity
ΔG	Gibbs free energy change
h	height
HLB	hydrophile–lipophile balance
k	Eötvös constant
k_L	liquid film mass transfer coefficient
l	slide length
L	length
m	mass of one drop
m_A, m_B	mass fractions
M	molecular weight
N	rotational speed
OTR	oxygen transfer rate
p	pressure
r	radius of capillary or wire
R	gas constant
R	radius or ring
S	surface area per unit volume

T	temperature
T_c	critical temperature
V	volume of one drop
W_{AA}	work of cohesion
W_{AB}	work of adhesion

6.15.1 Greek symbols

γ	surface tension
γ_{AB}	interfacial tension
γ_o	reduced surface tension
θ	contact angle
Π	surface pressure
ρ	density
τ	shear stress
τ_b	surface excess concentration

6.15.2 Dimensionless group

We	Weber number

7

Introduction to thermodynamic and thermal properties of foods

7.1 INTRODUCTION

Most methods of preserving foods involve the transfer of energy (heat), either to or from the food. Some examples are given in Table 7.1. This shows the wide variety of materials that are to be handled, the different heat transfer fluids and mechanisms, and the typical processing times and temperatures. An excellent review on heat transfer and temperatures obtained during processing has been given by Holmes and Woodburn (1981).

It is important to know how much energy is required for these processes, and to conserve energy wherever possible. Over the last decade (1975–1985), energy costs have increased more than five-fold. Consequently, there have been many analyses on energy utilization in food processing, with a view to reducing costs; these are reviewed in section 7.12. The expression used for the supply of energy is services, which include hot water, steam, electricity, compressed air and chilled water. Services used in food and catering have been discussed in more detail by Kirk and Milson (1982) and Unklesbay and Unklesbay (1982).

The recovery of materials from food-processing wastes can also be regarded as a form of energy conservation. In the production of chips, up to 50% of the potato is not used. It is now commonplace to recover the off-cuts in the form of reconstituted products. Other recovery processes include (a) protein recovered from cheese waste, (b) blood and (c) starch from potato-processing effluent. Sugars from many effluents can be either recovered or fermented to a wide variety of products including alcohol and microbial protein.

In most cases, it is desirable to be able to heat and cool foods as rapidly as possible. This improves the economics of the process by increasing the capacity and generally results in a better-quality product. Heat transfer rates

Table 7.1 — Operations involving energy transfer processes

Operation	Description	Examples	Heat transfer medium	Processing conditions
Pasteurization	Removal of pathogenic micro-organisms; increasing shelf-life of highly perishable products	Milk, fruit juice, beer, egg	Hot water, steam, electricity	63–85 °C 15 s to 30 min
Sterilization				
(i) In the container	Sterilization of solids and liquids; 6 months shelf-life	Meat, fish, soup, vegetables, fruit milk, cream, custard, desserts, soup	Hot water, steam	100–125 °C 15 min to 2 h
(ii) Continuous flow methods (UHT)	Sterilization of fluids followed by aseptic packaging		Hot water, steam direct or indirect	135–150 °C 1–10 s
Evaporation	Removal of water; production of liquid concentrate	Milk, fruit and vegetables, coffee, cheese whey	Steam	40–100 °C 2 s to 2 h
Dehydration	Removal of water, production of dried material with low water activity	Milk, potato, vegetables, fruit, meat, fish	Hot air, steam, hot water, electricity	150–250 °C
Cooking and baking	Cooking of foods, baking of cereal-based foods	Catering operations, bread, meat pies, cakes	Steam, hot air, microwaves	1 min to several hours
Frying	Immersion in hot oil	French fried potatoes, doughnuts, crisps	Hot oil	100–150 °C
Chilling and freezing	Reducing temperature to just above freezing point or to well below the freezing point	Dairy products, meat, fish, fruit, vegetables, frozen desserts	Cold air, indirect refrigeration (ammonia), cryogenic fluids (liquid nitrogen)	10–0 °C −18 °C to −30 °C 2 min to 4 days

and mechanisms are very important, with heat being transferred by conduction, convection or radiation, or by combinations of these. The heating or cooling rate will depend upon a variety of factors such as the shape, size and physical nature of a material (i.e. whether it is a solid, liquid or suspension), the thermal properties of the material (such as the specific heat and thermal conductivity), the mechanism of heat transfer and the temperature and nature of the heat transfer fluid. Thermal processes are primarily used to extend the shelf-life of perishable products and to remove pathogenic organisms. This is achieved by inactivating micro-organisms which are likely to cause food spoilage, food poisoning or disease. The two forms of heat treatment most widely encountered are pasteurization and sterilization (Table 7.1).

In thermal processing it is also necessary to consider the effects of heat on the chemical constitutents and to minimize reactions such as browning, vitamin losses and oxidation reactions; these may reduce both the nutritional value of the food and its acceptability in terms of colour, flavour and texture. Enzymes which may cause spoilage during subsequent storage also need to be inactivated.

Finally, food is the basic energy source for life. Therefore the energy content of foods, together with the knowledge of how the food is broken down and utilized by the body, is very important.

7.2 CONSERVATION AND CONVERSION OF ENERGY

Any substance which contains energy has the capacity to do work (see section 1.20). Therefore the units of energy are the same as those for work, namely joules (J). There are many different types of energy: the most important are mechanical (kinetic), potential, chemical, thermal, electrical and sound. A long-standing principle known as the conservation of energy states that energy can be neither created nor destroyed but can be converted from one form to another. To all intents and purposes, this principle still applies to most everyday situations. However, one case in which it is violated is in the destruction of matter and its conversion to energy, the amount of energy E produced being given by Einstein's equation

$$E = mc^2$$

where m (kg) is the mass destroyed and c ($= 3 \times 10^8$ m s^{-1}) is the velocity of light. Thus the destruction of small quantities of matter leads to the production of large quantities of energy.

Examples of energy conversion processes can be seen in the production of heat or electricity from primary energy sources, namely coal, gas, oil and wood. These contain chemical energy (energy locked in the bonding arrangement of the atoms), which is converted to thermal energy (steam) by combustion; this heat is converted to mechanical and then to electrical energy by a turbine and a generator, respectively. The electricity is distributed and used for a wide variety of industrial and domestic operations.

Sec. 7.3]   Thermal energy and thermal units

Electricity can also be generated from radioactive materials. These processes are summarized in Fig. 7.1.

Unfortunately the generation of electricity is not an efficient process, as not all the chemical energy is converted to electrical energy; much of it is lost as heat, up the chimney stack, in the cooling water rejected by steam turbine or as friction in the equipment.

Potential energy, which is energy due to position, can be converted to kinetic energy, e.g. a waterfall. Again the conversion efficiency is not 100%; as some of the potential energy is converted to heat, the temperature at the bottom of a waterfall would be slightly higher than the temperature at the top. This kinetic energy can also be converted to electrical energy (hydroelectric power).

It is difficult to store electrical energy. The electricity boards have the difficult task of ensuring that supply meets demand. This has been improved by schemes for storing electrical energy in terms of potential energy; the surplus electrical energy can be used to pump water from a reservoir at a low level to one at a higher level. Thus, when the demand suddenly increases, e.g. during the commercial break in certain television soap operas, the flow can be reversed and substantial quantities of electricity can be generated quickly. This is much simpler than stopping and starting steam turbines.

If the conversion efficiency were 100%, a mass flow rate of water of 100 kg s^{-1}, falling a distance of 50 m, should create a total energy production rate equal to *mgh*, i.e.

$$\text{mass flow rate} \times \text{distance fallen} \times \text{acceleration } g \text{ due to gravity}$$
$$= 100 \text{ kg s}^{-1} \times 50 \text{ m} \times 9.81 \text{ m s}^{-2}$$
$$= 49050 \text{ J s}^{-1} = 49.05 \text{ kW}$$

Turbine efficiencies are about 80% and so the amount of energy produced would be 0.8×49.05 kW (39.2 kW).

Much more could be said about energy conversion processes. The cost of primary energy sources are rising and there are limited reserves of many of them. Alternative energy sources such as solar heating, wind and tidal power are receiving considerable attention.

7.3 THERMAL ENERGY AND THERMAL UNITS

Thermal energy arises from the agitation of the molecules; any object above 0 K will possess thermal energy or heat and, as the temperature increases, the thermal energy increases. Thermal energy can be regarded as a form of kinetic energy.

Originally the quantity of heat was measured in terms of calories (cal). One thermal calorie (1 cal) was defined as the quantity of heat required to raise 1 g of water from 14.5 °C to 15.5 °C. Although this is unit temperature rise, the change in temperature is specified. It will take a slightly different

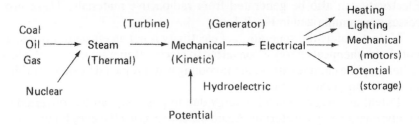

Fig. 7.1 — Some examples of energy conversion processes.

amount of energy to raise the same mass of water from 70 °C to 71 °C because of the temperature dependence of specific heat (see section 8.2).

Nutritionists also talk about the energy values of foods and express this in calories. However, one nutritionist's calorie is equivalent to one thousand thermal calories (1000 cal); to avoid confusion, the nutritionist's calorie is written as Cal or kcal to distinguish between them.

Joule showed experimentally a direct relationship between heat and work by means of a mechanically agitated turbine in a container of water. He was able to measure the work input by a system of weights and pulleys and to calculate the heat derived from the agitation by measuring the temperature rise; he determined that 4.18 J of mechanical work produced 1 cal of heat. This is termed Joule's equivalence.

$$4.18 \text{ J} = 1 \text{ cal}$$

It is possible to obtain 100% conversion efficiency of mechanical work to heat. (The conversion of heat to work, using a heat engine, can never reach 100% (Carnot efficiency).) Therefore, in all cases involving mechanical agitation (grinding, pumping, mixing and homogenizing), considerable mechanical energy will be absorbed by the material being processed and converted to heat. This is the principle behind extrusion cooking. Some form of cooling may be necessary in situations where temperature changes are undesirable.

The joule is now the recognized unit of heat in the SI system. It has also been used by nutritionists to express energy values; so an average daily intake of 2200 kcal can alternatively be expressed as 9.196 MJ. This is an improvement as it eliminates any of the ambiguities arising over the term calorie.

In the Imperial system, the unit of heat is the British Thermal unit (Btu). One British thermal unit (1 Btu) is the heat required to raise 1 lb of water from 59 °F to 60 °F. The conversion from British thermal units to calories or joules is as follows:

$$1 \text{ Btu} = 1\text{lb} \times 1°\text{C} = 454\text{g} \times \tfrac{9}{5}°\text{C} = 252.2 \text{ cal} = 1053 \text{ J}$$

1 therm, the unit still used by the British Gas board, is equal to 10^5 Btu:

$$1 \text{ therm} = 10^5 \text{ Btu} = 1.053 \times 10^8 \text{ J} = 29.25 \text{ kW h}$$

This conversion can be used to compare gas and electricity prices. Currently, in the UK, gas works out to be much cheaper than electricity for domestic purposes. The rates of heat transfer are expressed in watts (W) or British thermal units per hour (Btu h^{-1}). These are converted as follows:

$$\frac{1 \text{ Btu}}{1 \text{ h}} = \frac{1053 \text{ J}}{3600 \text{ s}} = 0.2925 \text{ W}$$

7.4 THERMODYNAMIC TERMS

Some of the terms used in describing properties which are necessary to understand food-processing operations are discussed below.

7.4.1 Systems and surroundings

In the analysis of many operations, the word system is often used. One common system encountered in thermodynamics is that of a gas enclosed in a cylinder by a piston. The gas itself is the system; anything external to the system is termed the surroundings. This implies the existence of a boundary, separating the system from the surroundings; in this case the boundary is the cylinder wall. The gas contained within the system will have certain fixed thermodynamic properties, i.e. temperature, pressure, volume, internal energy, etc. We can now perform certain operations on the system; we can compress or expand the gas, and we can heat or cool the gas. These operations will change the thermodynamic properties of the gas. In these cases, there is an interaction between the systems and surrounds and energy can cross the boundary. However, this is an example of a *closed system* because no mass or matter crosses the boundary (Fig. 7.2(a)). An *open*

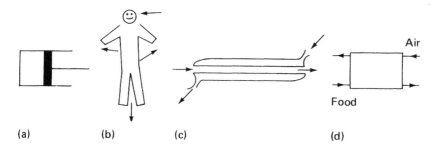

Fig. 7.2 — Systems and surrounds: (a) a closed system illustrated by a gas in a cylinder. Open systems illustrated by (b) a human being, (c) a continuous heat exchanger and (d) a food-drying plant.

system is one in which both matter and energy can cross the boundary. There are many examples of open systems in food-processing operations; some examples are shown in Figs. 7.2(b)–7.2(d). Mass balances in open systems are described in section 1.3.1.

7.4.2 Adiabatic and isothermal processes

If the temperature of the system is kept constant (consider the gas in the cylinder), the process is known as *isothermal*. If the gas is being compressed, i.e. gaining internal energy, then heat will need to be removed to keep the temperature constant. Thus, an isothermal compression will result in the transfer of energy across the boundary. If no energy crosses the boundary, the process is known as *adiabatic*. All the work of compression goes towards increasing the internal energy of the system. The isothermal and adiabatic compression represent the two extreme cases possible. Most practical compressions lie between the two and are known as *polytropic*. In a polytropic compression, there is both an increase in the internal energy and an increase in the temperature.

The pressure–volume–temperature (P–V–T) relationships for these situations are as follows:

$$\text{isothermal:} \quad pV = \text{constant}; \quad \Delta T = 0$$

$$\text{adiabatic:} \quad pV^\gamma = \text{constant}; \quad \frac{pV}{T} = \text{constant}$$

$$\text{polytropic:} \quad pV^n = \text{constant}; \quad \frac{pV}{T} = \text{constant}$$

where $\gamma = c_p/c_V$ (i.e. the specific heat c_p at constant presurse divided by the specific heat c_V at constant volume) and n is the polytropic index, where $1 < n < \gamma$. The specific heats of gases are discussed in more detail in section 8.4.

7.4.3 Reversible and irreversible processes

Most simple thermodynamics are concerned with reversible changes. A reversible change can be defined as a 'change taking place such that the properties hardly change by more than an infinitesimal amount from one instance to another'. This implies a very slow compression or expansion. In practice, most changes are irreversible. An alternative way of looking at a reversible change is by examining the expansion of a gas. If a gas expands reversibly, work is done on the surroundings. Therefore, it should be possible to put that same amount of work back into the system and to compress the gas back to its original condition. This will only be true if there is no frictional heat loss. Any energy lost that way will not be recoverable for useful work purposes. Any changes involving the loss of frictional heat are irreversible. In thermodynamics, it is usual to consider reversible processes and then to take into account the deviations from these processes.

7.5 THE FIRST LAW OF THERMODYNAMICS

The first law of thermodynamics is a statement of the law of conservation of energy. If we consider a gas enclosed by a piston in a cylinder, there is a relationship betwen the internal energy, the work done and the heat transferred to or from the system.

The internal energy of an ideal gas is a function of temperature. A change in internal energy takes place when there is a temperature change. Compressing a gas or heating a gas will normally increase the internal energy. The change ΔU in internal energy can be given from the first law:

$$\Delta U = q - W$$

Using this equation the sign convention is that the work done on the gas, i.e. compression, is negative and the work done by the gas, i.e. expansion, is positive; the heat added to the system is positive and the heat removed from the system is negative. Sometimes, this is written as

$$\Delta U = q - \int p \, dV$$

In an adiabatic process, $q = 0$. Therefore, $\Delta U = -\int p \, dV$. All the work is converted to internal energy. If a change takes place at constant volume, e.g. combustion of food products in oxygen in a bomb calorimeter, the work done is equal to zero. Therefore the heat liberated under these conditions gives a measure of the internal energy change for heat processes, i.e.

$$\Delta U = q$$

7.6 ENTHALPY

Enthalpy, normally denoted H, is a thermodynamic property; it is equal to the sum of the internal energy plus the product of the pressure and the volume, i.e.

$$H = U + pV$$

Changes in enthalpy during heating and cooling operations are important; the change in enthalpy ΔH is given by

$$\Delta H = \Delta U + \Delta(pV)$$
$$\Delta H = \Delta U + p \Delta V + V \Delta p$$

where Δ indicates the final minus the initial value. If a process takes place at constant pressure (i.e. $\Delta p = 0$), then

$$\Delta H = \Delta U + p \Delta V$$

From the first law of thermodynamics, $q = \Delta U + p\Delta V$, then

$$q = \Delta H$$

For a process taking place at constant pressure, the enthalpy change is equal to the heat absorbed or evolved. If the enthalpy change is positive, heat is absorbed and the reaction is *endothermic*. If the enthalpy change is negative, heat is evolved and the reaction is termed *exothermic*. For this reason, because many processes take place at a constant pressure, enthalpy is often equated with heat content. For example, when methane is burnt in air, the equation is as follows:

$$CH_4 + 2O_2 \rightarrow CO_2 + 2H_2O \qquad \Delta H = -890 \text{ kJ mol}^{-1}$$

Thus, when 16 g of methane burns at a constant pressure, 890 kJ of energy is released. Usually, we are not interested in absolute enthalpy values, and in many situations the enthalpy is expressed as a value above an arbitrarily chosen zero value. Often, in refrigeration, the enthalpy of a saturated liquid at −40 °C is given a value of zero (see section 11.7). With thermodynamic fluids it is common to talk about the specific enthalpy, i.e. the enthalpy associated with unit mass of the fluid (in kJ kg^{-1}) (kilojoules per kilogram)). Processes which take place at constant enthalpy ($\Delta H = 0$) are known as *isenthalpic*.

7.7 ENTROPY AND THE SECOND LAW OF THERMODYNAMICS

Entropy is probably one of the most difficult of the thermodynamic properties to understand. It can be regarded as a measure of the order or disorder of a system and is based on statistical mechanics. A system is regarded as being highly ordered if it is possible to predict the exact location of individual molecules comprising that system. As this becomes more difficult to do, the system becomes more disordered. As we go from the solid state through the liquid state to the vapour state, the disorder increases and the entropy is said to increase:

$$\text{solid} \rightarrow \text{liquid} \rightarrow \text{gas}$$

Most solids have a regular rigid structure and so there is a very high chance that one can predict the whereabouts of a particular molecule in that solid network. As the solid melts, the molecules become more mobile and their position less easily predictable. Nevertheless the liquid remains confined within its boundaries or surfaces. A further input of energy will convert the liquid into a gas; here there are no boundaries and it has become extremely difficult to predict the location of an individual molecule.

Sec. 7.8] Diagrammatic representation of thermodynamic changes

If we now consider an object which loses heat to the surroundings, T_2 the change in entropy of the object is defined as

$$\Delta S_{ob} = \frac{Q}{T_1}$$

where Q is the heat lost and T_1 is the absolute temperature of the object. The units of entropy are kilojoules per kelvin (kJ K^{-1}). Q is negative (sign convention). Therefore an object cooling is subjected to a decrease in entropy. However, the surroundings increase in entropy, by an amount

$$\Delta S_{surr} = \frac{Q}{T_2}$$

Therefore the total entropy change is given by

$$\Delta S_{surr} + \Delta S_{ob} = \frac{Q}{T_2} - \frac{Q}{T_1}$$

For a reversible process, $T_1 = T_2$ and thus $\Delta S = 0$. For an irreversible or spontaneous process where $T_1 > T_2$ (heat is always transferred from a high to a low temperature), the total entropy change will always be positive. Thus, in any spontaneous process, the total entropy always increases; only in a reversible change will the total entropy remain constant. This simple fact gives rise to one way of expressing *the second law of thermodynamics*: the total entropy of the Universe tends to a maximum.

In differential terms a change in entropy that takes place when an object loses a small amount of energy dq_{rev} is given by

$$\int dS = \int \frac{dq_{rev}}{T}$$

The total change in entropy is obtained by integrating this expression between the initial and final conditions:

$$dq_{rev} = \int_{S_1}^{S_2} T dS$$

A reversible adiabatic process ($dq_{rev} = 0$) is the same as an isentropic process ($dS = 0$). As was the case with enthalpy, the term specific entropy is used; this refers to the entropy of a unit mass of gas (in kilojoules per kilogram per kelvin (kJ kg^{-1} K^{-1})).

7.8 DIAGRAMMATIC REPRESENTATION OF THERMODYNAMIC CHANGES

Thermodynamic changes can be represented on pressure–volume (p–V) Diagrams (Fig. 7.3(a)) or temperature–entropy (T–S) diagrams

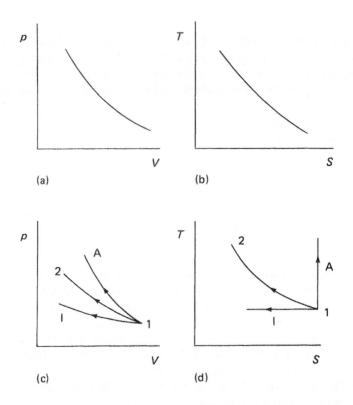

Fig. 7.3 — Thermodynamic changes represented on (a) a pressure–volume diagram and (b) a temperature–entropy diagram; (c) a compression between state 1 and state 2 on a pressure–volume diagram, together with an isothermal compression I and an adiabatic compression A; (d) the same compression shown on a temperature–entropy diagram, together with an isothermal compression I and an adiabatic compression A.

(Fig. 7.3(b)). A compression can be represented on a pressure–volume diagram, as shown in Fig. 7.3(c). In this case the gas is being compressed from condition 1 to condition 2. Such a process can also be represented on a temperature–entropy diagram (Fig. 7.3(d)). Also shown are the plots labelled I for an isothermal compression and the plots labelled A for an adiabatic compression ($\Delta S = 0$).

The total amount of heat liberated in a reversible process can be obtained from the area under the curve on the temperature–entropy curve for that process.

7.9 THE CARNOT CYCLE

Carnot considered that a gas is a cylinder being subjected to a series of reversible processes (Fig. 7.4). The gas initially at condition A is expanded

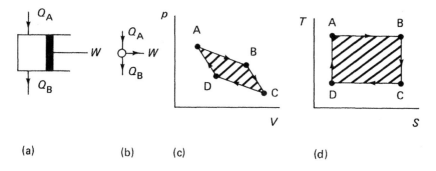

Fig. 7.4 — The Carnot cycle: (a) the conversion of heat to work; (b) the representation of a heat engine for converting heat to work; (c) the Carnot cycle represented as a pressure–volume diagram; (d) the Carnot cycle represented on a temperature–entropy diagram.

under isothermal conditions to condition B. This will require an input of heat Q_A. The gas is then expanded adiabatically to point C. During this expansion the work W_E done by the gas is equal to the area under ABC. The amount of heat supplied is equal to the area under ABC. The gas is then compressed isothermally forn C to D; during this process, heat Q_B is rejected from the gas. The gas is finally compressed adiabatically back to its original condition and the cycle is complete. During the compression the amount of work W_c done on the system is given by the area under ADC. Therefore

net work done = $W_E - W_c$ (shaded area in Fig. 7.4(c))

and

net heat supplied = $Q_A - Q_B$ (shaded area in Fig. 7.4(d))

Thus, during this cycle, heat is converted to work. For this reason, it is known as a power cycle. The efficiency of the conversion of heat to work is known as the Carnot efficiency CE and is given by

$$CE = \frac{\text{work done}}{\text{heat supplied}} \times 100$$

$$= \frac{Q_A - Q_B}{Q_A} \times 100$$

If the heat is supplied at a temperature T_A and is rejected at a temperature T_B, then the Carnot efficiency can be expressed as

$$CE = \frac{T_A - T_B}{T_A} \times 100$$

(Note that all temperatures are expressed in absolute terms.)

Effectively, heat is being taken in at a temperature T_A and rejected at a temperature T_B, and a quantity of work is done; this quantity of work is directly proportional to the temperature difference. Such machines that convert heat to work are known as heat engines. The Carnot cycle describes the workings of an ideal heat engine. Despite the fact that all the processes are reversible, the Carnot efficiency gives the maximum possible conversion efficiency of heat to work, when operating between two temperatures; in practice the efficiency will be less than the Carnot efficiency because of frictional and other losses. A steam turbine taking in steam at 400 °C and rejecting water from the condenser at 25 °C will have a Carnot efficiency given by

$$CE = \frac{673 - 298}{673} \times 100 = 55.7\%$$

Thus, 44.3% of the original energy in the steam is rejected, normally into the atmosphere or cooling water. This energy is often termed low-grade energy. The actual efficiency of a steam turbine will be less than this value; the low conversion efficiency explains why the production of electricity by any fuel involving steam production is a relatively inefficient process. One further consequence of this analysis is that the conversion efficiency from heat to mechanical work can only be 100% when the heat is rejected at absolute zero ($T_B = 0$). This is a practical impossibility. The second law of thermodynamics deals with the conversion of heat to work and can be expressed in several ways. In a spontaneous change, heat is transferred from a high to a low temperature; there can never be 100% conversion efficiency of heat to work and the maximumn possible conversion efficiency is given by the Carnot efficiency.

7.9.1 The reverse Carnot cycle

If the cycle is reversed, the result is that work is put into the system and the direction of energy flow is reversed (Fig. 7.5). Energy is now being transferred against the natural temperature difference. Energy is being removed from a low-temperature source and rejected at a higher temperature. This is the basis of the refrigeration cycle or a heat pump circuit. The efficiency of such a cycle is measured by the coefficient COP of performance where

$$COP = \frac{\text{heat removed}}{\text{work done}}$$

$$= \frac{T_1}{T_2 - T_1}$$

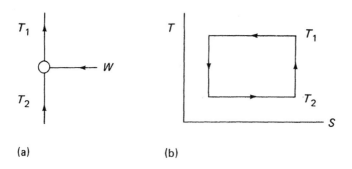

Fig. 7.5 — The reverse Carnot cycle: (a) represented by a heat pump; (b) represented on a temperature–entropy diagram.

Coefficients of performance greater than 1.0 are obtainable. It appears that an input of work reverses the direction of heat transfer. However, in a practical refrigeration cycle, this is not so, and the normal rules of heat transfer are obeyed. What is effectively happening is that energy is being removed from the media (water, air or food); this energy is then being upgraded and rejected to the atmosphere (air cooling on a refrigerator) or some other heating medium, by means of a heat transfer fluid, in this case termed a refrigerant. Maximum efficiency can be obtained from such a cycle if it is used for both heating and cooling purposes. Thus, in a supermarket, with a large number of deep-freezes, the energy derived from maintaining the food frozen, or from freezing the food, can be efficiently utilized for space heating or providing domestic hot water.

The third law of thermodynamics states that the entropy of a perfect crystal at 0 K is zero. At this temperature, there is no thermal motion, and therefore perfect order.

7.10 HEAT OR ENERGY BALANCES

In many of the examples that follow the law of conservation of energy will be used in the form of a heat or energy balance. If steam is being used to heat water, the energy balance states that

heat lost by the steam = heat gained by the water

Such balances enable the quantity of steam, hot water, chilled water, hot air or refrigerant to be evaluated. The equation is somewhat simplified because some heat is lost to the surroundings (heat losses). These should be minimized as far as possible, or attempts should be made to compensate for these if more accurate solutions are required.

Other thermal properties of importance, such as specific heat, latent heat, thermal conductivity and thermal diffusivity, are discussed in Chapters 8 and 9.

7.11 ENERGY VALUE OF FOOD

Food contains energy which is utilized by the body for performing useful work, keeping it warm and producing further cell and body tissue to replace that which is lost. This energy is released during the oxidation of food as it is broken down in the body. The energy content of food, which is currently of great concern to people, is expressed in terms of its calorific value (see section 7.3).

The *gross energy* value of a food can be determined using an adiabatic bomb calorimeter. The food is burnt in oxygen in a container of constant volume, and all the heat liberated is adsorbed by water; the heat losses to the surroundings are zero. The amount of heat q evolved is determined from the product of the mass of water, its temperature rise and its specific heat. This value is also called the heat of combustion or the internal energy change:

$$\Delta U = q - W$$
$$= q - \int p \, dV \qquad (dV = 0)$$
$$q = \Delta U$$

However, for routine use the adiabatic bomb calorimeter is tedious. Miller and Payne (1959) describe the use of a ballistic bomb calorimeter, which enables ten samples to be measured every hour. No attempt is made to prevent heat losses and the heat evolved is absorbed by the bomb itself, which rises in temperature fairly rapidly and then cools. The maximum temperature is used as a measure of the heat evolved and the bomb is calibrated with standards of known energy content, such as butter fat, benzoic acid, sucrose and urea. The amount of energy available for use by the body is known as the metabolizable energy. It is less than the gross energy value because certain substances are not adsorbed and are excreted in the faeces, e.g. some unavailable carbohydrate, or even when adsorbed are not completely converted to carbon dioxide and water and are excreted in the urine. Therefore, metabolizable energy is given by the following equation:

$$\begin{Bmatrix} \text{metabolizable} \\ \text{energy} \end{Bmatrix} = \begin{Bmatrix} \text{gross energy} \\ \text{intake} \end{Bmatrix} - \begin{Bmatrix} \text{energy excreted} \\ \text{in faeces} \end{Bmatrix} - \begin{Bmatrix} \text{energy excreted} \\ \text{in urine} \end{Bmatrix}$$

A reasonable estimate of metabolizable energy can be made by reference to the Artwater factors, which in SI units are as follows: protein, 17 MJ kg^{-1}; fat, 37 MJ kg^{-1}; carbohydrate, 16 MJ kg^{-1}; alcohol, 29 MJ kg^{-1}. Therefore, if compositional data are avilable, metabolizable energy can be determined; such data have been presented by Paul and Southgate (1978). In 1980, the contributions of energy dervied from protein, fat and carbohydrate were 13.0%, 42.8% and 44.4%, respectively, in the UK diet. The average contribution of alcohol to the total energy content of the diet was about 7% (Walker, 1984).

Muller and Tobin (1980) emphasized that accurate values for metabolizable energy can only be obtained from animal feeding trials. An individual energy requirement is governed by their basal metabolic rate and their level of activity.

The basal metabolic rate is a measure of the energy requirement of the body at rest and is the energy required for maintaining the body at 37 °C and for the activity of internal organs. Significant variations occur between individuals. It is interesting to note that basal metabolic rates for different species of animals are the same, when calculated on the basis of body surface area.

Over a 24 h period, the basal metabolic rate accounts for the greatest part of the body's total energy requirement. However, other daily activities require greater energy expenditure; some of these activities may be sustained for only very short periods whilst others may last over the working day. Energy expenditure during such activities can be measured, but with difficulty. Some values for everyday operations are given in Table 7.2.

Table 7.2 — Energy utilization during everyday operations

Energy utilization (kJ min^{-1})	Operation
10–20	Light assembly work, many types of housework, military drill, painting, golf, truck driving
20–31	Pick and shovel, soldier route marching, tennis, cycling at 10 miles h^{-1}
31–40	Coal mining, football
41	Lumberjack, furnaceman, swimming crawl, cross-country running

Adapted from the data of Hawthorn (1981).

Metabolic rates for a wide range of activities have been recorded by the American Society of Heating, Refrigerating and Air-Conditioning Engineers (1981) in terms of met. units, where

1 met. unit = 58.2 W m^{-2} or 18.4 Btu h^{-1} ft^{-2}

This takes into account the surface area of the individual.

The basic laws of energy conservation apply in nutrition. If energy intake exceeds energy utilization, the excess energy will accumulate and be stored as fat. For a person of stable weight, energy expenditure will equal energy supplied by the food.

7.12 ENERGY CONSERVATION IN FOOD PROCESSING

The costs of primary energy sources have increased significantly over the last 15 years worldwide. Singh (1984) reported that the cost of different energy sources increased between three and seven times over the period 1971–1980 in the USA. This has forced most sectors of the food industry to examine energy utilization, with a view to saving energy because most processing operations are energy intensive. Energy conservation starts even at the design stage in terms of both selecting the process and ensuring that energy recovery systems are incorporated.

There is usually an approximate inverse relationship between capital costs and energy costs, i.e. the more money invested, the lower are the energy requirements. One simple example is in the choice of pipe diameter for pumping applications (see section 3.10). For a given flow rate, the energy costs decrease as the pipe diameter increases but capital costs increase. The economic pipe diameter is calculated as the size which minimizes the total pumping costs (capital plus energy), over the plant operating life. Such economic balances are commonplace in most food engineering operations.

Equipment should be properly installed, all heating and cooling systems should be correctly lagged, and provision should be made to recycle steam condensate and to re-use waste heat. A well-planned maintenance schedule should ensure that these items are checked regularly and that leaking taps, steam valves and other items are quickly repaired. Personnel should also be educated in these matters. Some examples of energy conservation processes will now be considered.

In heating and cooling operations, it is far more energy efficient to use continuous rather than batch processes. Modern continuous heat exchangers operate at high regeneration efficiencies (up to 95%), resulting in savings in heating and refrigeration. However, one consequence of this is longer residence times, resulting in a greater loss of heat-sensitive components. Usually this is not too important in high-temperature short-time pasteurization processes, but it may be significant in UHT products (Lewis, 1986a; Reuter, 1984). The production of UHT products is much more energy favourable than in-container sterilization. However, it is possible to recover more energy using the indirect rather than the direct UHT heating system. In canning processes, considerable heat is lost as non-condensed steam during retorting. Savings can be made by using recirculating hot water as the heating medium and by insulating the retorts. Heat transfer rates may be slightly reduced compared with steam.

Evaporation is an energy-intensive process and considerable savings result from the use of multiple-effect evaporators (see section 9.16). If heat losses are neglected, 1 kg of steam will remove n kg of water vapour, where n is the number of effects. Seven-effect evaporators are now common in the dairy industry. An alternative way of conserving energy is to recompress the vapour produced during evaporation. Mechanical vapour recompressor evaporators are now available, which are very efficient but also expensive. Steam ejectors are also commonly used for recompressing the vapour. Often the feed is preheated using extracted vapour, and all condensate should be recycled to the boiler or used for space heating, wherever possible. Other methods for removing water, such as reverse osmosis and freeze concentration, may be advantageous. Reverse osmosis used much less energy than evaporation, but higher capital and maintenance (replacing membranes) are incurred (see section 13.11). However, it has been found to be useful for preconcentrating whey and skim-milk, up to 20% total solids. Some of these aspects of liquid concentration have been discussed in more detail by Schwartzberg (1977). The use of membrane techniques in the dairy industry has been reviewed by Glover *et al.* (1978) and for concentrating food proteins by Lewis (1982).

Dehydration is also very energy intensive. Most liquids and pastes are dehydrated by means of spray-drying methods, using hot air at an inlet temperature (150–250°C) and an outlet temperature of 90–95°C. Spray driers are usually inefficient, an approximate equation for the thermal efficiency TE being

$$\text{TE} = \frac{\theta_{in} - \theta_{out}}{\theta_{in} - \theta_a} \times 100$$

where θ_{in} is the inlet air temperature, θ_{out} is the outlet air temperature and θ_a is the ambient temperature of the air prior to heating.

The thermal efficiency is improved by increasing the inlet air temperature. However, this is limited as high temperatures may scorch the product. The outlet temperature is controlled at about 95°C. If it falls below this, the final powder becomes too wet. If it rises too much above this, energy is wasted. Therefore the exit air temperature should be as low as possible, commensurate with the product being sufficiently dry. Efficiency can be increased by recovering the heat from the air, either by an indirect exchanger (used to heat incoming air) or by wet scrubbing, where the air is contacted with the incoming concentrate. This serves two functions: it prewarms the concentrate and removes fine powder from the air stream. Microbiological problems may be encountered with scrubbing systems. The inlet air can be heated by direct methods (combustion gases) or indirect methods, with direct methods being more efficient; however, this may cause combustion gases such as SO_2 to contact the product, and a higher humidity in the inlet air.

Finally, to ensure that the output of powder from a spray drier is maximized, it is important that the feed concentration to the drier is as high as possible. In other hot-air drying processes some of the air leaving the drier is recycled as it very rarely absorbs as much moisture as it can in one pass.

Energy economy in low-temperature processes is attained by ensuring that the compression conditions are correct and by regular maintenance and inspection, e.g. checking the insulation, avoiding the build-up of ice in the evaporator. The cost of storing frozen foods for long periods should be taken into account when comparing the economics of freezing with canning or dehydration.

Heat pumps are now used widely for abstracting waste heat and provide a means for utilizing 'low-grade' energy. They are extremely useful when both heating and cooling are required. The energy removed when foods are chilled or frozen is used for space heating or producing hot water. Alternatively, warm chilled water leaving a cooler can be recooled and the energy removed used for heating purposes elsewhere. The energy costs involved in cleaning and sanitizing need attention with particular reference to operating temperatures, flow rates, times and detergent strengths.

Energy can be recovered from agricultural waste products by fermentation, e.g. the production of protein from a wide variety of low-grade carbohydrate sources, or by production of methane by anaerobic fermentation of organic matter. A wide variety of literature is available in food and agricultural biotechnology.

Alternative or 'free' energy sources have also been widely investigated; sun drying has been practised for centuries. Solar energy is now widely used for domestic heating, cooking, evaporation and dehydration proceses. Other 'free' energy sources include wind, tidal and hydroelectric power. However, the capital costs involved in exploiting these sources may be very high.

A vast amount of literature has appeared over the last 10 years about energy utilization. The use of services in catering operations have been described by Kirk and Milson (1982). An analysis of energy costs involved in food production has been given by Leach (1976). The texts of symposia in the following references should be consulted for analysis of energy utilization in food-processing industries: *Food Technology* (1979), International Dairy Federation (1977), Linko *et al.* (1980), McKenna (1984a), McKenna (1984b), Society of Dairy Technology (1982) and Unklesbay and Unklesbay (1982).

7.13 SYMBOLS

c velocity of light
c_p specific heat at constant pressure
c_V specific heat at constant volume
E energy
H enthalpy
m mass
n polytropic index

p pressure
q quantity of heat liberated
Q heat gained or lost
S entropy
T temperature
V volume
U internal energy
W work

7.13.1 Greek symbols
γ ratio of specific heats
Δ final value minus initial value
θ temperature

8

Sensible and latent heat changes

8.1 INTRODUCTION

The properties to be discussed in this chapter include specific heat, latent heat and specific enthalpy.

These properties play an important role in heat transfer problems when heating or cooling foods. It is necessary to know the specific heat to determine the quantity of energy that needs to be added or removed. This will give an indication of the energy costs involved and in a continuous process will have an influence on the size of the equipment.

Latent heat values, which are associated with phase changes, play an important role in freezing, crystallization, evaporation and dehydration processes.

8.2 SPECIFIC HEAT

The specific heat of a material is a measure of the amount of energy required to raise unit mass by unit temperature rise. As mentioned in Chapter 7, specific heat is temperature dependent. However, for the purpose of many engineering calculations, these variations are small and an average specific heat value is used for the temperature range considered.

The units of specific heat are kilojoules per kilogram per kelvin (kJ kg^{-1} K^{-1}), kilocalories per kilogram (kcal kg^{-1} K^{-1}) or British thermal units per pound per degree Fahrenheit (Btu lb^{-1} degF^{-1}). From the definitions of the different thermal units, the specific heat of water in the respective units is

$$1.0 \text{ kcal kg}^{-1} \text{ K}^{-1} \text{ or } 4.18 \text{ kJ kg}^{-1} \text{ K}^{-1} \text{ or } 1 \text{ Btu lb}^{-1} \text{ degF}^{-1}$$

In a batch heating or cooling process, the amount of heat (energy) Q required or removed is given by

Sec. 8.2] Specific heat 221

Q = mass × average specific heat × temperature change
= $MC\,\Delta T$ (J or kcal or Btu)

In a continuous process, the rate of heat transfer is given by

Q/t = mass flow rate × specific heat × temperature range

The units of Q/t are joules per second (J s^{-1}), i.e. watts (W), or British thermal units per hour (Btu h^{-1}). This is often termed the heating or cooling *duty* of the heat exchanger. If it is felt that this is not a sufficiently accurate procedure, the total energy requirement can be obtained by graphical integration. Specific heat is plotted against temperature; the total heat required to raise unit mass from T_1 to T_2 is given by $\int c_p\,dT$ or the area under the curve (Fig. 8.1) This should, in most cases, be not too far removed from

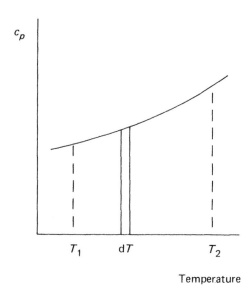

Fig. 8.1 — The relationship between specific heat and temperature and the evaluation of total heat change.

the value obtained by selecting a specific heat value at the average temperature $(T_1 + T_2)/2$. If the relationship between the specific heat and temperature is known in terms of temperature, then the integral $\int c_p\,dT$ can be evaluated directly. (see section 8.4.).

All these equations apply to what are known as *sensible* heat changes, i.e. energy changes that can be detected by a change in temperature. Latent heat changes which are involved in phase changes will be considered later; examples of these are the conversion of water to ice, other crystallization reactions particularly of fats, the vaporization of water or the condensation of steam, and the sublimation of ice, as in freeze-drying operations.

Table 8.1 shows some specific heat values for some foods and non-food materials.

It is pertinent to make some general observations about specific heat values at this point. Liquid water has an extremely high specific heat value, much higher than most other liquids. This is why it is so widely used as a cooling medium. The addition of ethylene glycol (antifreeze) will lower the specific heat and consequently the cooling efficiency. When water freezes, the specific heat capacity is drastically reduced, by a factor of approximately 2. Since water has a much higher specific heat than most other food constituents, the specific heat of a food is significantly affected by the amount of water present and the physical state of the water. Frozen foods with high water contents will have specific heat values approximately half that of their fresh counterparts.

Thus, considerably less energy is required to reduce food from −1°C to −30 °C than is required from 28 °C to −1 °C (most foods will commence freezing at about −1 °C); it should also be noted that freezing is not a sharp process, i.e. the water does not freeze at a constant temperature. During processes such as evaporation, and dehydration the specific heat of the food will fall. Water vapour has a specific heat value approximately equal to that of ice. Specific heat of vapours are considered in more detail in section 11.6.

Metals have very low specific heat values compared with those of foods. Oils and fats again have specific heats about half that for water. Dried grain and food powders also have very low specific heat values. Specific heats are temperature dependent; for most substances, there is a slight increase in the specific heat as the temperature rises. Since specific heats are dependent on moisture content and temperature, these are often recorded in more detail. Table 8.1 gives further specific heat values for a range of foods, together with a selection of data sources.

8.3 RELATIONSHIP BETWEEN SPECIFIC HEAT AND COMPOSITIONS

From these observations, one might expect that it would be possible to predict the specific heat of a food from a knowledge of its composition. For example the specific heat of skim-milk would be slightly lower than that of water because of the presence of milk solids. As the fat content increases, one would expect the specific heat to decrease (substitution of water or fat).

The simplest form of equation for estimating the approximate specific heat c of food is as follows:

$$c = m_w c_w + m_s c_s \text{ (kJ kg}^{-1}\text{ K}^{-1}\text{)}$$

where m_w is the mass fraction of water, $c_w = 4.18$ kJ kg^{-1} K^{-1} is the specific heat of water, m_s is the mass fraction of solids and $c_s = 1.46$ kJ kg^{-1} K^{-1} (Lamb, 1976) is the specific heat of solids. This reflects the major contribution due to water content.

An alternative form, given by Miles et al. (1983), distinguishes between fat and other solids. The equation given is

Sec. 8.3] Relationship between specific heat and composition

Table 8.1 — Specific heat of some foods and food processing materials.

Food	Temperature	Specific heat (kJ kg^{-1} K^{-1})	Specific heat (kcal kg^{-1} K^{-1}) or Btu lb^{-1} degF^{-1}
Water	59 °F	4.18	1.000
Ice	32 °F	2.04	0.487
Water vapour	212 °F	2.05	0.490
Air	−10 °F to +80 °F	1.00	0.240
Copper	20 °C	0.38	0.092
Aluminium	20 °C	0.89	0.214
Stainless steel	20 °C	0.46	0.110
Ethylene glycol	40 °C	2.21	0.528
Ethyl alcohol	0 °C	2.24	0.535
Glycerol	18–50 °C	2.43	
Oil, maize	20 °C	1.73	0.414
Oil, sunflower	0 °C	1.86	0.446
Oil, sunflower	20 °C	1.93	0.460
Apples (84.1% moisture content)	Above freezing point	3.59	0.860
Apples (84.1% moisture content)	Below freezing point	1.88	0.45
Potatoes (77.8% moisture content)	Above freezing point	3.43	0.82
Potatoes (77.8% moisture content)	Below freezing point	1.80	0.43
Potatoes, dried (10.9% moisture content)		1.85	0.443
Lamb (58.0% moisture content)	Above freezing point	2.80	0.67
Lamb (58.0% moisture content)	Below freezing point	1.25	0.30
Cod	Above freezing point	3.76	0.90
Cod	Below freezing point	2.05	0.49
Milk (87.5% moisture content)	Above freezing point	3.89	0.930
Milk (87.5% moisture content)	Below freezing point	2.05	0.490
Soya beans (8.7% moisture content)		1.85	0.442
Wheat (10.0% moisture content)		1.46–1.80	0.35–0.43

Compiled from a variety of sources.
Further data can be obtained from the following references: American Society of Heating, Refrigerating and Air-Conditioning Engineers (1981), Holmes and Woodburn (1981), Jowitt et al. (1983), Polley et al. (1980) and Singh (1982).

$$c = (0.5m_f + 0.3m_{snf} + m_w) \times 4.18 \text{ (kJ kg}^{-1}\text{ K}^{-1}\text{)}$$

where m_f, m_{snf} and m_w are the mass fractions of fat, solids non-fat and water, respectively.

If the approximate analysis for the material is available, the following equation can be used:

$$= m_w c_w + m_c c_c + m_p c_p + m_f c_f + m_a c_a$$
$$\text{(water)} \quad \text{(carbohydrate)} \quad \text{(protein)} \quad \text{(fat)} \quad \text{(ash)}$$

where m_w, m_c, m_p, m_f and m_a are the mass fractions of the respective components and c_w, c_c, c_p, c_f and c_a are the specific heats of the respective components.

Values for the specific heats of the various components are given in Table 8.2.

The specific heat of air and ice are given as 1.0 kJ kg^{-1} K^{-1} and 2.1 kJ kg^{-1} K^{-1}, respectively. The latter is extremely useful for estimating the specific heat of frozen foods, although it cannot be assumed that all the water is present in the frozen form until the temperature is reduced to below -40 °C. It is normal to state whether the specific heat of the component is measured above or below its freezing point; if this is not stated, it is normally accepted that it is above the freezing point.

The freezing of foods and crystallization of fats will be dealt with separately, as they involve the occurrence of simultaneous sensible and latent heat changes.

8.4 SPECIFIC HEAT OF GASES AND VAPOURS

Specific heats of gases and vapours are considered under two sets of conditions. A gas can be heated under conditions of constant volume or constant pressure. The specific heat c_V at a constant volume is the amount of heat required to raise unit mass by unit temperature rise under conditions of constant volume, whereas the specific heat c_p at constant pressure is the amount of heat required to raise unit mass by unit temperature rise at constant pressure. Additional energy is required in the latter case in order to expand the gas and to maintain a constant pressure (if the gas were not expanded, the pressure would increase). Therefore,

$$c_p > c_V$$

The ratio of c_p/c_V is given the symbol γ. Specific heats and γ values are given for some common gases in Table 8.3.

The γ value for gases decreases as the number of atoms in the molecule decreases; a common value for diatomic molecules is 1.4. The relationship between γ and the structure of molecules has been discussed by Nelkon (1970) and by Spalding and Cole (1973).

The specific heat of a gas also varies with temperature, the extent again being dependent upon the number of atoms in the molecule. There is hardly any variation for monatomic gases, a slight variation for diatomic molecules and much more considerable variation for triatomic molecules. Usually, specific heat increases as temperature increases. Table 8.4 shows how the specific heat of air varies with temperature.

For many gases a type of relationship between specific heat and temperature has been found to be applicable:

$$c_p = a + bT + cT^2 + dT^3$$

where a, b, c and d are constants and T is the absolute temperature.

Value for these constants are given for some common gases in Table 8.5, which allows the determination of specific heat over a wide temperature

Sec. 8.4] Specific heat of gases and vapours 225

Table 8.2 — Specific heat values for food components.

		(1)	(2)
Specific heat c_w of water	(kJ Kg^{-1} K^{-1})	4.18[a]	4.18[b]
Specific heat c_c of carbohydrate	(kJ Kg^{-1} K^{-1})	1.4[a]	1.22[b]
Specific heat c_p of protein	(kJ Kg^{-1} K^{-1})	1.6[a]	1.9[b]
Specific heat c_f of fat	(kJ Kg^{-1} K^{-1})	1.7[a]	1.9[b]
Specific heat c_a of ash	(kJ Kg^{-1} K^{-1})	0.8[a]	—

[a]From the data of Kessler (1981); recommended for dairy products.
[b]From the data of Miles et al. (1983).

Table 8.3 — γ and c_p values for some gases.

Gas	γ	c_p (kJ kg^{-1} K^{-1})
Air	1.4[a]	1.00[b]
Nitrogen	1.4[a]	1.04[b]
Oxygen	1.4[a]	0.92[b]
Hydrogen	1.4	1.42
Carbon monoxide	1.4	1.04
Carbon dioxide	1.3	0.84
Argon (monatomic)	1.67	0.52

[a]From the data of Spalding and Cole (1973); values at 15 °C.
[b]From the data of Batty and Folkman (1983); values at 27 °C.
Further data has been provided by Kaye and Laby (1973).

Table 8.4 — Relationship between the specific heat of air and temperature.

Temperature (°C)	−73	−23	27	127	227
Specific heat (kJ kg^{-1} K^{-1})	1.003	1.003	1.005	1.014	1.030

Table 8.5 — Values for the constants for specific heat c^p of gases.

Gas	a	$10^2 b$	$10^5 c$	$10^9 d$	Valid temperature range (K)
Ammonia	27.55	2.563	0.99	−6.686	273–1500
Carbon monoxide	27.11	0.655	−0.10	—	273–3800
Hydrogen	29.09	−0.1916	0.40	−0.870	273–1800
Nitrogen	27.32	0.6226	0.0950	—	273–3800
Oxygen	25.46	1.519	−0.7150	1.311	273–1800
Water	32.22	0.1920	1.054	−3.594	273–1800

Compiled from the data of Warn (1969).

range. The units of specific heat are joules per mole per kelvin (J mol^{-1} K^{-1}).

Therefore as an example, the specific heat of oxygen at 27 °C (300 K) is given by

$$c_p = a + bT + cT^2 + dT^3$$
$$= 25.46 + 1.519 \times 10^{-2}(300) - 0.7150 \times 10^{-5}(300)^2 + 1.311 \times 10^{-9}(300)^3$$
$$= 25.46 + 4.57 - 0.6435 + 0.0354$$
$$= 29.42 \text{ J mol}^{-1} \text{ K}^{-1}$$
$$= \frac{29.42}{32}$$
$$= 0.919 \text{ J g}^{-1} \text{ K}^{-1} \text{ (kJ kg}^{-1} \text{ K}^{-1})$$

This shows good agreement with the value in Table 8.3. Often the term dt^3 is ignored.

The total amount of energy Q required to raise the temperature of a gas (by dT) at a constant pressure is given by the equation

$$Q = \int c_p \, dT$$

If the value of c_p is known in terms of T, the integral can be easily evaluated. The amount of energy required to raise the gas from temperature T_1 to a higher temperature T_2 is

$$Q = \int_{T_1}^{T_2} (a + bT + cT^2 + dT^3) \, dT$$
$$= \left[aT + \frac{bT^2}{2} + \frac{cT^3}{3} + \frac{dT^4}{4} \right]_{T_1}^{T_2}$$
$$= a(T_2 - T_1) + \frac{b}{2}(T_2^2 - T_1^2) + \frac{c}{3}(T_2^3 - T_1^3) + \frac{d}{4}(T_2^4 - T_1^4)$$

8.5 DETERMINATION OF SPECIFIC HEAT OF MATERIALS (EXPERIMENTAL)

8.5.1 Method of mixtures

In the method of mixtures, the specimen of known mass and temperature is dropped into a fluid of known mass and temperature contained in a copper or aluminium metal container known as a calorimeter. The final temperature of the mixture is noted. To determine the specific heat of the specimen, it is assumed that the heat lost by the specimen is equal to the heat gained by the fluid and the calorimeter.

The following experimental readings are required: m_s, m_c and m_f are the masses of the specimen, calorimeter and fluid, respectively, and c_s, c_c and c_f are their respective specific heat values. The initial temperatures of the

Sec. 8.5] Determination of specific heat of materials (experimental)

specimen and fluid are T_s and T_f and the final temperature of the mixture is T_m.

Then the heat lost by the specimen is $m_s c_s (T_s - T_m)$, the heat gained by the fluid is $m_f c_f (T_m - T_f)$ and the heat gained by the calorimeter is $m_c c_c (T_m - T_f)$. Therefore, $m_s c_s (T_s - T_m) = m_f c_f (T_m - T_f) + m_c c_c (T_m - T_f)$ and

$$c_s = \frac{(T_m - T_f)(m_c c_f + m_c c_c)}{(T_s - T_m) m_s}$$

The greatest source of experimental error is involved in heat losses to the surroundings, particularly if the specific heat capacity is low. One way of reducing this is to insulate the calorimeter; alternatively, preliminary experiments with the materials in question can be performed to ascertain the temperature rise involved, and the initial temperature of the liquid adjusted below room temperature, so that the heat gained whilst the liquid was below room temperature balances the heat loss when the temperature rose above room temperature.

A more sophisticated modification is to avoid heat loss by ensuring that the surroundings are always at the same temperature as the calorimeter contents. Thus the temperature of the contents and surrounds are compared, and energy is put into the surroundings to keep them at the correct value. This principle is known as adiabatic calorimetry.

Care should be taken when heating biological materials to ensure that the moisture content is not unduly affected. The fluid usually chosen is water, although other reference substances with lower specific heat values (and hence which are more sensitive) have been used. An alternative approach to heating the fluid to too high a temperature would be to lower the fluid temperature. The method of mixtures is very suitable for most solids in the form of particulate matter grain or powder. Initial size reduction of large particles may be required to ensure that the heat is transferred relatively quickly.

8.5.2 Method of cooling

At a given temperature the rate of loss of heat from two liquids contained in calorimeters are equal. Newton's law of cooling states that the rate of loss of heat from a body is directly proportional to the excess temperature of the body above that of its surroundings. The sample liquid is heated and then placed in a calorimeter. The temperature is taken at regular intervals as the fluid cools. The procedure is repeated using the same calorimeter and the volume of a liquid of known specific heat. Cooling curves are obtained for the two liquids and the rates of cooling are determined at the same temperature (Fig. 8.2).

If m_1 and m_2 are the masses of the sample and the control liquid (often water), c_1 and c_2 are their respective specific heats, $(dT/dt)_1$ and $(dT/dt)_2$ are their respective cooling rates, and m_c and c_c are the mass and specific heat of the calorimeter, then

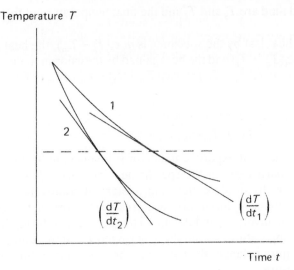

Fig. 8.2 — Cooling curves 1 and 2 for two liquids with different specific heats.

$$\left\{\begin{array}{c}\text{rate of heat loss}\\\text{from the sample}\end{array}\right\} = \left\{\begin{array}{c}\text{rate of heat loss from}\\\text{the reference material}\end{array}\right\}$$

$$(m_1 c_1 + m_c c_c)\left(\frac{dT}{dt}\right)_1 = (m_2 c_2 + m_c c_c)\left(\frac{dT}{dt}\right)_2$$

Since the specific heat of the sample is the only unknown, it can be determined from this equation. It should be noted that the rate of loss of heat is dependent upon a number of other factors such as the surface area of the sample, the temperature of the surroundings and the nature of the calorimeter surface. Care should be taken to ensure that these remain the same for the two liquids.

The cooling rate at a particular temperature is determined from the gradient of the tangent to the curve. This can be done either by drawing the tangent or by various mathematical methods available. A reasonable approximation can be made by taking the time for the temperature to fall over a specified temperature range, where the temperature at which the cooling rate is required is the mean value. For example, if the rate of cooling is required at 65 °C, the time required for the temperature to fall from 70 °C to 60 °C is determined from the cooling curve. If the time is 4 min, then the rate of cooling is 10/4, i.e. 2.5 degC min^{-1}

An alternative way of performing the experiment is to use a comparison calorimeter. This will speed up the operation because the two fluids are placed in identical calorimeters and the cooling curves are obtained simultaneously.

8.5.3 Electrical methods

It is possible to insert an electrical element in a calorimeter containing the test fluid and to measure the temperature rise brought about by the application of known voltage and current for a fixed time. Heat losses can be eliminated by cooling the liquid to below room temperature and applying electrical energy until the temperature reaches a value above room temperature such that the temperature differences above and below room temperature are the same.

In this case,

$$VIt = m_c c_c \Delta T + m_l c_l \Delta T$$

where V (V) is the voltage, I (A) is the current, t (s) is the time, ΔT (degC) is the temperature difference, m_l (kg) and m_c (kg) are the masses of liquid and calorimeter, respectively, and c_l and c_c (J kg^{-1} K^{-1}) are the specific heats of the liquid and calorimeter, respectively.

Continuous-flow calorimeters have been devised; a liquid passes through a tube, containing an electrical element, at a constant flow rate m' (kg s^{-1}) with the application of steady voltage V (V) and current I (A); the calorimeter is allowed to reach steady state (the inlet and outlet temperatures reach a constant value). At this point the heat gained by the fluid is equal to the electrical energy supplied. Therefore,

$$m' c_L \Delta T = VI$$

The flow rate can be measured and the only unknown is the specific heat c_L of the liquid.

Methods and precautions for specific heats of foods have been described in more detail by Mohsenin (1980) and Ohlsson (1983). Further specific heat data for foods and a list of compilations are given in Table 8.3.

8.6 LATENT HEAT

So far we have dealt with sensible heat changes, i.e. changes that can be detected by a rise or fall in temperature. However, in many food-processing operations, we encounter a change in phase; associated with these phase changes are energy changes. The phases involved are the solid, liquid and vapour phases. Water can exist as a solid, a liquid, a vapour or a combination of these phases in equilibrium. If the pressure and temperature are fixed, it is possible to predict what state the water will be in. The most usual form that phase diagrams take is pressure plotted against temperature.

Figure 8.3(a) represents such a phase diagram for water. At one particular pressure and temperature, the three phases solid (S), liquid (L) and vapour (V) are in equilibrium. This is referred to as the triple point T; the triple-point pressure and triple-point temperature for water are 4.6 τ and 0.01 °C, respectively. The line AT represents the pressure–temperature conditions for which the solid and the liquid are equilibrium, i.e. the melting-point line; therefore, it shows how the melting point varies with pressure.

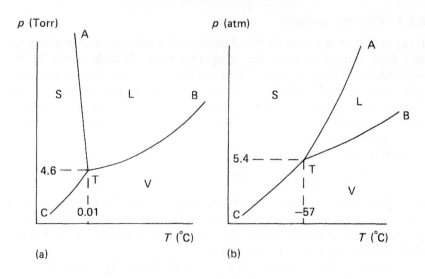

Fig.8.3 — Phase diagrams for (a) water and (b) carbon dioxide: S, solid; V, vapour; L, liquid; T, triple point.

The line TB represents the conditions where the vapour and the liquid are in thermal equilibrium; therefore, this predicts how the boiling point of a liquid varies with pressure.

The line TC represents the conditions for the solid and vapour to be in equilibrium. Thus, for water, if the watrer vapour pressure is maintained below 4.6 τ and the water is frozen, when energy is supplied, the solid will go directly to a vapour. This process is known as sublimation and is how water is removed in freeze drying; no melting or liquid phase is involved and problems such as shrinkage and case hardening, associated with hot-air drying, can be avoided. If the pressure goes above 4.6 τ, then ice will be converted to liquid water before it is removed as vapour.

In contrast with this, the phase diagram for carbon dioxide is shown in Fig. 8.3(b). The triple-point conditions are $-57\,°C$ and 5.4 atm. Therefore, solid carbon dioxide (Cardice) will sublime at atmospheric pressure.

If water is frozen down to about $-40\,°C$ and energy is supplied at a constant rate and the temperature is plotted against time, the plot shown in Fig. 8.4 results

The temperature steadily rises until it reaches B, which corresponds to the melting point of water. At this point, latent heat energy is required to melt the ice; this accounts for the plateau BC. The temperature remains at $0\,°C$ until all ice has melted at C. The temperature then rises steadily to the boiling point D. The slope CD for the water heating is about half that of AB for the ice, because of the different specific heat capacities. At point D, the water will start to boil; again, latent heat energy is required to cause vaporization. During this process the temperature remains constant until all the water is converted to vapour. Point D represents a saturated liquid at its

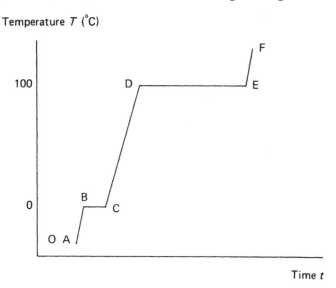

Fig 8.4 — Heating curve for water, during the transition from ice to superheated water at atmospheric pressure.

boiling point, point E a saturated vapour at its boiling point and anywhere between D and E represents the coexistence of vapour and liquid (a wet vapour), with droplets of water entrained in the vapour.

If the saturated vapour can be further heated, the time–temperature relationship follows EF. Again the slope EF is about twice that of DC because of the different specific heat capacities. In the region EF, the vapour is referred to as *superheated* (see section 11.6).

The lengths of the plateaus BC and DE are different, reflecting the different latent heat values for fusion and vaporization.

For water at atmospheric pressure,

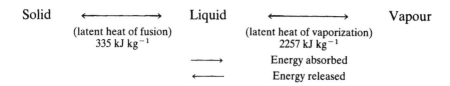

The latent heat of vaporization is approximately seven times higher than that for fusion. The latent heat of vaporization for water is extremely high. Thus, energy costs for evaporation and dehydration are potentially high in comparison with processes involving only changes in sensible heat. Steam is also a very useful heat transfer fluid because it gives out large quantities of energy when it condenses as well as having a high heat film coefficient value (see section 9.11). Furthermore, when a food freezes, it also gives out substantial amounts of energy which has to be removed by the refrigerant.

8.7 BEHAVIOUR OF WATER IN FOODS DURING FREEZING

When energy is removed from pure water, the temperature will fall and, at 0 °C, ice will start to separate. The temperature will remain constant as latent heat is removed until all the water has frozen; the temperature of the ice will then fall. This is depicted by AB in Fig. 8.5(a).

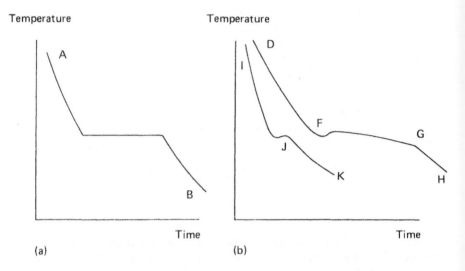

Fig. 8.5 — (a) Freezing curve for water; (b) freezing curve for food material, D–H representing slow freezing and I–K fast freezing.

The presence of dissolved solutes in foods tends to complicate this picture by depressing the freezing point. Most foods start to freeze at a temperature below −1 °C. Supercooling may also occur, followed by an increase in temperature, as the first ice crystals form and give out latent heat (Fig. 8.5(b)). This is followed by a thermal arrest period, FG, when most of the water is converted to ice. The food will then start to fall in temperature more rapidly, GH, and approach the temperature of the cooling medium.

Removal of ice has the effect of concentrating the solutes and further depressing the freezing point. Therefore, as more ice crystallizes, the concentration of solutes in the unfrozen material increases. this is known as freeze concentration and will occur in solid and liquid foods. This increase in concentration may result in an increase in reaction rates, despite the reduction in temperature. Therefore, as the temperature is reduced over the range from 0 °C to −20 °C, some chemical reactions may proceed at a faster rate, others may remain almost constant, whilst others will decrease, in the normal fashion. Some of the reactions have been discussed by Duckworth (1975).

Freeze concentration has received much attention as a unit operation for concentrating liquids. The liquid is frozen such that the ice crystals are large and of a uniform size. These are then separated from the liquid by centrifugation or other methods; the crystals may also need washing to remove any trapped concentrate. It is very useful for heat-labile liquids, such as grape juice; there is no loss of volatile compounds, and the theoretical energy requirements are low, compared with evaporation. Some of the technical problems have been discussed by Thijssen (1975).

Ice is removed until the eutectic temperature is reached and the entire mass freezes. Eutectic temperatures for foods are usually below −30 °C (see section 10.13). For practical purposes, most foods are considered to be 'frozen' at −15 °C, at which temperature between 90% and 95% of water present will be in the frozen form. Therefore the effective freezing time is defined as the time required to reduce the temperature at the slowest cooling point from the ambient temperature to −15 °C. This is very dependent on the size and nature of the product and the method of freezing used. Freezing times range from less than 1 min (e.g. peas in liquid nitrogen) to over 48 h for large carcases of meat using cold air. The relationship between temperature and time for small products, which are rapidly frozen, is shown by line IJK in Fig 8.5(b). For calculating refrigeration loads, it is necessary to estimate the amount of energy removed during chilling and freezing. A number of approaches have been adopted.

8.8 LATENT HEAT VALUES FOR FOODS (FUSION)

In this approach, it is assumed that all the water in the food freezes at a constant temperature, usually about −1 °C. The freezing process then involves two sensible and one latent heat change, i.e. bringing the product down to its freezing point, converting the water to ice and lowering the temperature of the frozen food to its final storage temperature. This requires knowledge of the specific heat of fresh and frozen food and the latent heat value of the food.

For example the amount of heat removed to reduce 200 kg of apples from +25 °C to −20 °C is as follows:

$$\text{Heat removed to reduce from } +25°C \text{ to } -1°C = mc\Delta T$$
$$= 200 \times 3.59 \times 26$$
$$= 18\,668 \text{ kJ}$$

$$\text{heat removed to freeze food} = mL$$
$$= 200 \times 2.815 \times 10^2$$
$$= 56\,300 \text{ kJ}$$

$$\text{heat removed to reduce from } -1°C \text{ to } -20°C = mc\,\Delta T$$
$$= 200 \times 1.88 \times 19$$
$$= 7144 \text{ kJ}$$

Therefore,
$$\text{total heat} = 18\,668 + 56\,300 + 7144 \text{ kJ}$$
$$= 82\,112 \text{ kJ}$$

It is required to remove 821 112 kJ of energy to effect this freezing process. Note that 68.6% of this is concerned with the removal of latent heat.

Latent heat values have been quoted for a wide variety of foods by Polley et al. (1980). Some values are given in Table 8.6.

Table 8.6 — Moisture contents and latent heat values for foods.

Food	Moisture content (%)	Latent heat (kJ kg^{-1})
Lettuce	94.8	316.3 (317.6)
Avocado pear	94.0	316.5 (314.9)
Strawberries	90.9	289.6 (304.5)
String beans	88.9	297.0 (297.8)
Apricots	85.4	284.0 (286.1)
Potatoes	77.8	258.0 (260.6)
Fresh lamb	58.0	194.0 (194.3)
Dried figs	24.0	79.0 (80.4)
Dried beans	12.5	41.9 (41.9)
Dried peas	9.5	32.6 (31.8)

The values in parentheses ( ) are calculated using the expression

$$L = 335 \, m_w \text{ (kJ kg}^{-1})$$

where m_w is the mass fraction of moisture.

It can be seen that the latent heat value is influenced by the moisture content of the food. Lamb (1976) has given the following equation for estimating the latent heat value:

$$L = 335 \, m_w \text{ (kJ kg}^{-1})$$

where m_w is the mass fraction of moisture. These values for various foods are shown in parentheses in Table 8.6 and show good agreement with the quoted values.

8.9 ENTHALPY–COMPOSITION DATA

For processes taking place at a constant pressure, heat changes can be associated with enthalpy changes. Therefore, if the enthalpy of the food is known at two temperatures, e.g. +25 °C and −20 °C, the amount of heat removed is simply obtained from the difference in values. Such enthalpy changes will account for both sensible and latent heat values. It is possible to measure enthalpy as a function experimentally. Loncin and Merson (1979) have listed some of the data sources; Rha (1975a) have provided a good table listing the enthalpy–temperature data for a wide variety of foods over

Sec. 8.9] Enthalpy–composition data 235

therange from −40 °F (−40 °C) to 40 °F (4.5 °C), as well as enthalpy–composition diagrams for bread, egg white, whole egg, fish and potato starch. When available, such data are extremely useful. Fig. 8.6 shows such a

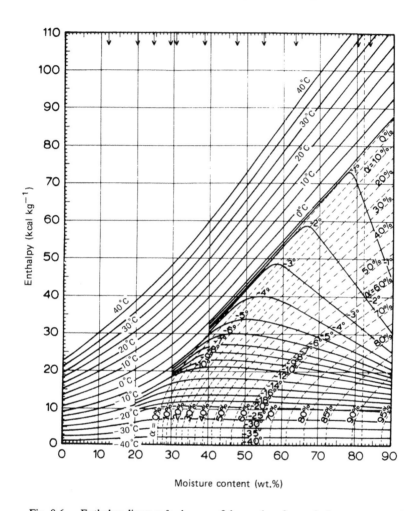

Fig. 8.6 — Enthalpy diagram for lean sea fish muscle, where α is the percentage of frozen water. (From Rha (1975a), with permission).

diagram for lean sea fish muscle. Enthalpy is plotted against moisture content; the other variables are temperature and percentage of frozen water. Enthalpy changes are evaluated as follows.

Fish with a moisture content of 80% at 25 °C would have an enthalpy value of 99.0 kcal kg^{-1}. If this is reduced to −20 °C, the new enthalpy value is 10 kcal kg^{-1} and the percentage of water frozen is 89%. Therefore,

$$\text{enthalpy change} = H_{-20°C} - H_{+25°C}$$
$$= 10 - 99.0$$

$$= -89.0 \text{ kcal kg}^{-1}$$

Therefore, 89.0 kcal kg^{-1} needs to be removed to bring about this temperature change. Such data are also useful for determining how the amount of frozen water changes with temperature; this is a useful way of presenting the behaviour of a food during freezing.

Figure 8.7 shows a graphical presentation of such data for two different

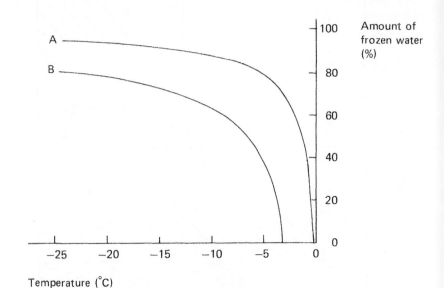

Fig. 7 — The relationship between the amount of frozen water and temperature for two different foods: line A, meat; line B, citrus fruit. (Adapted from the data of Jul (1984)).

foods. Most foods show similar trends; the vast majority of water freezes over the concentration range from −1 °C to −10 °C. Some data are also shown in Table 8.7. Microbial activity also ceases at temperatures below

Table 8.7 — Amounts of frozen water in various foods.

	Amount (%) of frozen water in the following foods		
Temperature (°C)	Lean meat	Haddock	Egg white
−2	48	55.6	75
−5	74	79.6	87
−10	83	86.7	92
−20	88	90.6	93
−30	89	92.0	94.0

Adapted from the data of Fennema (1975a).

−10 °C. This type of data is also useful for interpreting changes in the structure of frozen foods during conditions of storage at fluctuating temperatures. For example, if the temperature of a cabinet is fluctuating between −10 °C and −20 °C, because of poor temperature control, the amount of frozen water will also fluctuate, particularly at the surface of the food. Consequently, substantial proportions of the water will be changing state. These effects will be much more pronounced if the temperature is allowed to rise above −10 °C, and so storage temperatures above this must be avoided, to prevent deterioration of product quality. Thermodynamically, it can be shown that small ice crystals melt in preference to large ice crystals. During this process the small ice crystals will melt and form larger pools of water; when the temperature is subsequently reduced, these pools form larger ice crystals, which may disrupt the cell structure. This phenomenon is known as 'recrystallization'; it is manifested by the appearance of large ice crystals which will spoil the appearance and perhaps the texture of dessert products which are eaten frozen. To avoid the occurrence of this, it is important to keep frozen foods at a uniform temperature at all stages throughout the 'cold chain'. Probably the weak link in the cold chain is transportation from the retail outlet, e.g. from the supermarket to the home, where it is highly likely that some melting will occur. It has also been observed that chemical spoilage reactions proceed more rapidly in frozen foods as the rate and extent of temperature fluctuation increase.

Generally speaking, the shelf-life of most frozen products increases as the storage temperature decreases; the relationship between them is known as the time–temperature tolerance. It is illustrated in Fig. 8.8. Quality changes during cold storage have been discussed in more detail by Jul (1984).

Many foods can be stored almost indefinitely at temperatures below −30 °C without any noticeable loss in quality. Set against this are the high costs of producing and maintaining such low temperatures and storing materials for long times. Temperatures of −30 °C are commonly used for freezing and storing foods at the processing factory, whereas the storage temperature in most supermarket and domestic freezers is between −18 °C and −20 °C. However, this time–temperature relationship does not hold for all foods, particularly for those containing high quantities of fat or salt, e.g. herring, mackerel, cured bacon and butter. These products, which are subject to oxidation reactions, often exert a maximum reaction rate at a temperature of from −10 °C to −15 °C.

8.10 OILS AND FATS: SOLID–LIQUID TRANSITIONS

The simplest way to distinguish between an oil and fat is to see whether it appears to be a solid or a liquid at ambient temperature; lard, butter and hydrogenated vegetable fats are normally regarded as fats, while most other vegetable oils are regarded as oils. However, one common feature is that most oils and fats used in foods are completely liquid at temperatures above

50 °C. Therefore, above these temperatures, one needs only to consider sensible heat changes.

Oils are mixtures of triglycerides and, because of this, their melting

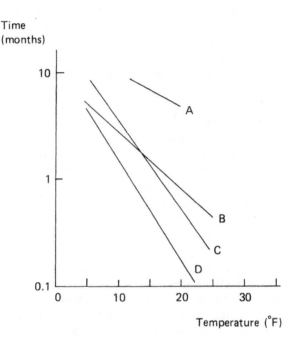

Fig. 8.8 — Time–temperature tolerance curves for storage of frozen foods: line A, packaged chicken; line B, peas, green beans; line C, raspberries; line D, peaches. (Adapted from the data of Fennema (1975a).

behaviour is quite complex. Oils and fats can be extracted from a wide variety of animal and vegetable materials; they will all have different triglyceride compositions. Techniques for determining the extent of crystallization, or solid–to–liquid fat ratio, include dilatation and nuclear magnetic resonance, differential thermal analysis and differential scanning calorimetry. Data are calculated in terms of the percentage of crystalline solids against temperature, and this is known as a melting characteristic (Table 8.8).

Data are also available for tallow, margarine fat, olive oil, rape seed oil and peanut oil. It can be seen that the fats have very different melting characteristics. Animal fats are regarded as being hard fats, i.e. most of the melting takes place at relatively high temperatures, whereas the vegetable fats are much softer and melt at much lower temperatures. Vegetable oils can be hardened by hydrogenation, a process which saturates the fatty acids in the triglycerides.

Such melting characteristics provide invaluable information about the

Sec. 8.1] Oils and fats: solid–liquid transitions 239

Table 8.8 — Melting characteristics and specific enthalpies of fats.

Temperature (°C)	Butter fat		Sunflower oil		Coconut oil		Lard	
	Enthalpy H (kcal kg^{-1})	Amount α of crystalline solids (%)	Enthalpy H (kcal kg^{-1})	Amount α of crystalline solids (%)	Enthalpy H (kcal kg^{-1})	Amount α of crystalline solids (%)	Enthalpy H (kcal kg^{-1})	Amount α of crystalline solids (%)
−40	3	100	4	100	4	100	3	100
−20	11	98	11	94	10	100	10	100
−10	17	90	25	18	15	98	15	94
0	24	75	37	0	20	87	22	82
10	32	56	42	0	25	69	33	59
20	45	20	47	0	45	35	39	50
30	54	10	52	0	59	0	50	33
40	60	0	57	0	65	0	60	10
50	65	0			70	0	67	2

Interpreted from the data of Rha (1975a).

fat, and how it will behave in formulated foods, e.g. ice-cream, chocolate, pastry and spreads. Cocoa butter is the fat mainly used in the manufacture of chocolate. It has a very sharp melting characteristics; at 25 °C, most of the fat is solid whereas, at 37 °C (body temperature), most of it is liquid, giving it a smooth mouth feel. Therefore, any fat which is to be used to replace cocoa butter must have a similar melting characteristic to avoid a noticeable change in the texture of the product.

Lowering the temperature of butter from 30 °C to −10 °C requires the removal of $17 - 54 = 37$ kcal kg^{-1}. (Note that this is 'butter fat'; butter contains approximately 16% moisture.) In practice, butter is frozen and stored below −10 °C

An alternative approach is to use an apparent specific heat value, which takes into account both sensible and latent heat changes. The apparent specific heat for butter over the temperature range from −40 °C to +50 °C is given in Table 8.9.

Table 8.9 — Variation in apparent specific heat for butter fat with temperature.

Temperature (°C)	−40	−20	−10	0	10	20	30	40	50
Apparent specific heat (kJ kg^{-1} K^{-1})	1.59	1.84	2.01	3.34	4.39	5.35	3.34	2.09	2.01

The highest apparent specific heat values are associated with temperatures at which the rate of crystallization or melting is highest. The enthalpy change involved between two temperatures can be obtained by plotting c_p against temperature over the range in question and evaluating the area under the curve (see section 8.4).

For products containing high quantities of fat, e.g. dairy and synthetic creams, considerable error may result if these crystallization effects are not taken into account. For example the apparent specific heat of 40% cream is about 3.34 kJ kg^{-1} K^{-1} at 0 °C, rising to 4.43 kJ kg^{-1} K^{-1} at 20 °C, before falling to 3.26 kJ kg^{-1} K^{-1} at 50 °C (Batty and Folkman 1983).

Correlations for predicting freezing points, percentages of frozen water and enthalpy values for fresh and frozen foods have been discussed in considerable detail by Miles et al. (1983).

8.11 DIFFERENTIAL THERMAL ANALYSIS AND DIFFERENTIAL SCANNING CALORIMETRY

Two useful techniques for analysing phase changes are differential thermal analysis and differential scanning calorimetry. Reactions which can lead to such phase changes are crystallization of fats and water (melting and freezing), evaporation of water, and certain chemical reactions, e.g. protein denaturation and starch gelatinization.

Sec. 8.11] Thermal analysis and differential scanning calorimetry   241

The two methods are fairly similar in principle but quantitative results can only be obtained from differential scanning calorimetry analysis. Both involve heating or cooling small quantities of experimental sample and a reference material, which should have a similar specific heat content to that of the sample and not undergo any phase change over the experimental temperature range.

In differential thermal analysis the same amounts of energy are put into the two materials during the heating cycle and their temperature difference is recorded. As this is very small, it needs to be amplified. Generally, results are presented as plots of differential temperature against the actual temperature of the sample; this is known as a thermogram. As soon as a phase change takes place in the sample, the temperature of the sample will be different from that of the cycle and a peak results. During the heating cycle, most phase changes adsorb heat (endothermic), and so the sample temperature lags behind the reference temperature.

Figure 8.9 shows how the differential temperature ΔT and temperature T

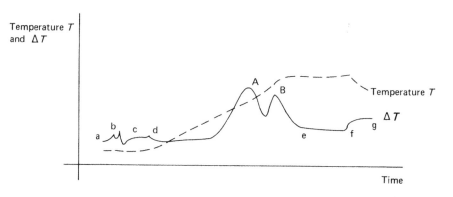

Fig. 8.9 — The relationship between sample temperature and time, and differential temperature and time during a differential thermal analysis cycle: ab, conditions before loading samples; b, sample and reference material are inserted into the machine; cd, conditions equilibrate; d, heating programme starts (ΔT changes slightly); de, heating cycle (this sample shows two peaks A and B during the heating cycle; these could be exothermic or endothermic reactions); ef, holding period; f, cooling cycle commences.

changes with time for a sample containing two endothermic peaks when taken through a heating cycle. Also included are the holding period and the beginning of the cooling cycle. This gives a qualitative picture of the heat changes involved.

In differential scanning calorimetry the temperatures of the sample and reference are maintained the same, and the amount of heat required to achieve this is recorded.

The thermogram is a plot of heat flow against temperature. Consequently the enthalpy change involved in the reaction can be determined from the area under the heat flow-time curve. Materials of known enthalpy can be used to calibrate the equipment. Such a thermogram is shown in Fig. 8.10.

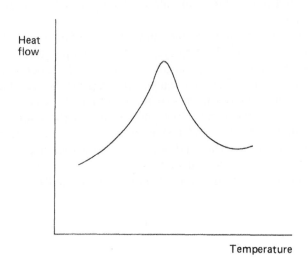

Fig. 8.10 — Differential scanning calorimetry peak, showing a plot of heat flow against temperature.

It is usually possible to vary the heating rate, and temperatures from −150 °C to 500 °C can readily be obtained. For many samples such as oils and fats, it is necessary to temper the sample beforehand. For some samples it is common practice to put the sample through a heating programme followed by a cooling cycle, to obtain a more complete picture. For samples containing water, evaporation is prevented using sealed containers. For examining frozen water in foods, results which are more reproducible are obtained whilst thawing frozen products than during the cooling and freezing of products, because the variable phenomenon of supercooling occurs during freezing.

Some of the uses will now be examined in more detail.

For proteins, the temperatrure corresponding to the maximum peak height is often used as an indication of the denaturation temperature, and the width of the peak (usually taken at 0.5 of the maximum peak height) as a measure of the complexity of the denaturation reaction. Using these assumptions, it is possible to observe how the denaturation temperature is affected by the changes in pH or the presence of other components, and the variation in enthalpy with denaturation temperature. Systems which have been investigated include meat, whey, egg and vegetable proteins. As a protein becomes more denatured, the size of the peak should decrease. Thus, it has been used to examine protein isolates with a view to assessing their level of denaturation; this will affect the solubility and other functional properties of the protein. Some of these aspects have been considered in more detail by Wright (1982).

With carbohydrates, differential scanning calorimetry has been used to measure heat-induced phase transitions or starch–water systems, particularly gelatinization. The behaviour of different starches can be compared, in addition to the effects of variations in moisture content and the addition of other components such as sugar, oil and protein (Lund, 1983; Biliaderis, 1983).

For fats, it has been used for observing crystalline behaviour and transitions in a single triglyceride and mixtures of triglycerides. It is particularly useful for determining the solid-to-liquid fat ratios in fats at different temperartures; the results compare favourably with those found from dilatation and wide-line nuclear magnetic resonance methods.

Finally, it has been used for observing the behaviour of water during freezing, and the water-binding capacities of different foods. one definition of bound water is water which is unfreezable. A material containing only unfreezable water will show no exothermic peak as its temperature is reduced to $-50\,°C$, in contrast with a food containing freezable water which will show such a peak. Therefore, as the moisture content of a food is reduced, a point will be reached where no peak is observed; for most foods this has been reported to occur at a moisture value 0.2–0.5 g water (g solid)$^{-1}$. The use of differential scanning calorimetry techniques for all these changes has been reviewed by Biliaderis (1983).

8.12 DILATATION

Dilatation is a technique involved for measuring the solid-to-liquid ratio (melting characteristic) of fats at different temperatures. it makes use of the experimental observation that 100 g of completely hydrogenated fat expands 10 ml on melting. A specially designed tube known as a dilatometer tube is used to measure this expansion (See Fig. 8.11).

About 1.5 ml of boiled-out dye solution is added to the bulb, followed by the melted fat. The melted fat should push the dye solution up the calibrated right-hand limb but no fat should enter the limb; the stopper is then inserted so that there is no air in the fat.

The expansion or contraction of the fat will now be measured by the level of dye. The tube is then placed in a refrigerator at 5 °C for several hours to allow the fat to solidify.

After storage the bulb is placed in a water bath at 5 °C and the dye level is noted when no further change is observed. The temperature of the water bath is increased, normally by intervals of 5 degC and the dye level is measured at each temperature, once equilibrium is attained. This procedure continues until the fat has completely melted. For most fats, this is below 50 °C. The expansion data are corrected for the thermal expansion, and the corrected data can be used to obtain the solid-to-liquid fat ratio over the

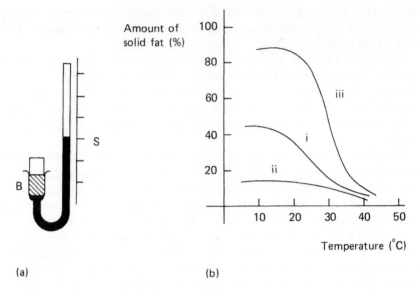

Fig. 8.11 — (a) Dilatometer tube showing the bulb B and the scale S; (b) the relationship between percentage solid fat and temperature for the three fats, butter (line i) fat from polyunsaturated margarine (line ii) and cocoa butter (line iii).

temperature range investigated. Typical data are shown for three fats in Fig. 8.11.

For more accurate work, intervals of 2 degC can be used. The method is quite time consuming but cheap. Many experiments can be performed at the same time. For reproducible results, attention should be paid to crystallization or tempering data. Dilatation is still used for calibrating more rapid and sophisticated methods, such as wide-line nuclear magnetic resronance (Mohsenin, 1984).

8.13 SYMBOLS

- c specific heat
- c_p specific heat at constant pressure
- c_V specific heat at constant volume
- I electric current
- H enthalpy
- L latent heat
- m mass
- m mass fraction
- m' mass flow rate
- Q quantity of heat liberated
- t time
- T temperature
- V voltage

8.13.1 Greek symbols

α   percentage of crystalline solids
γ   ratio of specific heat capacities

9

Heat transfer mechanisms

9.1 INTRODUCTION

The thermal conductivity and thermal diffusivity describe the heat transfer characteristics of food materials and will influence how quickly the product heats or cools during thermal processing. The diverse range of heat transfer processes occurring in food-processing operations have already been discussed. Heat is transferred by three different mechanisms, namely conduction, convection and radiation. Conduction is the principal mechanism involved in heat transfer processes into solid materials, whereas convection applies mainly to fluids. In radiant heat transfer processes, energy is transmitted by means of electromagnetic waves, e.g. by infrared, ultraviolet or microwaves. Many operations are not limited to one mechanism. In a heat exchanger, heat is transferred from one fluid to another through a metal tube or plate; both conduction and convection are involved. In infrared and microwave heating, energy is transferred to the surface or dissipated within the food by radiation. Any temperature gradients or localized heating effects are dissipated by conduction.

9.2 HEAT TRANSFER BY CONDUCTION

If one end of a metallic bar is heated, the kinetic energy of the molecules increases and the temperature rises. A temperature gradient will be set up and heat will be transferred along the bar. The exact mechanism of this energy transfer process is far from understood but it is believed that the two major contributory factors involved are the vibrational energy within the structure (vibrations are passed from molecule to molecule along the bar) and by means of a cloud of electrons which is free to move within the lattice structure. Materials differ in their capacity to conduct heat, metals being generally much better conductors than non-metals. The property used to quantify the ability of a substance to conduct heat is known as thermal conductivity. While it ought to be possible to predict the thermal conducti-

Sec. 9.3] Steady- and unsteady-state heat transfer 247

vity of a material from a knowledge of its molecular structure, it is not quite so straightforward.

9.3 STEADY- AND UNSTEADY-STATE HEAT TRANSFER

It is necessary to be able to quantify the rate of heat transfer. Heat transfer processes can occur under two sets of conditions namely steady-state and unsteady-state heat transfer.

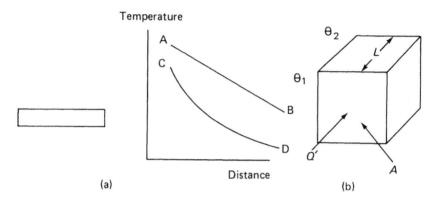

Fig. 9.1 — (a) Heat transfer and temperature gradient along a bar which is well lagged (line AB) and a bar with no lagging (line CD); (b) heat transfer through a cube of material

In a steady-state process, all the heat energy passes along the material, none being used to heat the material. Consider a bar of material (Fig. 9.1(a)) which is well heat insulated. Heat will be transferred along the bar from the high to the low temperature. When a steady state is achieved, all the heat entering the front face A will leave the back face B. Therefore, there is no accumulation of heat by the specimen and the temperature profile along the bar will remain constant (i.e. the temperature at any point will not appear to change). If the bar were not well insulated, a steady state would still be achieved, but the temperature profile across the bar would not be linear. Steady-state conditions prevail in most heat exchangers, cold stores and continuous food-processing operations, once the equipment has settled down.

If the bar (Fig. 9.1(a)) has reached a steady state and the temperature at one end is suddenly changed, the temperature at all points along the bar will change, until a new steady state is achieved. The period during which temperatures change is known as unsteady-state conditions. If a can of food is placed in a steam retort, the temperature of the can contents will rise. This is another example of unsteady-state heat transfer. A steady state is achieved in this situation when the contents reach the retort temperature. In many operations it is pertinent to ask how long it will take, for example, to chill or freeze a product, to dry a product or to heat or cook a product. These are the types of question that need to be solved by the application of unsteady-state heat transfer equations. The solution of unsteady-state heat transfer problems is more difficult than steady-state ones. Therefore, if the temperature at any location in any heat transfer application is changing with time, then unsteady-state conditions prevail. Some unsteady-state problems are discussed in Chapter 10, whereas the emphasis in this chapter will be placed on steady-state situations.

9.4 THERMAL CONDUCTIVITY

Thermal conductivity provides a means of quantifying the heat transfer properties of a solid material. Under steady-state conditions, the rate of heat transfer along a piece of solid material will depend upon the cross-sectional area of the surface, the temperature gradient and the thermal conductivity of the material (Fig. 9.1(b)). This can be expressed mathematically as

$$Q' = kA \frac{(\theta_1 - \theta_2)}{L}$$

where Q' ($J\,s^{-1}$ (W)) is the rate of heat transfer, A (m^2) is the surface area and $(\theta_1 - \theta_2)/L$ ($K\,m^{-1}$) is the temperature gradient. Thus the units of thermal conductivity k are joules per second per metre per kelvin ($J\,s^{-1}\,m^{-1}\,K^{-1}$), watts per metre per kelvin ($W\,m^{-1}\,K^{-1}$) or British therma units per hour per foot per degree Fahrenheit ($Btu\,h^{-1}\,ft^{-1}\,deg\,F^{-1}$). The conversion factor is $1\,Btu\,h^{-1}\,ft^{-1}\,deg\,F^{-1} = 1.731\,W\,m^{-1}\,K^{-1}$. The equation is a simplified form of the following steady-state heat conduction equation, known as Fourier's equation:

$$\frac{dQ}{dt} = -kA \frac{d\theta}{dL}$$

Thermal conductivity can be defined by considering the heat flow through unit cross-sectional area and unit temperature gradient. Under these conditions the thermal conductivity is equal to the rate of heat transfer through an area of 1 m² when a temperature difference of 1 K is maintained over a distance of 1 m, i.e.

$$Q' = k \times 1 \times \frac{1}{1}$$

Therefore,

$$Q' = k$$

The thermal conductivities of some common materials are given in Table 9.1. Further reference sources are given at the bottom of Table 8.1.

Silver is the best conductor of heat, shortly followed by copper. Copper

Table 9.1 — Thermal conductivity values

Material	Temperature (°C)	Thermal conductivity (W m^{-1} K^{-1})
Silver	0	428
Copper	0	403
Copper	100	395
Aluminium	20	218
Stainless steel	0	8–16
Glass	0	0.1–1.0
Ice	0	2.3
Water	0	0.573
Corn oil	0	0.17
Glycerol	30	0.135
Ethyl alcohol	20	0.24
Air	0	2.42×10^{-2}
Cellular polystyrene	0	3.5×10^{-2}
Freeze-dried peach (1 atm)	0	4.18×10^{-2}
Freeze-dried peach (10^{-2} Torr)	0	1.35×10^{-2}
Whole soya beans	0	0.097—0.133
Starch (compact powders)	0	0.15
Beef parallel to fibres	0	0.491
Frozen beef	−10	1.37
Fish		0.0324+0.3294m_w
Sorghum		0.564+0.0858m_w

m_w is the mass fraction of moisture.
1 Btu h^{-1} ft^{-1} degF^{-1} = 1.731 W m^{-1} K^{-1}.

was widely used in such industries as brewing and jam manufacture in the early part of the century but is now being replaced. Aluminium also has a high thermal conductivity value and aluminium pans are commonly used in the kitchen. Nearly all food-processing equipment is now manufactured from stainless steel. This is a much poorer conductor of heat than copper and therefore many seem a surprising choice. However, it will be seen later in section 9.12 that conductive heat trasnsfer is not very often the limiting resistance in a heat transfer situation, and changing copper to stainless steel would make very little difference to the performance of a heat exchanger. Stainless steel, like glass, varies in its composition and a range of values are quoted. Glass has a much lower thermal conductivity than metals; therefore, one may expect lower rates of heat transfer into material packaged in glass containers than those in cans.

Since thermal conductivity deals with the transfer of heat within solids, it may seem peculiar that the thermal conductivity of several fluids are quoted. In this case, values quoted are for when the fluids behave as solids and convection currents are eliminated. In many foods, water behaves like a solid and is immobilized within the solid matrix. The thermal conductivity of most foods is strongly influenced by the moisture content.

Air is a very poor conductor of heat. Many materials such as expanded polystyrene and foams have large porosities, and so the thermal conductivity of such materials approximates to that of air. Materials which have low thermal conductivities are termed insulators and are used as lagging material to prevent heat gains and losses. Steam pipes, hot-water tanks and high-temperature-holding tubes are lagged, as well as refrigerators, cold stores and chilled-water units.

The steady-state heat transfer equation can be used to evaluate the heat gain into a cold store.

9.5 HEAT TRANSFER THROUGH A COMPOSITE WALL

Consider a composite wall made of two materials A and B, with the respective thermal conductivities of k_A and k_B and the respective thicknesses L_A and L_B (Fig. 9.2). If the wall temperatures are θ_A and θ_B, what will be the

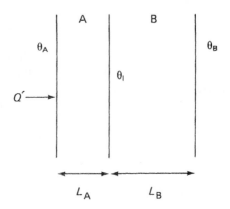

Fig. 9.2 — Heat transfer through a composite wall.

rate of heat transfer through the wall under steady-state conditions?

This problem is solved by assuming that the temperature at the interface is θ_i. Then, at steady state, the heat transferred through wall A is given by

$$Q'_A = k_A A \frac{\theta_A - \theta_I}{L_A} \tag{9.1}$$

and the rate of heat transfer through wall B is given by

$$Q'_B = k_B A \frac{\theta_I - \theta_B}{L_B} \tag{9.2}$$

However, at steady state

$$Q'_A = Q'_B$$
$$k_A A \frac{\theta_A - \theta_I}{L_A} = k_B A \frac{\theta_I - \theta_B}{L_B}$$

This equation can be solved for the interace temperature θ_I. This is then substituted into either equation (9.1) or equation (9.2) to determine the rate of heat transfer. Alternatively, these equations can be rewritten as

$$\theta_A - \theta_I = \frac{Q_A L_A}{k_A A}$$

$$\theta_I - \theta_B = \frac{Q'_B L_B}{k_B A}$$

θ_I is eliminated by summing the two sides of these equations, i.e.

$$\theta_A - \theta_B = \frac{Q'_A}{A}\left(\frac{L_A}{k_A} + \frac{L_B}{k_B}\right)$$

Therefore,

$$Q'_A = A(\theta_A - \theta_B) \bigg/ \left(\frac{L_A}{k_A} + \frac{L_B}{k_B}\right)$$

The rate of heat transfer through a thick-walled tube (Fig. 9.3) is given

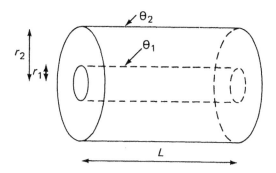

Fig. 9.3 — Heat transfer through a thick-walled tube.

by

$$Q' = \frac{2\pi L k(\theta_1 - \theta_2)}{\ln(r_2/r_1)}$$

9.6 THERMAL CONDUCTIVITY OF FOODS

Most foods are poor conductors of heat, and so heat transfer processes in which conduction is the predominant mechanism are slow. Heating and cooling times can be shortened by size reduction processes. In canning operations the term *conduction pack* is used for products where conduction is the major mechanism.

The thermal conductivity of a food is influenced by the composition of the food, in a similar manner to specific heat; water exerts the major influence. Other factors which can affect it are the pressure which is particularly important in freeze-drying operations and temperature.

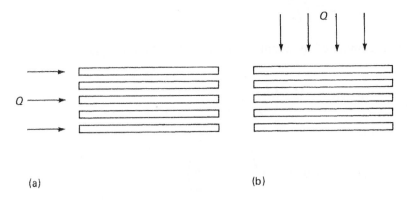

Fig. 9.4 — (a) Heat transmission parallel to the fibres; (b) Heat transmission across the fibres.

Some biological materials and fabricated foods have different conductivities in different directions; their properties are direction orientated or anisotropic. One example is meat and fish, where heat may be transferred better along the fibres than across the fibres (Fig. 9.4).

Thermal conductivities for a range of food materials are given in Table 9.1. There are many published equations relating thermal conductivity to the moisture content of the food, e.g. for fish

$$k = 0.0324 + 0.3294 m_w$$

where m_w is the mass fraction of water.

Further references have been given by Lamb (1976), Mohsenin (1980) and Miles *et al.* (1983). The thermal conductivity decreases as the food becomes drier. Freeze-dried materials, which are usually very porous, have extremely low thermal conductivities. During freeze drying, heat is normally transferred to the frozen material through the dried layer. Therefore, this is a slow process and the overall drying rate is limited by the rate of heat

Sec. 9.6] Thermal conductivity of foods 253

transfer. The conversion of water to ice increases the thermal conductivity approximately four-fold. Rates of heat transfer can be increased by heating through the frozen layer. It is also important to ensure a good contact between the heating surface and the food. The application of pressure to push the food against the surface and the use of a porous metal mesh to allow the vapour to escape have been adopted to accelerate the process (accelerated freeze drying).

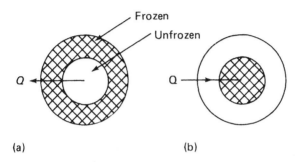

Fig. 9.5 — Comparison of (a) the freezing and (b) the thawing mechanisms.

Since frozen foods are better conductors than fresh foods, the rate of heat transfer during freezing (where heat is removed through the frozen layer) is greater than during thawing (where heat is added through the defrosted material); this is particularly so if conduction offers the major resistance to heat transfer (Fig. 9.5).

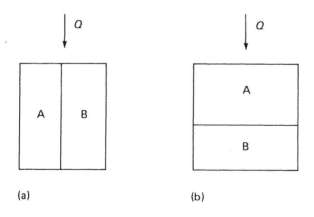

Fig. 9.6 — The thermal conductivity of two-component systems: (a) the parallel model; (b) the perpendicular model.

9.6.1 Compositional factors

Compositional data can also be used to obtain a more accurate thermal conductivity value. Two models have been proposed, namely the parallel model and the perpendicular model. These are shown for a two-component system in Fig. 9.6. For such a system containing solids and water, the equations for the thermal conductivity are given, for the parallel model, by

$$k = V_s k_s + V_w k_w$$

and, for the perpendicular model, by

$$\frac{1}{k} = \frac{V_s}{k_s} + \frac{V_w}{k_w}$$

where V_s and V_w are the volume fractions of the solids and water respectively, and k_s and k_w are the thermal conductivities of the solids and water respectively. It is important to note that volumetric fractions are used rather than mass fractions.

The thermal conductivities of various components have been given by Miles *et al.* (1983) as follows: k_a (air) = 0.025; k_p (protein) = 0.20; k_c (carbohydrate) = 0.245; k_s (solids) = 0.26; k_f (fat) = 0.18; k_w (water = 0.6; k_i (ice) = 2.24. Thus, for an n-component system, using the parallel model (the same principle would apply to the perpendicular model)

$$k = V_1 k_1 + V_2 k_2 + \ldots + V_n k_n$$

The composition of an apple is 0.844 water anmd 0.156 solid (mass fractions) and the densities of water and solids are 1000 kg m^{-3} and 1590 kg m^{-3} respectively.

The volume fraction of solids is given by

$$V_s = \frac{m_s/\rho_s}{m_s/\rho_s + m_w/\rho_w} = 0.104$$

and the volume fraction of water is given by

$$V_w = 1 - V_s = 0.896$$

Using the parallel model,

$$k_{\text{apple}} = V_s k_s + V_w k_w = 0.104 \times 0.26 + 0.896 \times 0.6$$
$$= 0.565 \text{ W m}^{-1} \text{ K}^{-1}$$

The value for the perpendicular model is $0.528 \text{ W m}^{-1} \text{ K}^{-1}$. The average value would be $0.540 \text{ W m}^{-1} \text{ K}^{-1}$. The thermal conductivity of apple is quoted as 0.422 for green apples and 0.513 W m^{-1} K^{-1} for red apples (Mohsenin, 1980).

One difficulty of this approach is that it ignores any air within the food. If the apple contains 20% air (porosity, 0.2), the thermal conductivity calculated from the parallel model is

$$k = 0.8 \times 0.540 + 0.2 \times 0.025 = 0.437$$

and from the perpendicular model

$$\frac{1}{k} = \frac{0.8}{0.540} + \frac{0.2}{0.025}$$

$$k = 0.105$$

In this example the parallel model appears to give closer agreement to the value in the literature.

In drying operations, energy is transferred by conduction and also as latent heat by water vapour diffusion from wet to drier areas. The use of practical methods involving long time periods for equilibrium to be established should be avoided because such moisture migration may occur and affect the results. Heat transfer by moisture migration is known as the 'heat pipe' effect.

9.6.2 Temperature effects

For many structrual materials (particularly metals), the thermal conductivity of the material changes considerably with temperature. However, in most food materials the temperature effect is not so pronounced; the conductivity is more affected by cellular structure and moisture content. Note that the obvious exception to this is during the transition from water to ice or vice versa. Furthermore, the range of temperatures to which foods are subjected is not as large as some other materials. Mohsenin (1980) suggests that, if k varies linearly with temperature, the temperature effect can be taken into consideration by taking the value at the average temperature. There is very little published information about the thermal conductivity of foods at higher temperatures.

9.6.3 Pressure effects

In most freeze-drying applications, the operating pressure within the chamber is maintained below 4.6 τ (613 Pa). Often the pressure is considerably lower than this. Furthermore, most freeze-drying processes are heat transfer controlled, i.e. the drying rate is controlled by the rate of heat transfer to the frozen material.

It has been found that the thermal conductivity of a material increases as the pressure increases, over the pressure range 10–10^4 Pa, as shown in Fig. 9.7. Over this range the increase is almost linear when k is plotted against log p and the thermal conductivity can be written as

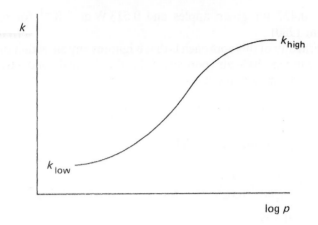

Fig. 9.7 — The relationship between thermal conductivity and pressure of materials.

$$\frac{k_e - k_{low}}{k_{high} - k_{low}} = \frac{1}{1 + C/p}$$

where k_{high} and k_{low} are depicted in Fig. 9.7, C is a constant, p is the pressure and k_e is the effective thermal conductivity at pressure p. This is known as Harper's equation.

Mellor (1978) gives values of k_{low}, k_{high} and C for freeze-dried beef, apples and peach, together with variations in thermal conductivity with pressure for freeze-dried sucrose and gelatin solutions.

This equation is based on the suggestion that the effective thermal conductivity of any gas-filled porous material has two components: one associated with a solid network and the other with the gas that fills the pores. The thermal conductivity can be significantly improved by raising the pressure or by introducing a gas such as helium into the pores of the material. Increasing thermal conductivity values by increasing the pressure results in shorter drying times and is the basis of the cyclic pressure freeze-drying process which has been described in more detail by Mellor (1978).

9.7 DETERMINATION OF THERMAL CONDUCTIVITY

Most of the simpler methods for determining thermal conductivity rely upon passing heat through the sample, achieving steady-state conditions and measuring the rate of heat transfer and the temperature gradient over the sample. If the geometry of the sample is known, the values can be substituted in the steady-state heat transfer equation to determine k:

$$Q' = kA \frac{(\theta_2 - \theta_1)}{L}$$

Good conductors of heat can be prepared in the form of a bar or rod (Fig. 9.8(a)), and the heat transferred along the bar can be measured by controlling the energy input into the high-temperature end or by removing the heat from the low-temperature end by passage of a known flow rate of water m through the coils. The classical experiment with Searle's bar (Tyler, 1972) uses the following equation:

electrical energy = rate of heat transfer = water

$$VI = kA\frac{(\theta_2 - \theta_1)}{L} = mc\,\Delta\theta$$

For a poor conductor, the sample needs to have a much larger surface area, so that the rate of heat transfer can be reasonably measured. The sample is often prepared in the form of a disk, as this is the simplest form available (Fig. 9.8(c)). However, glass can be prepared in the form of a tube (Fig. 9.8(b)). Water flows through the inside of the tube at a constant flow rate m' and is heated by condensing steam at a temperature θ_3 on the outside of the tube. The rate Q' of heat transfer through the tube wall to the water is given by

$$Q' = \frac{2\pi Lk(\theta_3 - \theta_4)}{\ln(r_2/r_1)}$$

where $\theta_4 = (\theta_1 + \theta_2)/2$ is the average water temperature.

This is equal to the heat gained by the water. Therefore,

$$\frac{2\pi Lk(\theta_3 - \theta_4)}{\ln(r_2/r_1)} = m' \times 4.18 \times 10^3(\theta_2 - \theta_1)$$

where r_2 is the outer tube radius and r_1 is the inner tube radius.

Preparing the sample in the form of a slab is the basis of the most common technique for poor conductors, namely the guarded ring method. This has been described in more detail by Mellor (1978) and Mohsenin (1980). Mellor has also described a modification of the equipment for measuring freeze-dried materials under vacuum conditions.

Unsteady-state methods, which involve the application of heat to the material which is in thermal equilibrium and measuring the change in temperature at some other location within the material, have been described in more detail by Mohsenin (1980). On the whole, these methods are quicker but not as accurate as steady-state methods.

9.8 THERMAL DIFFUSIVITY

The thermal diffusivity a is the ratio of the thermal conductivity to the specific heat of the product multiplied by its density.

Thus,

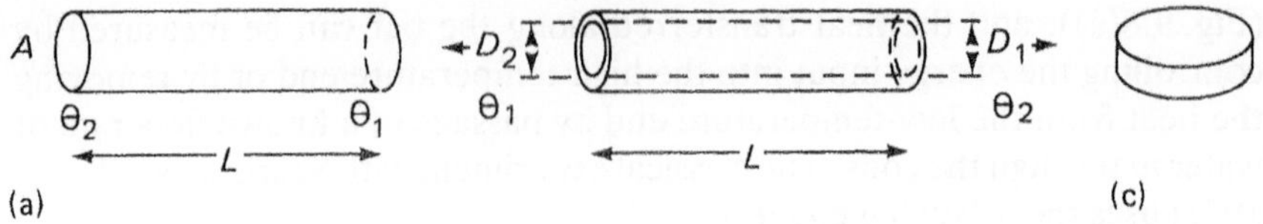

Fig. 9.8 — Sample preparation for thermal conductivity measurement: (a) bar; (b) tube; (c) disk.

$$a = \frac{k}{\rho c} \frac{(\text{J s}^{-1}\text{m}^{-1}\text{K}^{-1})}{(\text{kg m}^{-3})(\text{J kg}^{-1}\text{K}^{-1})}$$

Therefore the units of thermal diffusivity are square metres per second ($m^2\,s^{-1}$) or square feet per second ($ft^2\,s^{-1}$).

In physical terms, thermal diffusivity gives a measure of how quickly the temperature will change when it is heated or cooled. Materials with a high thermal diffusivity will heat or cool quickly; conversely, substances with a low thermal diffusivity will heat or cool slowly. Thus, thermal diffusivity is an important property when considering unsteady-state heat transfer situations. Thermal diffusivities can be determined experimentally (see references covering thermal conductivity) or evaluated from a knowledge of the individual properties (Singh, 1982). Thermal diffusivity will be further encountered in unsteady-state heat transfer problems (see section 10.3).

9.9 PARTICULATE AND GRANULAR MATERIAL

When dealing with particulate or granular material (grain, peas, legumes, etc.) there are two possible approaches. Firstly, one could consider the thermal conductivity of an individual unit, or one could consider the thermal conductivity of a bulk sample of the material. The former quantity should be less ambiguous, but the value obtained may be of limited usage when attempting to predict heat transfer rates in the bulk material which are not fluidized. Mathematical models used to predict heat transfer rates in packed solids required the knowledge of such factors as pore size and porosity, thermal conductivity of solids and contact area between solids; assumptions such as spherical shape and a known pressure in the gas space are made to simplify the solution. Mohsenin (1980) has reviewed the various types of thermal conductivity experiments performed on particulate matter.

9.10 HEAT TRANSFER BY CONVECTION (INTRODUCTION)

Heat is transferred in liquids and gases by the bulk movement of molecules; this process is termed convection. Heat distribution and elimination of temperature gradients are brought about by molecular motion. Convection itself can be further subdivided into natural and forced convection. If a beaker of water is heated (Fig. 9.9(a)), the water in the vicinity of the heat source becomes warm and decreases in density. This warm water rises and is replaced by cold water. A circulation pattern is set up within the fluid because of natural density differences; such heat transfer processes are referred to as natural convection. If the water is stirred with a spoon, external energy is put into the system, turbulence is artificially induced and we have moved to a system of forced convection; heat transfer rates are also increased. Forced convection applies whenever external energy is put into the systen. Fig. 9.9(b) represents a batch circulation evaporator. If heat is supplied in the form of steam to a liquid in a tube, natural circulation patterns are set up; as shown, the liquid will boil in the tube, and a mixture of liquid and vapour will climb up the tube. The vapour is separated from the liquid in the separator and removed from the system, while the liquid is recirculated for further evaporation. As the solids concentration increases, the viscosity will increase and the circulation rate will fall. There comes a point at which the natural convection currents will not overcome the viscous forces. The liquid stops circulating and fouling of the heat exchange surface may occur. If further evaporation is required, it will be necessary to incorporate a pump to circulate the liquid; the situation is now one of forced convection. Similarly, low-viscosity liquids can be heated in cans under static conditions, and the circulation patterns set up are shown in Fig. 9.9(c). With high-viscosity products there may be a need to agitate the cans, in order to prevent excessively long processing times and consequent overcooking.

9.11 HEAT FILM COEFFICIENT

As for conduction it is important to quantify the rate of heat transfer within fluids. If we imagine a fluid flowing over a solid surface, a boundary layer exists near the surface (see section 3.4). This boundary layer offers the major resistance to the transfer of heat to or from the solid (Fig. 9.10).

The rate Q' of heat transfer to the solid surface is given by

$$Q' = hA(\theta_b - \theta_{surf})$$

where h is the heat film coefficient, A is the surface area, θ_{surf} is the surface temperature on the solid and θ_b is the bulk fluid temperature. It is assumed that all the temperature gradient exists across the film.

The units of the heat film coefficient are watts per square metre per kelvin (W m^{-2} K^{-1}) or British thermal units per hour per square foot per degree Fahrenheit (Btu h^{-1} ft^{-2} deg F^{-1}). The conversion factor is 1 Btu h^{-1} ft^{-2} deg F^{-1} = 5.68 J s^{-1} m^{-2} K^{-1}. Table 9.2 shows average heat film coefficient values for some flow situations.

The heat film coefficient is equivalent to the term k/L in the heat conduction equation. As the flow situation becomes more turbulent, the boundary layer thickness decreases and the heat film coefficient increases.

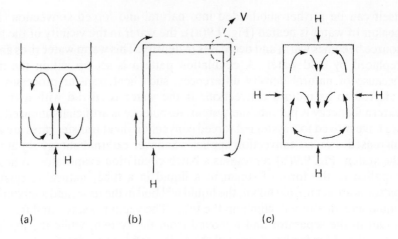

Fig. 9.9 — Convection currents in (a) a beaker of water, (b) a closed circular tube and (c) a tin can: H, energy source; ↑, convection currents.

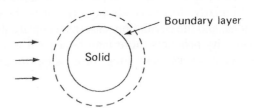

Fig. 9.10 — The boundary layer and heat film coefficient.

Table 9.2 — Heat film coefficients for some flow situations

Flow situation	Heat film coefficient (Btu h^{-1} ft^{-2} degF^{-1})	Examples
Boiling liquids	400–4000	Evaporation
Condensing vapours	300–3000	Steam, ammonia
Still air	1	Refrigerator, deep-freeze cabinet
Moving air	10	Blast freezer, spraying drying, hot-air drying
Liquids through pipes, low viscosity	100–200	Milk, water, fruit juices, dilute solution salts, sugar
Liquids through pipes, high viscosity	10–100	Starchy products, pastes, suspensions

The best types of heat transfer situation are obtained when boiling liquids or condensing vapours. However, still air is a very poor convector of heat. Circulating the air can increase the heat film coefficient by a factor of approximately 5–10. Heat transfer into liquids is an intermediate situation. Two consequences of this are that condensing steam at 100 °C is a better heat transfer fluid than hot water at the same temperature and that any heat transfer situation involving air or other gases are relatively inefficient and will require larger surface areas than liquid transfer situations. Air heat exchangers are required for heating and cooling large volumes of air for drying and freezing applications.

Loncin and Merson (1979) have presented some data on the effect of the presence of air on the heat at film coefficient for saturated steam. Saturated steam has a value of 11 000 W m^{-2} K^{-1} but, when accompanied by 3 vol. % and 6 vol. % of air, the value falls to 3400 W m^{-2} K^{-1} and 1100 W m^{-2} K^{-1}, respectively. Therefore, the presence of air in steam has a serious effect on its heat transfer properties. Consequently the first process in in-container sterilization is to ensure that all the air is removed from the retort when the steam is first introduced. This process is known as venting.

9.11.1 Evaluation of heat film coefficients

The technique of dimensional analysis has been used to evaluate heat transfer coefficients. This approach results in equations which relate the heat transfer coefficient to other physical properties of the fluid. These can be used for a wide range of fluids. For example, for a fluid flowing which is being heated as it flows along a tube, the following relationship has been found to apply:

$$\underbrace{\frac{hD}{K}}_{\substack{Nusselt \\ number}} = 0.023 \underbrace{\left(\frac{vD\rho}{\mu}\right)^{0.8}}_{\substack{Reynolds \\ number}} \underbrace{\left(\frac{c\mu}{k}\right)^{0.4}}_{\substack{Prandtl \\ number}}$$

The Nusselt number, Reynolds number and Prandtl number are all dimensionless groups (see section 1.22).

One parameter used in natural convection calculations is the Grashof number Gr; this is the ratio of the buoyancy forces to the viscous forces and is represented by

$$Gr = \frac{D^3 \rho^2 g \beta \Delta T}{\mu^2}$$

where β is the coefficient of volumetric expansion. It takes the place of the Reynolds number, when no turbulence is present. An alternative form described by Loncin and Merson (1979) is

$$\mathrm{Gr} = \frac{D^3 \rho \Delta \rho g}{\mu^2}$$

where $\Delta\rho$ is the density difference between the fluid in contact with the wall and the bulk of the fluid. Natural convection occurs in the canning of low-viscosity fluids, and some natural circulation evaporators. In most cases, turbulence is induced to increase the overall heat transfer rates.

A wide range of correlations for other flow situations have been given by Loncin and Merson (1979), by Milson and Kirk (1980) and in most books dealing with chemical engineering unit operations. Other methods for determining heat film coefficients haved been discussed by Mohsenin (1980).

9.12 COMBINATION OF HEAT TRANSFER BY CONDUCTION AND CONVECTION

In most practical heat transfer situations, both conduction and convection will be employed; most indirect processes involve three resistances. If we

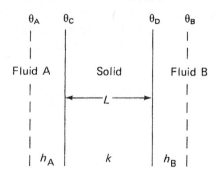

Fig. 9.11 — Heat transfer involving conduction and convection.

imagine the transfer of heat from one fluid through a wall to second fluid (Fig. 9.11), there are two convective heat transfer resistances h_A and h_B and a resistance due to the thermal conductivity k of the wall of thickness L. If the bulk fluid temperatures are θ_A and θ_B, respectively, and steady-state conditions prevail, the rate of heat transfer through the boundary layer of fluid A is equal to the rate through the wall, which in turn is equal to the rate through the boundary layer of fluid B. The wall surface temperatures are not known but can be designated by θ_C and θ_D. If we consider a cross-sectional heat transfer area of A (m^2) and the rate of heat transfer to be Q', then we have the following:

Sec. 9.12] Heat transfer by conduction and convection

$$Q' = h_A A(\theta_A - \theta_C) \quad \text{or} \quad \theta_A - \theta_C = \frac{Q'}{h_A A}$$

$$Q' = \frac{kA}{L}(\theta_C - \theta_D) \quad \text{or} \quad \theta_C - \theta_D = \frac{Q'L}{Ak}$$

$$Q' = h_B A(\theta_D - \theta_B) \quad \text{or} \quad \theta_D - \theta_B = \frac{Q'}{h_B}A$$

If the left-hand sides and the right-hand sides of these three equations are added together and equated, the following equation results:

$$\theta_A - \theta_B = \frac{Q'}{A}\left(\frac{1}{h_A} + \frac{L}{k} + \frac{1}{h_B}\right) \tag{9.3}$$

The important aspects of this equation are that the unknown wall temperatures have been eliminated and the rate of heat transfer is expressed in easily measured values.

9.12.1 Overall heat transfer coefficient

A new constant called the overall heat transfer coefficient U is now introduced where

$$\frac{1}{U} = \frac{1}{h_A} + \frac{L}{k} + \frac{1}{h_B} \tag{9.4}$$

where U is the overall heat transfer coefficient. Therefore, combining equations (9.3) and (9.4) gives

$$Q' = UA(\theta_A - \theta_B)$$

Therefore the rate of heat transfer from the bulk of one fluid to the bulk of the other will depend upon the overall heat transfer coefficient, the surface area and the temperature driving force $\theta_A - \theta_B$. It is desirable to maximize the overall heat transfer situation.

Some average values for overall heat transfer situations are given in Table 9.3 for different heat exchange situations.

The same analysis can be applied to many other situations.

Fig. 9.12(a) shows a more complex situation in which heat is being transferred through a composite wall (more than one material), extra resistance being added. The usual effect of the addition of extra resistances is to decrease the overall heat transfer rate. One situation in which a composite wall situation might be set up is in the fouling of heat exchanger surfaces. The overall heat transfer coefficient and rate of heat transfer are given, respectively, by

Table 9.3 — Overall heat transfer coefficients

Heat transfer fluids		Example	Overall heat transfer coefficient (W m^{-2} K^{-1})
Hot water	air	Air heater	10–50
Viscous liquid	hot water	Jacketed vessel	100
Viscous liquid	hot water	Jacketed vessel with agitation	500
Flue gas	water	Steam boiler	5–50
Evaporating ammonia	water	Chilled water plant	500
Viscous liquid	steam	Evaporator	500
Non-viscous liquid	steam	Evaporator	1000–3000

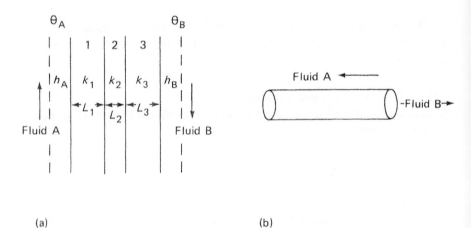

Fig. 9.12 — (a) Heat transfer through a composite wall; (b) heat transfer through a thin-walled tube.

$$\frac{1}{U} = \frac{1}{h_A} + \frac{L_1}{k_1} + \frac{L_2}{k_2} + \frac{L_3}{k_3} + \frac{1}{h_B}$$

$$Q' = UA(\theta_A - \theta_B)$$

Deposits either from the food or from the water build up on the heat exchanger surface and will alter its performance. Usually, these deposits have a low thermal conductivity and lead to a decrease in heat transfer rates.

Another flow situation is a simple tubular heat exchanger (similar in design to a laboratory condenser (Fig. 9.12(b)). The fluid to be treated passes through the centre pipe and the heat transfer fluid passes normally in a counter-current direction in the annulus surrounding the pipe (counter-

Sec. 9.12] Heat transfer by conduction and convection

current flow normally gives a better performance to co-current). In this case, if the pipe wall is thin, the overall heat transfer coefficient is given by

$$\frac{1}{U} = \frac{1}{h_A} + \frac{1}{h_B} + \frac{L}{k}$$

If the temperature profile along the tube is examined when one fluid is being heated from θ_1 to θ_2 by the heat transfer fluid flowing in the opposite

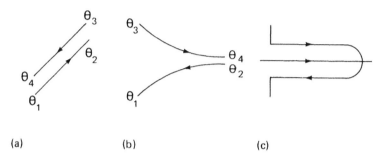

Fig. 9.13 — (a) Temperature profile through a tube in counter-current flow; (b) co-current flow; (c) flow arrangement involving both co-current and counter-current flow.

direction, which in turn cools from θ_3 to θ_4, it can be seen that the temperature driving force at different locations along the heat exchanger is not necessarily constant. To overcome this problem an average value is used, known as the logarithmic mean $\Delta\theta_m$ of the temperature difference.

With reference to the temperatures θ_1, θ_2, θ_3 and θ_4 (Fig. 9.13(a)), the temperature difference at one end is given by

$$\Delta\theta_1 = \theta_4 - \theta_1$$

the temperature difference at the other is given by

$$\Delta\theta_2 = \theta_3 - \theta_2$$

and the log arithmic mean of the temperature difference is given by

$$\Delta\theta_m = \frac{\Delta\theta_1 - \Delta\theta_2}{\ln(\Delta\theta_1/\Delta\theta_2)}$$

(Note that, to avoid the use of negative logarithms, the larger of the two end temperature differences should be made equal to $\Delta\theta_1$.)

Thus, if a fluid is being heated from 30 °C to 60 °C by a fluid cooling from

90 °C to 50 °C, with reference to Fig. 9.13(a) the logarithmic mean of the temperature difference equals 24.1 degC. What this means in practical terms is that one would obtain the same rate of heat transfer in this situation as one would from the same heat exchanger if a constant temperature difference of 24.1 degC had been maintained along its entire length. As a first approximation, an arithmetical mean value will suffice, i.e. 25 degC in this example. However, it can be shown theoretically that the logarithmic mean of the temperature difference is the correct one to apply.

Figs 9.13(a) and 9.13(b) illustrate the difference between parallel or co-current and counter-current flow. Counter-current heat exchange is preferred to co-current heat exchange for two main reasons, the first being that it usually results in a higher mean temperature driving force and the second that it allows the fluid being heated or cooled to approach the heating or cooling medium temperature more closely.

Most heat exchangers are not of such simple flow design, e.g. Fig. 9.13(c). However, there are extensive correlations for correcting the logarithmic mean of the temperature difference values for these more complex situations (Coulson and Richardson, 1977).

If the tube wall is not thin, then the heat transfer considerations have to be made on the basis of either the inside of the tube or the outside of the tube (because the areas will be different).

The equation

$$Q' = UA\Delta\theta_m$$

is the basic design equation for heat exchangers and evaporators where Q' ($J\,s^{-1}$) is the duty, A (m^2) is the surface area, U ($J\,s^{-1}\,m^{-2}\,K^{-1}$) is the overall heat transfer coefficient and $\Delta\theta_m$ (K) logarithmic mean of the temperature difference.

9.12.2 Limiting resistances

In a situation involving heat transfer by conduction and convection, it may be necessary to improve the rate of heat transfer. In most situations, this can be done by paying attention to one particular resistance, namely the limiting resistance. In the expression

$$\frac{1}{U} = \frac{1}{h_1} + \frac{1}{h_2} + \frac{L}{k}$$

the limiting resistance to heat transfer is normally given by the largest of the terms in the expression. Thus, if air is being heated by steam in a heat exchanger made of stainlesss steel ($k = 16$) of wall thickness 2 mm, and if the heat film coefficient for water and air, are, respectively, 500 W m^{-2} K^{-1} and 25 W m^{-2} K^{-1}, the overall heat transfer coefficient would be given by

$$\frac{1}{U} = \frac{1}{25} + \frac{1}{500} + \frac{2 \times 10^{-3}}{16}$$

$$= 0.04 + 0.002 + 0.000\,125$$

$$= 0.042\,125$$

Therefore,

$$U = 23.7\,\text{W m}^{-2}\,\text{K}^{-1}$$

The largest term (0.04) is associated with the heat film coefficient for air. Thus, increasing the heat film coefficient for the water by increasing turbulence, or by substituting copper for the stainless steel or making it thinner will have little effect on the performance.

Some of the changes are summarized in Table 9.4. By examining the

Table 9.4 — Effect of changing individual resistances on the overall heat transfer coefficient

Change	h_1, air	h_2, water	k, metal	U	Increase in overall heat transfer coefficient (%)
Normal	25	500	16	23.7	
Change the metal	25	500	400	23.8	0.4
Change the water	25	1000	16	24.3	2.5
Change the air	50	500	16	43.0	81.4

individual terms the limiting resistance can easily be determined; any attempts to improve the heat transfer efficiency should be directed towards improving this resistance. If one of the terms is much larger than the others, then the overall heat transfer coefficient will approximate to the reciprocal of that term, i.e.

$$\frac{1}{U} = \frac{1}{h_{\text{limiting}}}$$

or

$$\frac{1}{U} = \frac{L}{k}$$

if the conductive heat transfer resistance is limiting.

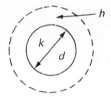

Fig. 9.14 — Limiting resistances in fluid–solid heat transfer processes: the Biot number Bi = hL/k, where $L = d/2$.

9.12.3 Biot number

When transferring or removing heat from soilds (freezing, heating, etc.) the ratio hL/k of the convection and conduction coefficients is termed the Biot number (Fig. 9.14); it gives an indication of which mechanism is controlling the heat transfer rate (see section 10.4).

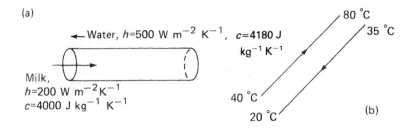

Fig. 9.15 — (a) Diagram of a heat exchanger; (b) temperature profile using counter-current flow.

9.12.4 Example of heat exchanger design

Consider a heat exchanger, designed to cool 3000 kg h^{-1} of milk from 80 °C to 40 °C, using cooling water at 20 °C, with a maximum allowable exit temperature of 35 °C (Fig. 9.15(b)). What surface area is required for this duty? The heat film coefficients for water and milk are shown in Fig. 9.15(a), together with other data.

The duty Q' is given by

$$Q' = \frac{3000}{3600} \times 4.0 \times 40$$

$$= 133.3 \text{ kJ s}^{-1}$$

and $\Delta\theta_m$ is given by

$$\Delta\theta_m = \frac{\Delta\theta_1 - \Delta\theta_2}{\ln(\Delta\theta_1/\Delta\theta_2)}$$

$$= \frac{45 - 20}{\ln(45/20)}$$

$$= 30.8 \text{ K}$$

If the heat film coefficients for milk and water are $200 \text{ W m}^{-2} \text{ K}^{-1}$ and $500 \text{ W m}^{-2} \text{ K}^{-1}$, respectively, and the wall thickness is 2 mm ($k = 20 \text{ W m}^{-1} \text{ K}^{-1}$),

$$\frac{1}{U} = \frac{1}{200} + \frac{1}{500} + \frac{2 \times 10^{-3}}{20}$$

$$= 0.071$$

Therefore,

$$U = 140.8 \text{ W m}^{-2} \text{ K}^{-1}$$

Thus,

$$A = \frac{Q'}{U \Delta\theta_m}$$

$$= \frac{1.33 \times 10^5}{140.8 \times 30.8}$$

$$= 30.7 \ m^2$$

The surface area required for counter-current flow is 30.7 m². For parallel or co-current flow, $\Delta\theta_m = 22.2$ K and the surface area equals 42.5 m²., For a tubular heater of diameter 2.54 cm, a surface area of 30.7 m² would require a tube length of 384 m.

9.13 APPLICATION TO HEAT EXCHANGERS

Heat exchangers have a wide variety of applications in the food industry and a number of designs are available. They can be batch or continuous. Batch processes are favoured for smaller operations; they are more labour intensive, not so easily automated but more flexible in operation. Heating and cooling times are long and they are more suitable for processes employing long holding times. They are more energy intensive, as there is no scope for regeneration. A typical batch process is the pasteurization of milk by the holder process (63 °C at 30 min). The heating and cooling times may increase the overall processing time significantly.

The simplest type of heat exchanger for batch heating and cooling is a tank with either a jacket or internal coils; the same vessel can be used for heating and cooling processes, by supplying steam and chilled-water services. Heat transfer rates are improved by agitation. Often the same vessels are used for mixing ingredients followed by subsequent heat treatment.

The heating time in a batch processing operation is given by

$$t = \frac{mc}{UA} \ln\left(\frac{\theta_H - \theta_I}{\theta_H - \theta_F}\right)$$

where m is the mass, c is the specific heat, θ_H is the temperature of the heating medium, θ_I is the initial temperature, θ_F is the final temperature, U is the overall heat transfer coefficient and A is the surface area.

This equation is derived in section 10.2. The design and operation of batch pasteurizers have been described in more detail by the Society of Dairy Technology (1983).

Other factors which need consideration are hygienic aspects, cleaning and sanitation. Heat exchangers should be totally enclosed to prevent airborne contamination of the product. Wherever possible, they should be capable of being cleaned by circulating caustic or acid detergents and sanitized by the use of hot water, steam or chemical sterilants. It is also advantageous to be able to inspect or clean manually the heat exchange surfaces, should the need arise. It may also be necessary to control and record the product temperatures and for continuous processes a flow controller may be required.

In contrast, continuous processes are more easily automated and are more suitable for larger throughputs. They result in more rapid heating and cooling and permit short residence times and consequently higher temperature. The conditions for the high-temperature short-time HTST process for pasteurized milk are 72 °C for 15 s, and the milk passes through the entire process in a few minutes. An additional advantage is the opportunity to employ regeneration, whereby the hot fluid is used to heat the incoming fluid, during which time it also cools. Both heating and refrigeration costs are reduced.

9.13.1 Regeneration efficiency

The energy in the hot fluid is used to heat the incoming fluid and, in so doing, cools, thereby saving heating and refrigeration costs.

Fig. 9.16 shows a schematic layout for a high-temperature short-time pasteurizer employing regeneration. The cold fluid enters the regeneration section, followed by mains-water and chilled-water cooling. The regeneration efficiency is a measure of how much of the total energy required is provided by regeneration.

The regeneration efficiency RE is defined as

$$RE = \frac{\text{amount of energy supplied by regeneration}}{\text{total amount required assuming no regeneration}} \times 100$$

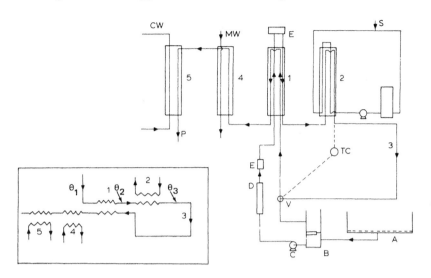

Fig. 9.16 — Layout of high-temperature short time pasteurizer (the insert shows a schematic diagram of the heat exchange sections): A, feed tank; B, balance tank; C, feed pump; D, flow controller; E, filter; P, product; S, steam injection (hot-water section); V, flow diversion valve; MW, mains-water cooling; CW, chilled water; TC, temperature controller; 1, regeneration; 2, hot-water section; 3, holding tube;, 4, mains-water cooling; 5, chilled-water cooling. (From Lewis (1986a), with permission.)

It can be expressed in terms of the mass flow rate m', specific heat c and temperature of the product at various locations (assuming no heat losses) (Fig. 9.16) as

$$\mathrm{RE} = \frac{m'c(\theta_2 - \theta_1)}{m'c(\theta_3 - \theta_1)} \times 100$$

$$= \frac{\theta_2 - \theta_1}{\theta_3 - \theta_1} \times 100$$

High regeneration efficiencies are obtained by increasing the surface area of the regeneration section, i.e. increasing the capital cost. Regeneration efficiencies in excess of 95% can now be achieved. However, this increases the residence time within the heat exchanger and for some UHT products may result in a poorer-quality product (Lewis, 1986a; Reuter, 1984).

9.13.2 Continuous heat exchangers

The simplest form of continuous heat exchanger is the tubular heat exchanger. The processed fluid is usually in the inner tube and the heating medium (or coolant) in the outer jacket. Again, flow is counter-current and the material of construction is generally stainless steel. Heat transfer is improved by inducing turbulent flow, taking into account which is the limiting resistance (see section 9.12.2). Such heat exchangers are used for small duties such as chilling of milk on the farm, For larger duties, *shell-and-tube* heat exchangers are used. The feed is divided between many tubes; all the tubes are fixed in a shell and the heat transfer fluid is passed through the shell over the outside of the tubes. Baffles are built into the shell to direct the fluid over the tubes. These types of heat exchanger, which are widely used in the chemical industry, are not suitable for food materials as the insides of the tubes are not accessible for inspection or cleaning.

The most popular type of heat exchanger for heating and cooling low-viscosity liquids is the plate heat exchanger. Plate heat exchangers were introduced in 1923 and the basic design concept has changed little over this period.

The constructions of various types of plate heat exchanger are shown in Fig. 9.17. Groups of stainless steel plates are clamped vertically in a robust frame. The frame can easily be dismantled, and the plates pulled apart for inspection or cleaning. The plates are separated by gaskets securely fixed in grooves. The gaskets are made from materials that can withstand temperatures up to 150 °C, as well as hot solutions of acid and caustic detergent. The gaskets form seals at the outer edges of the plate and around two of the four ports, so that the fluid enters from one end and flows over the plate, leaving the other end.

The plates are specially designed to induce turbulence but to minimize pressure drops. The gap between the plates is narrow, which means that they are not suitable for viscous liquids or liquids containing particulate matter. The layout for a typical plate pasteurizer is shown in Fig. 9.16. The plates are grouped in sections for regeneration, heating and cooling. Flow control is provided by using a positive-displacement pump or a centrifugal pump, with an additional flow controller. The temperature in the holding tube is recorded and, if it falls below the set fixed point temperature (e.g. 72 °C for milk), the under-processed milk is automatically diverted back to the balance tank for reprocessing. Such pasteurizers are used for milk, cream, beer, fruit juices and other low-viscosity fluids. High-viscosity fluids or fluids containing particulate matter cannot be processed in a plate heat exchanger because of the high pressure drop and blockage of the flow passages. They have to be treated using scraped-surface heat exchangers. The material is pumped through a cylinder in which an agitation device scrapes the fluid off the surface of the cylinder wall; this prevents the material from burning onto the hot surface and redistributes the energy by forced convection. This type of equipment can be used not only for heating and cooling but also for freezing; for ice-cream, air needs to be incorporated into the mix during the freezing process.

9.14 DIRECT STEAM INJECTION

Mixing saturated steam with a liquid has been found to provide a very rapid method of heat transfer, and such methods have been used for liquid foods for a considerable time for both batch and continuous processes. Either the

Sec. 9.14] Direct steam injection 273

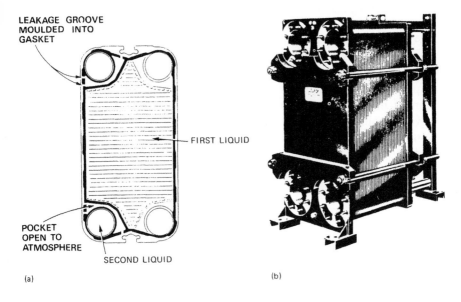

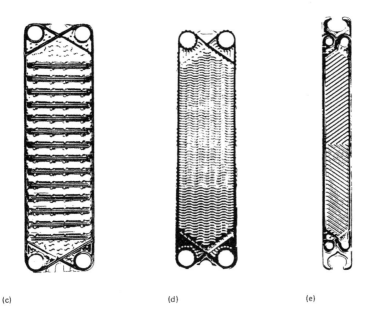

Fig. 9.17 — (a) Typical plate; (b) typical plate heat exchanger; (c) plate with corrugations and support pips; (d) plate with wavy groove type of corrugations; (e) plate with herring-bone type of corrugations. (The APV Company Ltd., Crawley.)

steam can be injected into the product (injection) or the product can be injected into steam (infusion). The steam condenses and dilution occurs. The steam must also be free of extraneous matter (e.g. water droplets, oil and rust).

A typical example is the production of UHT milk by direct steam injection. Milk is pre-heated to 75 °C and then brought into contact with saturated steam, resulting in almost instantaneous heating to the UHT temperature, It then passes through a holding tube and is passed into a chamber held under vacuum conditions. At this pressure the milk is well above its boiling point; this results in rapid cooling. This energy provides latent heat of vaporization to remove water equivalent in quantity to that condensing from the heating steam (Fig. 9.18).

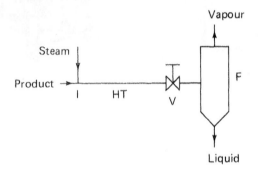

Fig. 9.18 — Steam injection and vapour removal (flash cooling): I, steam injection point; V, back-pressure valve; F, flash cooling chamber; HT, holding tube.

The heat balance for the process has been described by Perkin and Burton (1970). As a first approximation to ensure that the correct amount of water vapour is removed, the temperature in the separator is about 2 degC higher than that of the milk, prior to steam injection. This flash cooling process also removes other important volatile compounds, namely oxygen and low-molecular-weight sulphur components, e.g. H_2S and CH_3SH_3, which are derived from sulphur-containing amino-acids and are responsible for the cooked flavour in milk. Milk processed by this method has a better flavour and is less susceptible to oxidation reactions during storage. The main advantage of the proces is instantaneous heating and cooling. One of the main disadvantages is that energy lost in the flash cooling process is not easy to recover, and so regeneration efficiencies are lower (Mehta, 1980).

9.15 FOULING

During the operation of a heat exchanger, materials can be deposited from the processing fluids or water onto the surface of the heat exchanger. These deposits generally have a low thermal conductivity and offer an additional resistance to heat transfer. Consequently the rate of heat transfer may fall and the performance of the heat exchanger deteriorates; normally, this results in a fall in the outlet temperature. Occasionally, deposit formation may be severe enough to block the flow passage in plate heat exchangers. In such cases, processing stops and deposits are removed by cleaning-in-place techniques (see section 6.10).

There are three major types of deposit.

(1) Salts deposited as a result of using hard water. In many cases, water is softened or even demineralized.
(2) Deposits from the fluid being processed. One common example is deposits from heating milk. The deposits themselves are complex, containing fat, protein and minerals. The extent of deposit formation will increase as the processing temperature and acidity of the milk increase. Pasteurization at 71.8 °C usually produces minimum deposits, whereas UHT treatment at 140–145 °C will produce extensive deposits in a short time. An alcohol stability test is used for assessing whether milks are stable to heat, prior to UHT treatment. In many cases the indirect process has been found to be not suitable for UHT treatment of poor-quality (high-acidity) milks.
(3) Microbiological fouling can occur when using untreated river water for cooling purposes. Film-forming bacteria grow on the warm surface of heat exchangers. Similarly, in fermentation vesels, the microorganisms may accumulate on heat transfer surfaces and reduce heat transfer efficiencies.

A moderate amount of fouling is allowed for at the design stage, in terms of fouling factors:

$$\frac{1}{U} = \frac{1}{h_1} + \frac{1}{h_2} + \frac{L}{k} + \text{FF}$$

where FF is the fouling factor.

9.16 EVAPORATOR DESIGN

In an evaporator, steam is used to bring a liquid to its boiling point and then to vaporize the water. The layout of a simple evaporator is shown in Fig. 9.19. The heat exchange section is known as the calandria A. The mixture of liquid and vapour passes into a separator B where water vapour is withdrawn at the top and condensed. The evaporation may well operate under vacuum conditions, and the concentrate is either removed or evaporated further in the same or another unit. The total heat requirement Q is given by

$$Q = \left\{ \begin{array}{c} \text{heat required to bring feed} \\ \text{to boiling point} \end{array} \right\} + \left\{ \begin{array}{c} \text{heat required to cause} \\ \text{evaporation} \end{array} \right\}$$

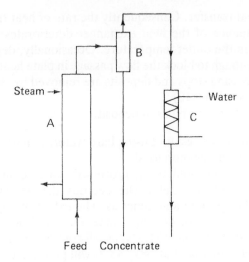

Fig. 9.19 — Layout of a single effect continuous evaporator: A, calandria; B, separator; C, water condenser.

$$= m'_1 c \Delta\theta + m' h_{fg}$$

where m'_1 is the feed rate, c is the specific heat, m' is the evaporation rate, $\Delta\theta$ is the evaporation temperature minus the feed temperature and h_{fg} is the latent heat of vaporization.

In the example cited in Chapter 1, 100 kg h^{-1} of liquid with a total solids content of 12% was fed to an evaporator, where 62.5 kg h^{-1} of water were removed, producing a concentrate containing 32% total solids. Therefore the feed rate would equal 100 kg h^{-1}, and the evaporation rate 62.5 kg h^{-1}.

The sensible heat change is generally low compared with the latent heat change, and particularly so on modern evaporators where the feed is generally pre-heated, almost to its boiling point, prior to entering the main evaporation section.

Heat is provided by the steam described by the heat balance:

heat lost by steam = heat required for evaporation

Therefore a simple evaporator requires approximately 1 kg of steam for every 1 kg of water removed. Energy economy is achieved by multiple-effect evaporation and recompression of the exhaust steam, by either steam ejectors or mechanical compression (see section 7.12). The properties of steam are discussed in more detail in section 11.4.

Other examples involving evaporation or condensation are in the use of refrigerants for abstracting heat from foods.

One problem encountered when concentrating fruit juices is the loss of volatile components which contribute to the flavour. These volatile compo-

Sec. 9.17] Heat transfer by radiation 277

nents would be lost if only a water condenser is used. They can be recovered by a refrigerated condensing system operating at much lower temperatures; they form the basis of natural fruit flavours and can be added back to the concentrate.

An alternative technique, to reduce the loss of volatile components, is to evaporate part of the juice to a higher concentration than is required and to use the raw juice to bring it down to the desired solids content. Membrane techniques such as reverse osmosis are also proving popular for concentrating such juices, one of the main advantages being that no phase change is involved.

There are many types of evaporator design available. The quality of the product will depend upon the evaporation temperature and the residence time. Temperatures range from 40 °C to 100 °C whereas residence times can be as short as several seconds to several hours for some batch processes. Evaporation has been described in more detail by Loncin and Merson (1979) and Coulson and Richardson (1978).

9.17 HEAT TRANSFER BY RADIATION

All objects above a temperature of 0 K give off electromagnetic radiation; the nature of this radiation will depend upon the nature of the substance and its temperature. This radiation will travel through a vacuum at the speed of

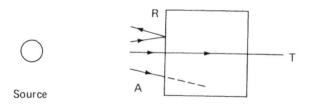

Fig. 9.20 — Electromagnetic radiation and its interaction with matter: R, reflected; T, transmitted; A, absorbed.

light; on contacting a further object, this radiation will be reflected, transmitted or absorbed (Fig. 9.20). Again, this will depend upon the property of this object. Only radiation that is absorbed will impart its energy and cause a temperature change.

9.17.1 Characteristics of electromagnetic radiation

Electromagnetic radiation consists of an electric and magnetic field mutually at right angles; these are normally in phase although they may be out of phase at the source. Both fields vary sinusoidally with time and distance. All electromagnetic radiation travels at the speed of light in a vacuum. The type of radiation is characterized by its wavelength λ or its frequency f. The relationship between these is

frequency × wavelength = velocity of light

$$f\lambda = c$$

where the frequency is equal to the number of cycles passing a fixed observer in 1 s; it is measured in hertz (Hz).

Fig. 9.21 shows the respective wavelengths and frequencies of the major

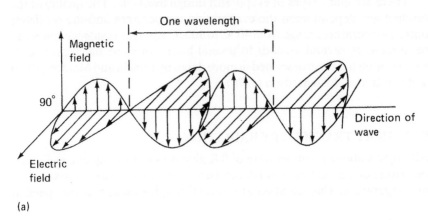

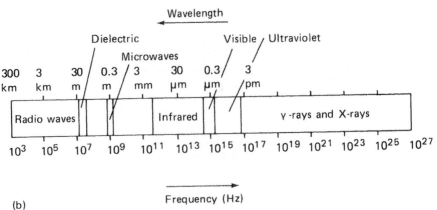

Fig. 9.21 — (a) Electromagnetic radiation considered as a waveform; (b) types of electromagnetic radiation. (From Milson and Kirk (1980), with permission.)

types of electromagnetic radiation. Radio waves have the longest wavelengths and shortest frequencies, whereas γ-rays have the shortest wavelengths and highest frequencies.

All the major types of electromagnetic radiation find use in food-

Radiation emitted from heated surfaces

processing operations or food-analytical techniques. Many examples from the field of food analysis have been described by Mohsenin (1984). The velocity of propagation may be less than the speed of light in some media. The amount that the wave is slowed down gives a measure of the dielectric constant for that medium:

$$v_p = \frac{c}{\sqrt{\varepsilon'}}$$

where v_p is the propagation velocity, c is the velocity of light and ε' is the dielectric constant. There is also some dispute about the exact nature of electromagnetic radiation; it is useful to imagine that it exists as discrete packets of energy or photons. The amount of energy per photon is given by the equation

$$E = h_{Pl} f$$

where E (J) is the energy, h_{Pl} (= 6.626×10^{-34} J s) is Planck's constant and f (Hz) is the frequency of radiation.

As the frequency increases, the energy per photon increases and the radiation becomes more harmful and dangerous to health. Therefore, electromagnetic radiation consists of a stream of photons, which are considerd to have zero rest mass but contain energy and momentum. These photons interact with matter and may possess sufficient energy to be lethal to living tissue, to break chemical bonds, to give rise to fluorescence or to eject electrons.

The energy in joules can be converted to electronvolts (eV), where one electronvolt (1 eV) is the energy acquired by an electron when accelerated by a potenetial difference of one volt (1 V), and

$$1 \text{ eV} = 1.6 \times 10^{-19} \text{ J}$$

When discussing high-energy radiation it is customary to describe the radiation in terms of its type and energy (in electronvolts) (see section 9.22).

9.18 RADIATION EMITTED FROM HEATED SURFACES

If an object is heated, e.g. to 1000 K, and the amount of energy emitted at a whole series of discrete wavelengths is measured, the relationship shown in Fig. 9.22 is found.

In the diagram the total energy emitted at a particular temperature is given by the area under the curve. As the temperature is raised, two important things happen: firstly, more energy is emitted and, secondly, the distribution of energy about its wavelength changes, with a shift of the maximum f_{max} towards the ultraviolet. The energy emitted at higher temperatures e.g. 2000–6000 K is also shown in Fig. 9.22. This is also the temperature on the surface of the Sun with the Sun providing us with our

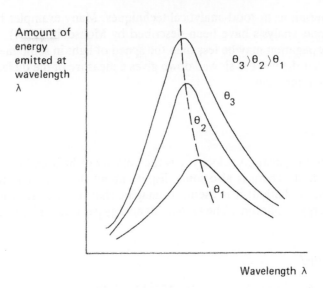

Fig. 9.22 — Radiation emitted from a surface at different temperatures.

major source of radiant energy. The radiant energy from the Sun consists approximately of 48% infrared, 40% visible and 12% ultraviolet. Much of the ultraviolet is filtered out in the ozone layer in the upper atmosphere. Solar heating is now becoming a very important source of 'free' energy. It is estimated that the energy in solar radiation amounts to $0.75\,\text{kW}\,\text{yard}^{-2}$ ($1.3\,\text{kW}\,\text{m}^{-2}$). In most applications, solar energy is converted to thermal energy (hot water) by means of a collector. Collection efficiencies greater than 50% are now commonplace. Solar energy is also used for evaporation, drying, cooking and many other operations.

9.19 STEFAN'S LAW

Stefan considered the role of energy emitted from a heated object. He found that the amount of energy depended upon the surface area, the temperature and the nature of the surface. Surfaces that gave the maximum rate of emission at a particular temperature were black and so the name *black body* is now given to surfaces which give a standard distribution of energy among the wavelengths as well as the maximum quantity of energy at a fixed temperature. The expression *full radiator* is also encountered.

Stefan's law for black bodies can be summarized as

$$Q' = \sigma A T_1^4$$

where Q' (J s^{-1}) is the rate of emission, $\sigma\,(= 5.7 \times 10^{-8}\,\text{J s}^{-1}\,\text{m}^{-2}\,\text{K}^{-4})$ is Stefan's constant, A (m^2) is the area and T_1 (K) is the absolute temperature.

Thus the rate of energy emission is proportional to the fourth power of the absolute temperature T. A correction factor is used for non-black bodies; it is known as the emissivity ε and is dimensionless. Therefore,

$$Q' = \varepsilon A T_1^4 \sigma$$

Emissivity values range from 0 (highly polished metal surfaces) to 1.0. Some emissivity values are shown in Table 9.5.

Table 9.5 — Emissivity values

Material	Emissivity	Material	Emissivity
Black body	1.00	Ice	0.97
White paper	0.9	Lean beef	0.74
Polished surfaces	0.05	Beef fat	0.78
Water	0.955	Aluminium foil	0.09
		Stainless steel	0.44

As well as emitting radiation, materials also absorb it. Substances that are good emitters of radiation are also good absorbers. The amount of radiation absorbed is given by

$$Q' = \alpha A T_1^4 \sigma$$

where α is the absorptivity of the surface and T_1 is the temperature of incident radiation source. The absorptivity value for a surface is numerically equal to the emissivity value, i.e. $\alpha = \varepsilon$ (see Table 9.5). Thus, for a food such as a loaf passing through an oven, the net rate of heat transfer by radiation to the loaf when the oven wall temperature and loaf temperature are T_1 and T_2, respectively, is

$$Q' = \left\{ \begin{array}{c} \text{rate of} \\ \text{absorption} \end{array} \right\} - \left\{ \begin{array}{c} \text{rate of} \\ \text{emission} \end{array} \right\}$$

$$= \alpha A T_1^4 \sigma - \varepsilon A T_2^4 \sigma$$

$$= \varepsilon A (T_1^4 - T_2^4) \sigma$$

As the temperature of the loaf rises, the rate of heat transfer by radiation will fall. Convection will also play an important part in most baking operations. Both radiated heat and convected heat are absorbed at the surface and redistributed into the loaf by conduction.

In practice the radiant heat transfer process is more complicated than the simple example presented here, and the geometries of the food and ovens walls will have a significant effect. Radiation exchange between a number of surfaces has been described in more detail by Milson and Kirk (1980).

9.20 INFRARED RADIATION

Infrared radiation can be used for heating purposes. The design of the heater should be such that the optical absorption properties should coincide with those of the source. Preferably the food should be capable of absorbing greater than 75% of the energy emitted from the source at the selected operating conditions, and these considerations will fix the operating temperature of the source.

The near-infrared region is defined as having a wavelength betwen $0.7\ \mu m$ and $3.0\ \mu m$, but the infrared range extends to $300\ \mu m$. A variety of heating elements are available, including tubular or flat metallic heaters, ceramic heaters, quartz tubes or halogen heaters. Quartz tubes are heated by an electric filament and emit radiation in the range 2.5–$5.0\ \mu m$. As the temperature of the source rises, the mean wavelength decreases; the wave has more penetration power (penetration depth increases) but, if the samples are thin, it may not be all absorbed. If surface heating is required, a source with a longer mean wavelength may be required. For water, radiation of wavelength longer than $1.4\ \mu m$ is almost completely adsorbed by a 3 mm layer, while absorption drops sharply for wavelengths below $1.4\ \mu m$. If carbon dioxide or water vapour is presented in the transfer medium, they will also absorb infrared radiation. Therefore the main factors governing the temperature rise of the sample are the incident power, its absorption characteristics and the sample thickness. Infrared radiation can also be used for drying applications.

Certain components absorb infrared radiation of particular wavelengths, and infrared absorption and reflectance techniques are used for measuring water, fat and protein concentrations (Mohsenin, 1984).

9.21 RADIO-FREQUENCY WAVES

9.21.1 Microwave and dielectric heating

This section covers the heating effects brought about by microwaves and macrowaves. Both these are types of electromagnetic radiation of long wavelength, in the radio-frequency range. To avoid interfering with the radio network, it has been agreed that manufacturers of commercial equipment are permitted to use the frequencies given in Table 9.6.

The distinction between macrowave and dielectric heating is solely in the wavelength of the incident radiation (note that the frequency multiplied by the wavelength equals the velocity of light). For microwave heating, the most common frequencies used are 915 MHz (896 MHz in Europe) and 2450 MHz.

Radio-frequency waves are generated in a device known as an applicator by means of a magnetron.

Radio-frequency waves

Table 9.6 — Permitted frequencies for radio-frequency heating

Frequency (Hz)	Wavelength (cm)	
13.56×10^6	2220.0	macrowave or dielectric heating
27.12×10^6	1100.0	
40.68×10^6	735.0	
915×10^6	32.8	microwave heating
2450×10^6	12.24	
5800×10^6	5.17	
22125×10^6	1.36	

The waves are transferred to the food which is placed in a cavity (oven) by means of waveguides. These waveguides are often aluminium tubes along which the waves are internally reflected (Fig. 9.23(a)). The food is placed within the cavity, and energy that is actually adsorbed by the food brings about a rise in temperature. If the electric field within the cavity is not uniform, uneven heating may occur. A more even distribution of the field is achieved either by using a metal fan which reflects the radiation randomly or by moving the food through the variable field by putting it on a turntable. Radiation is reflected by the walls of the cavity which do not rise in temperature, making the oven very efficient. Since microwaves are very effective at heating biological tissue, leaking radiation can be hazardous to the operators; paticularly vulnerable are the eyes since the blood supply is too low to provide sufficient cooling. Microwave ovens should be checked periodically to ensure that there is no leakage, although microwave ovens which are currently available are safe and reliable.

9.21.2 Absorption of microwave energy

There are some popular misconceptions about microwave heaters, and statements such as the food heats up from the centre are not uncommon. Therefore, it is necessary to examine how microwave radiation is adsorbed by a food material. For simplicity, let us examine microwave radiation which is incident on one surface only (Fig. 9.24(a)). As the wave passes through the food, it is attenuated, i.e. it loses energy. It is this energy which is converted to heat, at the point where the energy is lost. Any temperature gradients that develop within the food are then eliminated by normal processes of conduction and/or convection.

If the power in the incident radiation is P_0 (Fig. 9.24(b)) and the power after a penetration distance of d has been reduced to P, then the following two extreme situations could have occurred. Firstly, hardly any of the power is absorbed, with no consequent heating affect (this is almost typical of the reaction between microwaves and frozen food). Conversely, all the energy is absorbed very near the surface. In this situation, surface heating effects will predominate and the wave will hardly penetrate the food; this is the situation

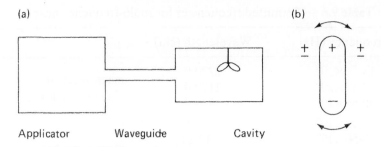

Fig. 9.23 — (a) Transfer of microwaves; (b) dipole rotation mechanism.

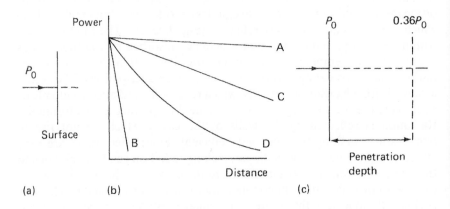

Fig. 9.24 — (a) Microwave energy of power P_0 incident on the surface of a food; (b) attenuation of power by different materials; (c) penetration depth evaluation.

with most infrared heaters and conventional cooking processes. For uniform even heating throughout the food, there should be a linear decrease in power as the wave penetrates into the food. However, in practice, what usually happens is that an exponential power loss occurs, i.e. equal fractions of the incident power are lost over the same distances. These four situations are depicted in Fig. 9.24(b). The situation for the loss of power in water and ice is shown in Fig. 9.25(a). Microwave energy is poorly attenuated in ice. The attenuation is also affected by the temperature of the water, with attenuation decreasing as temperature increases. One way of characterizing this exponential loss in power is to use the penetration depth. This is defined as distance taken for the power in incident radiation to fall to 36% of its value on the surface (that is 1/e of its surface value where e is the exponential function equal to 2.718) (Fig. 9.24.(c)). A large penetration depth indicates that radiation is poorly absorbed; a short penetration depth means that the heating will be predominantly surface heating. Fig. 9.25(b) shows how the penetration depth for cooked beef is affected by its temperature and frequency of the radiation. Again, frozen beef gives much higher penetration depths than unfrozen beef does. With exponential power loss, heating is not entirely even; proportionally less energy is adsorbed per unit volume (or mass) as one moves away from the surface; temperature gradients are set up which are then dissipated by other heat transfer mechanisms. However, it is true to say that energy is generated within the food itself.

The mechanism by which energy is produced depends very much on the presence of polar molecules, particularly water, within the food. The water molecule contains a dipole moment, i.e. one end is positively charged and the other end negatively charged. Thus, in a rapid oscillating field, the molecule will vibrate in an attempt to align itself with the field (Fig. 9.23(b)); the dipole rotation results in the formation of frictional heat which, together with some electrical resistance heating, produces a rise in temperature. Therefore the presence of liquid water within the food will be conducive towards such a heating effect.

The rate of heating for any substance will depend upon the amount of energy absorbed and its specific heat value. The amount of power P_0 absorbed by a substance can be expressed as

$$P_0 = 55.61 \times 10^{-14} f E_f^2 \varepsilon''$$

where P_0 (W cm^{-3}) is the power absorbed, f (Hz) is the frequency of radiation, E_f (V cm^{-1}) is the field strength and ε'' is the dielectric loss factor.

Thus the amount of power absorbed by a food will increase as the frequency of the radiation, the field strength and the loss factor increase. Normally the field strength and the frequency are fixed in a microwave heater. The dielectric loss factor is the important physical property of the food which will affect the amount of radiation adsorbed. Materials with a high dielectric loss factor are termed lossy materials and are very suitable for microwave heating. The loss factor has in turn been found to be dependent upon the frequency of the radiation and the temperature.

Fig. 9.26(a) shows the dielectric loss factor for three foods at different temperatures. Above the freezing point the loss factor for cooked ham increases as temperature increases, for raw beef there is no change, whereas the value for distilled water decreases as temperature increases. There is no obvious set pattern. Fig. 9.26(b) shows how the dielectric loss factor is affected by moisture content. This will be important when using microwaves in dehydration processes. The dielectric properties of foods are considered in more detail in section 12.3, in Figs. 12.24 and 12.25 and in Tables 12.7–12.12.

Frozen foods absorb microwave energy poorly and, when one defrosts food too quickly, run-away heating may occur. This happens when parts of the food which melt first then preferentially absorb energy and get very hot, whilst other areas remain frozen. To a limited extent, this can be

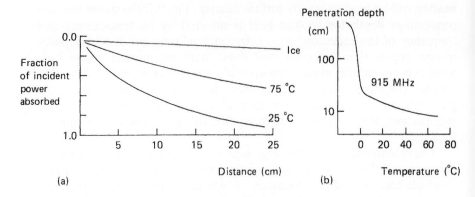

Fig. 9.25 — (a) Fraction of incident power adsorbed plotted aginst distance for water at three different temperatures; (b) penetration depth plotted against temperature at 915 MHz for cooked beef.

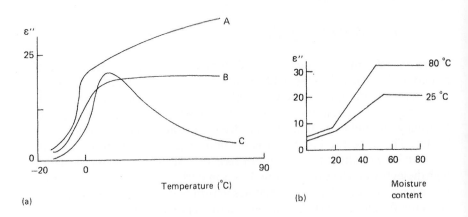

Fig. 9.26 — (a) Dielectric loss factor plotted against temperature for three different materials, cooked ham (line A), raw beef (line B) and distilled water (line C); (b) dielectric loss factor plotted against moisture content at two different temperatures for ground beef, at 915 MHz.

overcome by subjecting the food to intermittent doses of energy, thereby giving the food time to equilibrate during the rest period.

Microwaves offer a rapid means of heating and cooking materials and their use has been suggested in a wide range of food-processing operations, being extremely useful for catering operations. The principles and uses of microwaves and their effects on micro-organisms and food quality have been described in a series of articles in the *Journal of Food Protection* (1980).

9.22 IRRADIATION

One way of preserving the shelf-life of foods is to subject the food to high-frequency radiation, namely X-rays and γ-rays or an electron beam. As well as killing organisms, such radiation will kill insects, prevent over-ripening of fruits and inhibit sprouting. Unfortunately, undesirable side reactions may take place, such as free radical reactions, oxidation, reduction and the darkening poly(vinyl chloride) and Cellophane films. Some of these reactions can lead to the production of substances that may be toxic or carcinogenic, and for this reason the safety of foods sterilized by irradiation has been questioned. Furthermore, there may well be consumer resistance to any food that is associated with radiation treatment, particularly from the thought that the food might still be radioactive after treatment.

In practice, γ-radiation is most commonly used. The most popular sources being cobalt-60 (^{60}Co) and caesium-137 (^{137}Cs). Desirable properties are a long half-life, ease of handling and low cost. ^{60}Co has a fairly long half-life of 5.26 years and emits γ-rays very efficiently. There are two waves, with energies of 1.17 MeV and 1.33 MeV; two α-waves are also emitted.

Activity is now measured in becquerels (Bq), where one becquerel (1 Bq) is equivalent to one disintegration per second. The former unit, the Curie (Ci), is equivalent to 3.7×10^{10} disintegrations per second (or 3.7×10^{10} Bq). Thus a plant formally rated at 10^6 Ci would now be rated at 3.7×10^{16} Bq:

$$1 \text{ Ci} = 3.7 \times 10^{10} \text{ Bq}$$

The specific activity of ^{60}Co is between 60 Ci g^{-1} and 100 Ci g^{-1}, and radiation energy is generated at the rate of 15 kW MCi^{-1}. Not all the energy is absorbed by the sample, because γ-rays are given off in all directions. The absorption efficiency will range between 10% and 30%. ^{60}Co requires the addition of 12.3% of the original activity each year to maintain a constant capacity. The current cost (1984) of ^{60}Co is approximately £1 per curie.

The severity of the radiation treatment is expressed in terms of the number of radiological units ('rads') that the food is exposed to. One radiological unit is that quantity of radiation which results in the absorption of 100 erg g^{-1} at the point of interest. Therefore,

$$1 \text{ rad} = 100 \times \frac{10^{-7} \text{ J}}{10^{-3} \text{ kg}} = 10^{-2} \text{ J kg}^{-1}$$

The SI unit is termed the gray (Gy), where one gray (1 Gy) results in the absorption of 1 J kg^{-1}. Therefore,

$$1 \text{ rad} = 10^{-2} \text{ Gy} \quad \text{or} \quad 100 \text{ rad} = 1 \text{ Gy}$$

Table 9.7 gives an indication of the potential uses for food irradiation, together with the dosage involved.

As with heat treatment, the reduction in organisms follows first-order reaction kinetics, and so irradiation treatment is not an absolute form of sterilization. A dosage is selected to give a suitable reduction of the principle spoilage organisms (usually 12 decimal reductions for *Clostridium botulinum*). The potential advantage of irradiation sterlization of food is that the heating effect is negligible.

A dose of 6×10^6 rads will produce a temperature rise $\Delta\theta$ for water determined by

$$\frac{6 \times 10^6 \times 10^{-2}}{\text{(dose)}} = \frac{1 \times 4.18 \times 10^3 \, \Delta\theta}{\text{(m)} \quad \text{(c)} \quad (\Delta\theta)}$$

Therefore,

$$\Delta\theta = \frac{6 \times 10^4}{4.18 \times 10^3} = 14.4 \text{ degC}$$

Irradiation can be used to treat packaged materials, e.g. cans and flexible pouches, and there should be none of the heat transfer problems associated with viscous materials.

Clostridium botulinum is very resistant to irradiation conditions. Although it is not the most heat resistant of the food spoilage organisms, it has been claimed to be one of the most resistant to radiation. A dosage of 4.8 Mrad has been recommended for achieving 12 decimal reductions for this organism. Such a dosage would adversely affect the flavour of the food and make it organoleptically unacceptable.

In 1981, after analysing experimental data collected by scientists over a period of 10 years, the Joint Expert Committee of the World Health Organization concluded that food subjected to an average dose of 1.0 Mrad or less resulted in no toxiological or nutritional problems. Such dosages and the energy levels involved are not sufficient to make the food radioactive. This is below the 4.8 Mrad required to reduce *Clostridium botulinum* spores to a safe level but is sufficient to inactivate most food-poisoning organisms associated with food, including *Salmonella*. The effects of ionizing radiation in micro-organisms have been reviewed in more detail by Hersom and Hulland (1980).

Therefore, it appears that irradiation treatment is unlikely to be useful for producing a commercially sterile product that can be stored at ambient temperature. However, a list of applications involving doses less than 1.0 Mrad have been proposed (Table 9.7). Irradiation methods have been approved in USA, The Netherlands and South Africa but not yet in the UK, where it is being reviewed.

Currently, UK regulations (Jukes, 1984) pertaining to irradiation state that no person shall prepare food using ionizing radiation, except as follows.

Table 9.7 — Applications for irradiation treatment

Dosage (Mrad)	Applications and examples
2–6	Sterilization of foods in sealed containers
1	Decontamination of food ingredients, e.g. spices and powders
0.3–1.0	Pasteurization, extension of shelf-life, destruction of surface moulds, e.g. meat, poultry, fish and egg products, often combined with refrigeration
0.01–0.2	Extended storage of fruit, prevention of over-ripening, e.g. strawberries, mangoes and papaya
0.01–0.2	Control of insects, e.g. disinfection of grain
0.01–0.3	Inhibition of sprouting or growth, e.g. onions, potatoes, and carrots

(a) Up to 50 rad of ionizing radiation, where the energy of the radiation delivered does not exceed 5 MeV.
(b) Where food is intended for patients who require a sterile diet for their treatment.

In addition, no person shall sell or import food which is prohibited by the above regulations. Consequently, all irradiation techniques listed in Table 9.7 are prohibited in the UK. However, it is likely that irradiated food could be imported, because it is almost impossible to detect it when applied at the lower dosages described in Table 9.7.

Often, materials for irradiation are containerized. It is important that all areas of the container get the same dosage. This can be achieved by ensuring that the containers are not too large and that all sides of the container are exposed to the source for equal times. Irradiating the source from two sides will increase the thickness of the sample that can be processed.

The amount of radiation absorbed can be assessed both qualitatively and quantitatively by counters placed on the box. Quantitative estimates are accurate to within ±2%. Again, this may not fully represent the amount absorbed by the material because the radiation is attenuated within the container. The rate of attenuation will depend upon the packing density and energy of the radiation; electron beams are attenuated much more rapidly than are γ-rays. A sample thickness of less than 8 cm is recommended for electron beams of energy or 10 MeV, whereas γ-rays can penetrate much larger distances. Some of these factors have been discussed in more detail by Ley (1984).

Operating personnel need to be shielded from the radiation. For γ-rays, this will require approximately 2 m of concrete or 6 m of water to attenuate the wave completely.

Irradiation methods have been discussed in more detail in recent texts by Elias and Cohen (1983) and Josephson and Peterson (1982, 1983a, 1983b).

9.23 SYMBOLS

a	thermal diffusivity
A	surface area
c	specific heat
c	velocity of light (section 9.17)
C	constant
D	diameter
E	energy
E_f	electric field strength
f	frequency
FF	fouling factor
g	acceleration due to gravity
h	heat film coefficient
h_{Pl}	Planck's constant
h_{fg}	latent heat of vaporization
I	electric current
k	thermal conductivity
L	length
m	mass
m	mass fraction
m'	mass flow rate
p	pressure
P_0	power
Q	heat required
Q'	rate of heat transfer
r	radius
RE	regeneration efficiency
T	temperature
U	overall heat transfer coefficient
v	velocity
v_p	propagation velocity
V	volume fraction and voltage

9.23.1 Greek symbols

α	absorptivity
β	coefficient of volumetric expansion
ε	emissivity
ε'	dielectric constant
ε''	dielectric loss factor
θ	temperature
λ	wavelength
μ	coefficient of viscosity
ρ	density
σ	Stefan's constant

9.23.2 Dimensionless groups

- Bi   Biot number
- Gr   Grashof number
- Nu   Nusselt number
- Pr   Prandtl number
- Re   Reynolds number

10

Unsteady-state Heat Transfer

10.1 INTRODUCTION

In many heat transfer situations, one is interested in the time taken to heat, cool, freeze, cook or dry food. Solutions to these problems may be found by considering unsteady-state heat transfer equations which describe how the temperature at a fixed location changes with time.

In the first part of this chapter, unsteady-state heat transfer involving only sensible heat changes will be considered, followed by a discussion of latent heat changes, involved in freezing and thawing time calculations. Drying operations, in which both heat and mass transfer processes occur, will be considered in section 13.9.

10.2 HEAT TRANSFER TO A WELL-MIXED LIQUID

One relatively simple case to handle mathematically is concerned with the batch heating or cooling of a liquid in a stirred tank. It is assumed that the liquid is well mixed so that there are no temperature gradients within the liquid.

Consider heat being transferred from steam or hot water at a temperature θ_{st} to a liquid at a temperature θ. In a small time interval dt, the rate dQ/dt of heat transfer is as follows:

$$\frac{dQ}{dt} = UA(\theta_{st} - \theta) \qquad (10.1)$$

Assume that this brings about a small temperature change $d\theta$; therefore,

$$dQ = mc \, d\theta \qquad (10.2)$$

Substituting equation (10.2) in equation (10.1) gives

$$mc\frac{d\theta}{dt} = UA(\theta_{st} - \theta)$$

$$\frac{d\theta}{\theta_{st} - \theta} = \frac{UA}{mc}dt$$

On integration between the initial temperature θ_I and the final temperature θ_F the total heating time t is given by

$$t = \frac{mc}{UA} \ln\left(\frac{\theta_{st} - \theta_I}{\theta_{st} - \theta_F}\right) \tag{10.3}$$

As an example, how long will it take to heat 90 l of milk from 10 °C to 90 °C, in a jacketed heating pan, using saturated steam at 1 bar (121 °C)? The overall heat transfer coefficient and surface area are 400 W m^{-2}K^{-1} and 1.8 m^2, respectively.

The milk density is 1030 kg m^{-3} and the specific heat is 3.9 kJ kg^{-1} K^{-1}. Thus,

$$\text{mass of milk} = 90 \times 10^{-3} \times 1030$$
$$= 92.7 \text{ kg}$$

Therefore,

$$t = \frac{92.7 \times 3900}{400 \times 1.8} \ln\left(\frac{121 - 10}{121 - 90}\right)$$

$$= 502.1 \times 1.275$$

$$= 640.2 \text{ s or } 10.66 \text{ min}$$

The equation is useful for examining the factors affecting the heating time.

A similar situation exists when the fluid is being cooled. Newton's law of cooling states that the rate of loss of heat is proportional to the excess temperature. Therefore the time required to cool a liquid from θ_I to θ_F using a cooling medium at θ_c is given by

$$t = \frac{mc}{UA} \ln\left(\frac{\theta_I - \theta_c}{\theta_F - \theta_c}\right)$$

10.3 UNSTEADY-STATE HEAT TRANSFER BY CONDUCTION

When a potato or an egg is being heated in boiling water, experience indicates that the temperature of the food will depend upon the time of heating and the position within the food. Considerable temperature gradients will exist within the food, particularly during the initial period of heating.

Factors that are likely to affect the temperature rise are the temperature of the heating medium, the thermal conductivity of the food and the specific heat value.

The basic equation, which defines heat transfer by conduction in one direction x only, is second order (contains a second-derivative term) and is as follows:

$$\frac{d\theta}{dt} = \frac{k}{c\rho} \frac{d^2\theta}{dx^2}$$

The term $k/c\rho$ is called the thermal diffusivity a. This is a measure of how quickly a substance will change its temperature during a heating or cooling process (see section 9.8).

The solution to this equation is complex and is discussed for the situation involving bringing the two longest sides of the slab of material initially at a uniform temperature θ_0 immediately to a higher temperature θ_1, considering heat transfer only in a direction perpendicular to these faces. The temperature θ at a point a distance x from the surface after a time t is given by

$$\frac{\theta_1 - \theta_1}{\theta_0 - \theta_1} = \sum_{n=0}^{\infty} \frac{4}{(2n+1)\pi} \frac{\exp[-(2n+1)^2\pi^2 at]}{L^2} \frac{\sin[(2n+1)\pi x]}{L}$$

where n is an integer, a is the thermal diffusivity and L is the slab thickness. When the computed temperature is not too close to the initial value, a reasonable approximation can be obtained from the first expression in the series obtained by substituting $n = 0$.

This will simplify to

$$\frac{\theta - \theta_1}{\theta_0 - \theta_1} = \frac{4}{\pi} \exp\left(\frac{-\pi^2 at}{L^2}\right) \sin\left(\frac{\pi x}{L}\right)$$

An example of the use of this equation has been given by Jackson and Lamb (1981) for determining the temperature at the centre of a slab, i.e. $x = L/2$. Analysis of this solution shows that the heating or cooling time is proportional to the square of the slab thickness. Milson and Kirk (1980) have described the use of finite-difference methods for solving this equation. These involve dividing the material up into a number of small elements and considering the accumulation (or loss) of energy in each element over a short time period. This is repeated over the total time period under consideration. Considerable computation is required, and so these techniques are best solved by computer.

Both these approaches assume that there is no resistance due to convection, i.e. the value for the heat film coefficient is infinity. Therefore, they are likely to give an underestimate of the time in cases where convection may be significant.

10.4 HEAT TRANSFER INVOLVING CONDUCTION AND CONVECTION

When solid foods are being heated or cooled, using either air or a liquid,

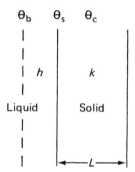

Fig. 10.1 — Heat transfer involving conduction and convection.

there are two resistances to heat transfer; these are the heat film coefficient h and the resistance due to conduction within the solid (see section 9.12). It is very important to know which is the limiting resistance (Fig. 10.1).

If a long flat slab of thickness L is being cooled by air or water, consider the rate of heat removal of heat when the temperature of the centre is θ_c and the temperature of the surface is θ_{surf} and the bulk fluid temperature is θ_b.

The rate of heat transfer from the surface is given by

$$\frac{dQ}{dt} = hA(\theta_{surf} - \theta_b)$$

This amount of energy must be transferred from the solid to the surface by conduction. Hence,

$$\frac{dQ}{dt} = kA \frac{\theta_c - \theta_{surf}}{L/2}$$

Equating the two rates gives

$$hA(\theta_{surf} - \theta_b) = kA \frac{\theta_c - \theta_{surf}}{L/2}$$

Rearranging in a way which gives two dimensionless groups, we obtain

$$\frac{hL/2}{k} = \frac{\theta_c - \theta_{surf}}{\theta_{surf} - \theta_b}$$

Note that this is the ratio of the temperature drop over the solid to that of the fluid film. The group $(hL/2)/k$ is termed the *Biot number* (see Section 9.12.3). It is effectively a measure of convection and conduction. At small values of Biot number (less than 0.2) the major resistance to heat transfer is due to convection and takes place over the film, i.e. most of the temperature gradient takes place over the film, and the solid can be assumed to be at a uniform temperature (no temperature gradients exist). In this case the time required to heat the solid from an initial temperature θ_I to a final temperature θ_F (θ_H is the temperature of the heating medium) is given by a modification of equation (10.3), where the overall heat transfer coefficient U is replaced by the heat film coefficient.

Unfortunately, the Biot number for most foodstuffs of any reasonable thickness or dimension is greater than 0.2 and heat transfer by conduction becomes the predominant mechanism; the object will heat or cool more slowly than predicted by the equation. In these situations the solution of the relevant unsteady-state heat transfer equations has been represented graphically and the problem can be solved using appropriate charts. The factors which affect the temperature at the geometric centre can be accounted for by the following dimensionless groups. The *temperature factor*, which is the fraction of temperature change which is unaccomplished, is given by

$$\frac{\theta_H - \theta_F}{\theta_H - \theta_I} \text{ or } \frac{\theta_F - \theta_H}{\theta_I - \theta_H}$$

The *Fourier number* Fo concerning the physical properties and dimensions of solid and the time is given by

$$\frac{k}{c\rho} \frac{t}{\delta^2}$$

The *Biot number* Bi, the ratio of convective and conductive heat transfer rates, is given by

Sec. 10.4] Heat transfer involving conduction and convection   297

$$\frac{h\delta}{k}$$

δ is a characteristic half-dimension; δ is the radius for a cylinder or sphere and it is the half-thickness for a slab.

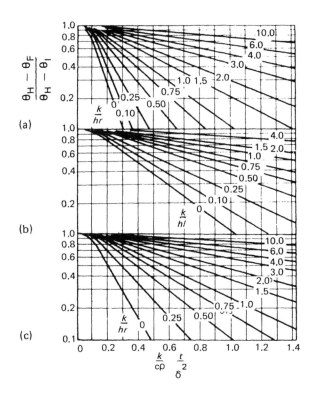

Fig. 10.2 — Heat transfer data into (a) a sphere, (b) a slab and (c) a cylinder. (From Henderson and Perry, 1955, with permission.)

The charts giving the relationship between these factors are shown in Fig. 10.2(a), Fig. 10.2(b) and Fig. 10.2(c) for a sphere, an infinite slab and an infinite cylinder, respectively, over the temperature factor range 1.0–0.1. The data are normally presented as a plot of the temperature factor against the Fourier number for different values of the Biot number (inverse).

As an example, consider peas initially at 20 °C which are blanched in hot water at 100 °C. How long will it take for the temperature at the centre of the peas to reach 80 °C?

The following data are used: diameter of the peas, 0.6 cm; specific heat, 3.31 kJ kg^{-1} K^{-1}; density of the peas, 950 kg m^{-3}; thermal conductivity, 0.3 W m^{-1} K^{-1}; surface film coefficient, 500 W m^{-2} K^{-1}.

The Biot number is

$$\frac{h\delta}{k} = \frac{500 \times 0.3 \times 10^{-2}}{0.3} = 5$$

Therefore,

$$\frac{k}{h\delta} = 0.2$$

The temperature factor is

$$\frac{\theta_H - \theta_F}{\theta_H - \theta_I} = \frac{100 - 80}{100 - 20}$$

$$= \frac{20}{80}$$

$$= 0.25$$

Using the chart for the sphere (Fig. 10.2(a)), the Fourier number equals 0.31:

$$\text{Fo} = \frac{k}{c\rho} \frac{t}{\delta^2}$$

$$= 0.31$$

Substituting all the data into this equation (SI units) gives

$$t = \frac{0.31 \times 3310 \times 950 \times (0.3 \times 10^{-2})^2}{0.3}$$

$$= 29.24 \text{ s}$$

Therefore the centre of the peas will reach a temperature of 80 °C after about 29.2 s. The infinite slab and infinite cylinder are not usually encountered in practice. Two geometries of importance are a finite cylinder, e.g. tin cans and sausage-shaped objects (Fig. 10.3(a)) and a rectangular block (Fig. 10.3(b)).

It can be shown mathematically that a finite cylinder can be considered as the product of an infinite cylinder of radius r and a slab of thickness h, i.e. $\delta = h/2$.

Therefore the ratio for the cylinder becomes

Sec. 10.4] Heat transfer involving conduction and convection

$$\left(\frac{\theta_F - \theta_H}{\theta_I - \theta_H}\right)_{\text{cylinder}} = \left(\frac{\theta_f - \theta_H}{\theta_I - \theta_H}\right)_{\text{infinite slab, thickness } h} \left(\frac{\theta_F - \theta_H}{\theta_I - \theta_H}\right)_{\text{infinite cylinder, radius } r}$$

and, in a similar fashion, the ratio for the block becomes

$$\left(\frac{\theta_F - \theta_H}{\theta_I - \theta_H}\right)_{\text{block}} = \left(\frac{\theta_F - \theta_H}{\theta_I - \theta_H}\right)_{\text{infinite slab, thickness } L_1} \left(\frac{\theta_F - \theta_H}{\theta_I - \theta_H}\right)_{\text{infinite slab, thickness } L_2}$$

These equations are corrected for heat entering the side of the cylinder or block.

In the case in which one of the dimensions is much greater than the other, this approach will give a similar answer to that obtained using the model for the infinite slab or cylinder if, in each case the smallest of the dimensions is used for the calculation, i.e. r and L_2 in Fig. 10.3.

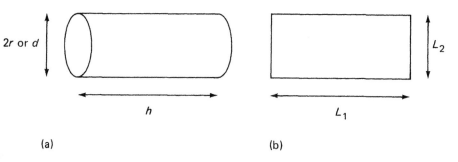

Fig. 10.3 — (a) Formation of a finite cylinder from an infinite cylinder and slab; (b) a block from two infinite slabs.

The following procedure can be used for calculating the temperature at the centre of a solid cylinder of height h and radius r after a given heating time.

(1) Determine the Fourier number using the heating time and physical properties of the food.
(2) Determine the Biot number for an infinite cylinder of radius r. Use the chart for the infinite cylinder to determine the temperature factor θ_{IC}.
(3) Determine the Biot number for an infinite slab of thickness h. Use the chart for the infinite slab to determine the temperature factor θ_{IS}.

The temperature factor θ_C for the cylinder is then determined from the product $\theta_{IS}\theta_{IC}$ and the final temperature is thus determined from

$$\theta_C = \theta_{IC}\theta_{IS}$$

This could be used for predicting heat penetration into a conduction pack using the thermal conductivity of the product (any resistance due to the can is ignored) or other cylindrical objects.

These correlations are extremely useful for estimating heating and cooling times and determining how changes in processing variables will affect these times.

The same charts can be used for cooling foods; in this case the temperature factor is

$$\frac{\theta_F - \theta_c}{\theta_I - \theta_c}$$

where θ_I is the initial temperature, θ_F is the final temperature and θ_c is the coolant temperature. The solution can be extended to determine the temperature at locations other than the centre by introducing a further dimensionless parameter n.

For a slab, cylinder and sphere, $n = x/\delta$, where x is the distance from centre. Therefore, $n = 0$ corresponds to the centre, and $n = 1$ corresponds to the surface.

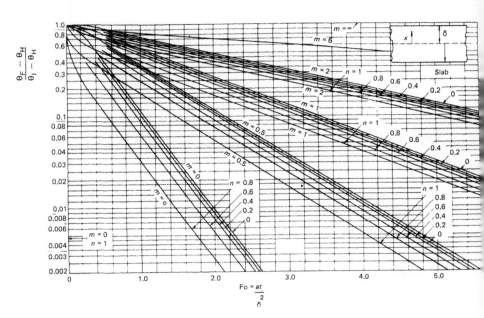

Fig. 10.4 — Conduction in an infinite slab: $n = x/\delta$; $m = k/h\delta$. (From American Society of Heating, Refrigerating and Air-Conditioning Engineers (1981), with permission.)

Fig. 10.4, Fig. 10.5 and 10.6 show these charts for an infinite slab, a sphere and an infinite cylinder, respectively, over a much wider range of

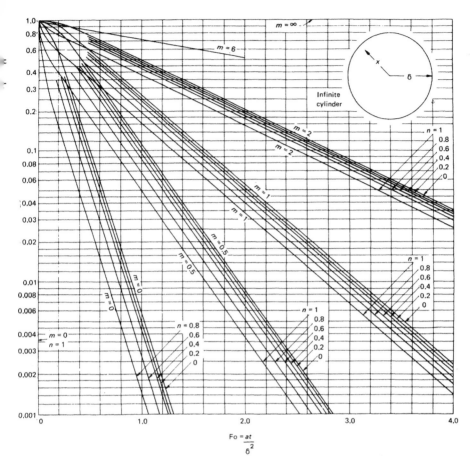

Fig. 10.5 — Conduction in an infinite cylinder: $n = x/\delta$; $m = k/h\delta$. (From American Society of Heating, Refrigerating and Air-Conditioning Engineers (1981), with permission.)

temperature factors (1.0–0.001). Note again that an inverse Biot number is used, i.e. $k/h\delta$. These charts are used in a similar manner to that described earlier, with the addition of the parameter n.

In contrast with the correlations described above, Holmes and Woodburn (1981) have provided a compilation of experimental heat penetration data for a wide variety of foods. Comprehensive tables of cooking methods, times, temperatures and food dimensions together with internal food temperatures are given. Similar experimental data have been compiled for chilling and freezing processes by Mohsenin (1980).

10.5 THERMAL PROCESSING

Heat treatment can be considered to be one of the more traditional forms of food preservation. The food is either placed in a container which is sealed

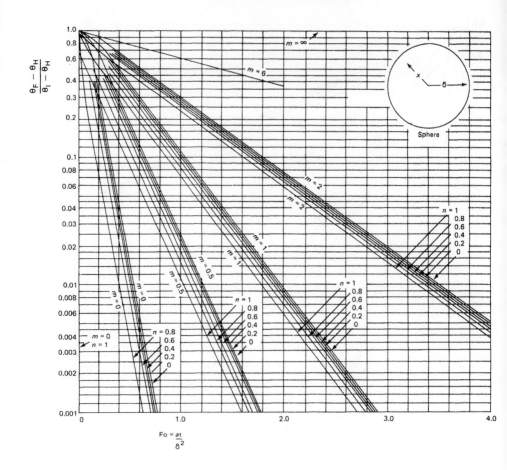

Fig. 10.6 — Conduction in a sphere: $n = x/\delta$; $m = k/h\delta$. (From American Society of Heating, Refrigerating and Air-Conditioning Engineers (1981), with permission.)

and then heated in that container or heated in a heat exchanger before being filled into a container. Heat is used to inactivate vegetative micro-organisms such as moulds, yeast and bacteria together with their spores.

Pasteurization is a relatively mild form of heat treatment, designed to remove pathogenic bacteria (e.g. *Brucella* and *Salmonella*) and the vast majority of vegetative organisms, thereby making it safe and increasing its shelf-life which is further increased by refrigerated storage. Milk is pasteurized at 63 °C for 30 min or 72 °C for 15 s. Other liquids which are pasteurized are cream, ice-cream formulations, fruit juices and other beverages and eggs.

However, heat-resistant spores will not be inactivated under these conditions and, to achieve a longer shelf-life, a more drastic form of heat treatment is required.

10.5.1 D and Z values

The thermal inactivation of heat-resistant spores normally follows first-order reaction kinetics, i.e. the rate of destruction at a constant temperature is directly proportional to the number of surviving spores present.

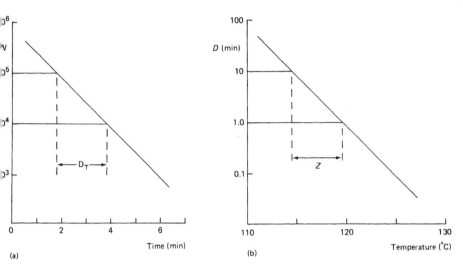

Fig. 10.7 — (a) Relationship between the population of organisms and time at a constant heating temperature; (b) relationship between the heating temperature and time to achieve the same lethal effect.

Fig. 10.7(a) shows how the population changes with time, with a straight-line relationship resulting when the data are plotted on semilogarithmic coordinates. The heat resistance of micro-organisms or spores, at a constant temperature, is measured by the decimal reduction time D_T, where D_T is defined as the time required to reduce the population by 90% (one logarithmic cycle). Therefore, if a population of spores are heated at a constant temperature T for a time t, the number N of spores surviving is given by

$$\log_{10}\left(\frac{N_0}{N}\right) = \frac{t}{D_T}$$

where N_0 is the initial population, N is the final population, t (min) is the heating time and D_T (min) is the decimal reduction time. The D_{121} °C values for *Clostridium botulinum* and *Bacillus stearothermophilus* are 0.21 min and 4.0 min, respectively. *Bacillus stearothermophilus* is one of the most heat-resistant spores found in foods.

In practical situations, most foods are not heated at a constant temperature but are subjected to a change in temperature. As the temperature increases, the rate of inactivation increases, resulting in a decrease in the decimal reduction time. For most vegetative organisms and spores, there is a reasonable straight-line relationship between the logarithm of the decimal reduction time, and the temperature. Fig. 10.7(b) shows such a relationship for one type of spore. From this a new parameter for the spore, known as the Z value is defined as follows. The Z value is the temperature change which results in a ten-fold change in the decimal reduction time.

The line shown in Fig. 10.7(b) is known as a thermal resistance line, and all combinations of time and temperature along that line have the same lethal effect on that particular spore.

The equation describing the thermal resistance line is (see Fig. 10.7(b))

$$\log\left(\frac{t_1}{t_2}\right) = \frac{(\theta_2 - \theta_1)}{Z}$$

The Z value for most heat-resistant spores is 10 °C (18 °F). D and Z values for a variety of organisms have been given by Lund (1975).

Physical chemists measure reaction rates in terms of a reaction velocity constant k and activation energy E. The relationship between the reaction velocity constant and decimal reduction time and that between the Z value and activation energy have been discussed by Lund (1975) and Lewis (1986a).

It is also important to minimize the loss of nutrients during food processing. Reaction kinetic data for a wide range of chemical reactions including browning reactions, protein denaturation, enzyme inactivation and vitamin destruction have been reviewed by Lund (1975) and Thompson (1982).

10.6 COMMERCIAL STERILITY AND F_0 EVALUATION

Canned foods and UHT products are termed commercially sterile and usually have a shelf-life of at least 6 months. Commercial sterility describes the situation in which viable micro-organisms may be found in the product, but conditions are not favourable for growth and the micro-organisms present will not cause spoilage or disease or have a detrimental effect on the product quality during its stated shelf-life.

The most dangerous of the food-poisoning organisms is *Clostridium botulinum* the spores of which are very resistant to heat and can produce a very powerful toxin in packaged foods. However, the spores will not usually germinate below pH 4.5. For this reason, foods are classified into low-acid (pH > 4.5) and acid foods.

Low-acid foods are subjected to a heat treatment that will ensure at least 12 decimal reductions (12D) for *Clostridium botulinum* spores. Such a process is referred to as the 'minimum *Clostridium botulinum* cook' and is achieved by heating the food to a temperature of 121 °C for 3 min. For

Sec. 10.6] Commercial sterility and F_0 evaluation

canned foods the temperature is monitored at the slowest heating point in the can, which is normally the geometric centre. A temperature of 121 °C (250°F) is often used as a reference temperature for comparing thermal processes.

Acid products require a less drastic process as the low pH prevents the growth of pathogenic and spoilage bacteria. Thermal processes for low-acid foods are evaluated and compared in terms of F_0 values. F_0 is defined as the total integrated lethal effect expressed in terms of minutes at 121 °C; it is derived for micro-organisms with a Z value equal to 10 °C.

It has been shown that the lethalities accumulated at different temperatures are additive. The lethality L is derived from the equation

$$\log L = \frac{T - \theta}{Z}$$

where L is the lethality, i.e. the number of minutes at the reference temperature equivalent to 1 min at the experimental temperature, T is the experimental temperature and θ is the reference temperature. For F_0 evaluation, this equation becomes

$$\log L = \frac{T - 121}{10} \text{ or } L = 10^{\{(T - 121)/10\}} \quad (10.4)$$

and

$$F_0 = \int L \, dt$$

Lethality tables are derived from equation (10.3). Some lethality values for experimental temperatures between 110 °C and 130 °C are shown in Table 10.1. It is important to know that there is no significant lethal effect until a

Table 10.1 — Lethality tables (defined for a reference temperature of 121 °C and $Z = 10$ °C).

Experimental temperature (°C)	Lethality L	Experimental temperature (°C)	Lethality L
110	0.079	121	1.000
111	0.100	122	1.259
112	0.126	123	1.585
113	0.159	124	1.995
114	0.199	125	2.512
115	0.251	126	3.162
116	0.316	127	3.981
117	0.398	128	5.012
118	0.501	129	6.310
119	0.631	130	7.943
120	0.794		

temperature of 100 °C is reached, where L is still below 0.01.

F_0 values are evaluated by placing a thermocouple at the slowest heating point in the container and recording how the temperature changes with time, during processing, which includes the heating, holding and cooling periods (see Fig. 10.8). The temperature–time data obtained can be transformed to lethality–time data, and F_0 is determined from the area under the lethality–time (min) curve. This is known as the general method. More detail has been provided by Pflug and Esselen (1979). F_0 values used for products lie between 3 and 18. Some examples are given in Table 10.2.

Note that some processes require F_0 values considerably in excess of 3, the main reason being that some spores which may cause spoilage are more heat resistant than *Clostridium botulinum* and that some types of material may be highly contaminated with spores.

10.7 UHT PROCESSES

Products produced using UHT processes, which involve higher temperatures for shorter times, by means of continuous heat exchangers, require to be packed aseptically in order to attain commercial sterility. Compared with in-container sterilized products, these are subject to less chemical damage such as browning reactions, protein denaturation and vitamin loss, because chemical reactions are less sensitive to changes in temperature than spore destruction is (Mehta, 1980). Consequently the quality of UHT products is claimed to be superior.

If it is assumed that the heating and cooling periods make no contribution to the total lethality, the F_0 value for a UHT process can be evaluated from

$$F_0 = 10^{\{(T-121)/10\}} \frac{t}{60}$$

where T (°C) is the UHT temperature and t (s) is the holding time. This assumption is reasonable for direct UHT plants but may give an underestimate for indirect UHT plants, where heating and cooling times may be considerable. This has been discussed in more detail by Reuter (1984).

Kessler (1981) has recommended an alternative procedure for evaluating UHT processes, based on using a reference temperature of 135 °C and a Z value of 10.1 °C for estimating the microbiological parameter and a Z value of 31.4 °C to indicate thiamin (chemical) damage. This higher reference temperature is used because it is closer to the UHT temperatures (135–150 °C) used in practice.

10.8 HEAT PENETRATION INTO CANNED FOODS (f_h AND f_c VALUES)

In canning operations, heat penetration experiments are performed by positioning thermocouples at the slowest heating point in the container, and

Sec. 10.8] Heat penetration into canned foods (f_h and f_c values)

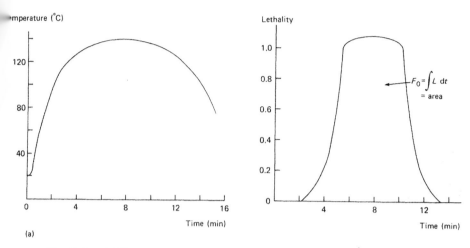

Fig. 10.8 — (a) Heat penetration into a food material; (b) corresponding lethality time relationships for F_0 evaluation.

Table 10.2 — F_0 values that have been successfully used commercially for products in the UK market.

Product	F_0	Product	F_0
Beans in tomato sauce	4–6	Milk puddings	4–10
Carrots	3–4	Herrings in tomato sauce	6–8
Meats in gravy	12–15	Pet food	15–18

Adapted from the data of Brennan et al. (1976).

the temperature monitored during the retorting process. For most products, during heating there is a straight-line relationship, between the logarithm of the approach temperature $\theta_s - \theta$ and time, where θ is the temperature and θ_s is the retort temperature.

The gradient of this straight line has been used to characterize the heat transfer rate for that product. One of the simplest ways of presenting these data is to turn the semilogarithmic plot upside down and to plot $\theta_s - \theta$ against time directly. Fig. 10.9 shows data presented in this way for baked beans in tomato sauce in A1 cans. The retort temperature was 238 °F (114.4 °C), and the filling temperature 72 °F (22 °C); the cans were not agitated. The approach temperature is plotted in both degrees Fahrenheit and degrees Celsius.

The f_h value for the product is defined as the time taken for the approach temperature to pass through one logarithmic cycle. Therefore, as f_h decreases, heat penetration rates increase; for this product it equals 34.5 min.

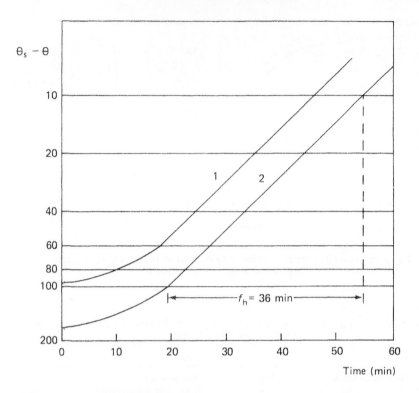

Fig. 10.9 — Heat penetration into a can of beans in tomato sauce: line 1, temperature difference in degrees Celsius; line 2, temperature difference in degrees Fahrenheit.

For conduction packs the f_h value is related to the thermal diffusivity of the product and the can size and can be estimated as follows:

$$f_h = \frac{6.63 \times 10^{-7}}{(1/r_c^2 + 1.708/h_c^2)/a} \text{ (min)}$$

where a (m² s⁻¹) is the thermal diffusivity, r_c is the can radius and h_c (cm) is the can height. Conduction packs are normally characterized by high f_h values of 50–200 min (Jackson and Lamb, 1981).

Convection packs have much lower f_h values than conduction packs (5–20 min). There is no simple method relating the f_h value to the physical properties, such as viscosity. However, for low-viscosity fluids, where convection is sufficient to give no internal temperature gradients, the f_h value is inversely proportional to the surface-area-to-volume ratio. This provides a useful method for estimating f_h values in cans 1 and 2 of different sizes using data from one can size. Thus,

$$\frac{(f_h)_1}{(f_h)_2} = \frac{r_1 h_1 (r_2 + h_2)}{(r_1 + h_1)(r_2 h_2)}$$

Similar data can be obtained during the cooling cycle, the approach temperature in this case being $\theta - \theta_c$ where θ_c is the coolant temperature. A cooling characteristic f_c is obtained in a similar fashion to f_h.

These heating and cooling characteristics can be used in two ways: firstly, to evaluate the F_0 value for a selected processing time and, secondly, to determine the processing time required to achieve a particular F_0 value. Both procedures are based on a mathematical method developed by Ball. The procedure has been described by Stumbo (1973), Hersom and Hulland (1980) and Jackson and Lamb (1981).

More recently, Stumbo et al. (1983a, 1983b) have produced detailed tables of processing times required to achieve a safe product and a commercially sterile product, for conduction and convection packs. A safe product is defined on the basis of finding one surviving *Clostridium botulinum* spore in every 10^{12} cans, whilst a commercially sterile product is based on finding one surviving spore of *Clostridium sporogenes* in every 10^4 cans.

Factors which are taken into account include filling temperature, can size, retort temperature and the Z value of the major spoilage organism (normally, $Z = 10$ °C). The f_h value of the product must be known before the tables can be used.

10.9 FREEZING AND THAWING TIMES

If a slab of food is being frozen, the surface of the food is subjected to a low

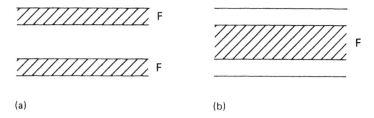

Fig. 10.10 — (a) Freezing and (b) thawing times; F = frozen zone.

temperature and heat is removed from the food to the cold source. The food will start to freeze at the surface. Fig. 10.10(a) shows the situation part of the way through the freezing process as the frozen zone advances into the food and energy is being removed through the frozen layer. When a food which is partially frozen is dissected, such a sharp interface between the two layers is observed. This provides an alternative method for measuring freezing rates, namely the speed of advancement of the frozen layer into the food. In contrast, thawing or defrosting a frozen food is an intrinsically slower process than freezing. The surface of the food is subjected to the thawing medium and melting proceeds from the surface (Fig. 10.10(b)). During this process, heat is transferred through the melted (fresh) food, which has a much lower thermal conductivity than the frozen food. In addition the temperature of the thawing medium is limited as the food surface will be close to this temperature for the duration of the process. This may cause microbiological or other problems when thawing times are long. This also puts a limit on the temperature driving force available.

The rate of heat transfer during freezing and thawing foods of different geometries was first considered by Planck, who derived the following equation for predicting the freezing time (Planck's equation):

$$t = \frac{L_f \rho}{\Delta \theta} \left(P \frac{a}{h} + R \frac{a^2}{k} \right)$$

where L_f (kJ kg^{-1}) is the latent heat of fusion for food, ρ (kg m^{-3}) is the food density, $\Delta\theta$ is the freezing point of the food minus the temperature of the coolant, a is the smallest dimension of the food (e.g. thickness), h is the heat film coefficient, k is the thermal conductivity of the frozen food, and P and R are shape factors given in Table 10.3.

Table 10.3 — The values of the shape factors P and R.

Shape	P	R
Infinite cylinder	$\frac{1}{4}$	$\frac{1}{16}$
Infinite slab	$\frac{1}{2}$	$\frac{1}{8}$
Sphere or cube	$\frac{1}{6}$	$\frac{1}{24}$

Fig. 10.11 shows the values of P and R for various values of β_1 and β_2, where β_1 and β_2 are the ratios of largest and second-largest sides to the shortest. Thus a pack measuring 20 cm × 14 cm × 2 cm would have β_1 and β_2 values of 10 and 7; from the chart the respective P and R values are 0.40 and 0.108. The P and R values for bricks range between those for an infinite slab and a cube.

The following were amongst the most important assumptions made by Planck.

(1) Foods have a sharp freezing point.
(2) Heat transfer from the freezing front in the food to the cooling medium is in steady state, i.e. freezing takes place relatively slowly.
(3) Freezing starts with all the food at the freezing temperature; the pre-freezing and tempering stages are not accounted for.

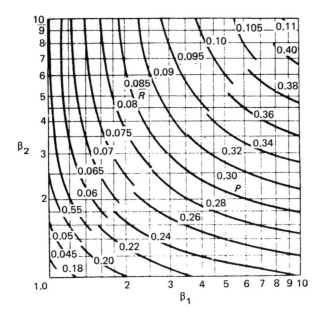

Fig. 10.11 — The constants P and R as a function of brick size for use in Planck's equation. (From Ede (1949), with permission.)

(4) The freezing medium remains at a constant temperature, and the thermal conductivity and density of the food are independent of temperature.

If packaging material is used, the heat film coefficient, h_p is modified to

$$\frac{1}{h_p} = \frac{1}{h} + \frac{L}{k}$$

where L is the thickness of the package, k is the thermal conductivity of the package and h is the film heat coefficient.

In order to account for the cooling of food before and after freezing, the term for the latent heat can be substituted by a term giving the total enthalpy change involved (see section 8.9).

Because of the nature of the assumptions made, calculated freezing times are not always in agreement with experimentally determined times. Nevertheless, the equation is useful because it shows how freezing time depends upon product-related properties such as density, thermal conductivity, shape and enthalpy change, where scope for change is limited, and upon process-related variables, such as type of packaging, refrigeration temperature and type of refrigeration system, where there may be more scope for change.

As an example, consider a tray of fish fillets, measuring 1 m × 80 cm ×

2.5 cm which is frozen in an air blast freezer using air at -30 °C. The initial freezing point of the fish is -2 °C, and the other relevant properties are as follows: thermal conductivity of frozen fish, 1.69 W m^{-1} K^{-1}; heat film coefficient for air, 20 W m^{-2} K^{-1}; density of fish, 980 kg m^{-3}; latent heat, 260 kJ kg^{-1}. β_1 and β_2 are calculated as 100/2.5 and 80/2.5, respectively. Therefore, this shape is approximately that of an infinite slab, i.e. $P = 0.5$ and $R = 0.125$.

Substituting these values in Planck's equation gives

$$t = \frac{L_f \rho}{\Delta \theta} \left(P \frac{a}{h} + R \frac{a^2}{k} \right)$$

$$= \frac{260 \times 10^3 \times 980}{28} \left(\frac{0.5 \times 2.5 \times 10^{-2}}{20} + \frac{0.125 \times (2.5 \times 10^{-2})^2}{1.69} \right)$$

$$= 6045 \; s \; or \; 1.68 \; h$$

Thawing times can be evaluated by using the thermal conductivity of the fresh food ($k = 0.56$ for fresh fish). Sensible heat changes can be accounted for by substituting the total enthalpy change from start to finish (see section 8.9).

The accuracy of all these calculations depends upon the accuracy of the physical property data and correlations for predicting heat film coefficients. Perhaps one of the major problems of all these correlations is that they do not account for changes in physical properties as the temperature changes.

Methods for evaluation of freezing time have been described by Jackson and Lamb (1981), Milson and Kirk (1980) and Singh and Heldman (1984). More recently, Cleland and Earle (1982) have developed a simple method based on Planck's equation that is capable of predicting freezing times for regular and some irregular objects. It takes into account the cooling period to the freezing point, the tempering period and the fact that heat is transferred through only partially frozen material. They claim that their freezing times are within 10% of published experimental results.

10.10 REFRIGERATION METHODS

Refrigeration involves the production of temperatures below the ambient temperature, the main areas of interest being freezing and chilling. Chilling extends from just below the ambient temperature down to -1 °C and only sensible heat changes are involved. Freezing commences at -1 °C and involves the conversion of water to ice, with the removal of the latent heat associated with this phase change. Virtually all microbial activity ceases below -10 °C. Most domestic food freezers opeate at -18 °C, whereas some commercial cold-storage units are much lower (between -25 °C and -30 °C). Some chemical and enzymatic reactions may still take place below -18 °C, albeit at a very low rate.

It is very important to ensure that chilled and frozen foods are maintained at their optimum storage temperature at all points between their production and point of sale (and preferably afterwards). The distribution network for refrigerated products is known as the cold chain.

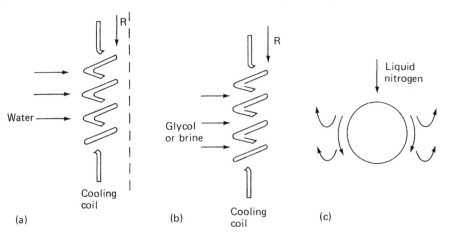

Fig. 10.12 — Some methods for chilling and freezing foods: (a) chilling of water; (b) chilling of brine, glycol or air for freezing purposes; (c) direct contact of liquid nitrogen or Freon with foods: R, refrigerant.

Freezing and chilling processes can be divided into direct and indirect processes in a similar fashion to heating processes. However, the distinction is not so clearcut, because the heat transfer fluid is not always the refrigerant. Fig. 10.12 shows the distinction between them. For example, cold air or chilled brine is usually cooled by a refrigerant before being used to freeze the food. Here, direct processes refer to situations in which the heat transfer fluid or the refrigerant is in actual contact with the food. The use of refrigerants for obtaining low temperatures in a vapour compression refrigeration cycle is covered in section 11.9. Some methods for freezing foods will now be described.

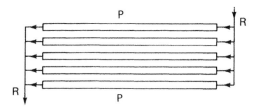

Fig. 10.13 — Plate freezer: P, plates; R, refrigerant.

10.11 PLATE FREEZERS

Food such as fish, meat and vegetables is arranged in trays to a maximum thickness of 4–5 cm. The trays are loaded between metal plates (usually aluminium alloy) which are hollow and through which passes an evaporating refrigerant at a temperature between − 30 °C and − 40 °C (Fig. 10.13). The flow rate of the refrigerant is several times in excess of that corresponding to the refrigeration duty. It is essential to get good contact between the food and the plates to improve conductive heat transfer; to ensure this, a slight pressure is applied. Freezing times range between 30 min and 60 min and are affected by the thickness of the product and the thermal properties of the food (see calculations for flat slabs).

At the end of the process the trays are removed and the products are 'knocked out' as a slab. This is then quickly broken into pieces or cut to the desired shape. For example, fish is sawed into fish-shaped pieces or fish fingers. In some cases, such products are not as attractive as 'individually quick frozen' foods, particularly for vegetables, e.g. peas and beans. Such freezers are labour intensive but are also extremely useful for freezing food in thin regularly shaped packages, such as hamburgers and pre-cooked meals, or for hardening ice-cream blocks. Note that the packaging material will offer an additional resistance to heat transfer.

10.12 COLD-AIR FREEZING

Air is reduced in temperature by passing it over coils containing evaporating refrigerant. In a deep-freeze cabinet, the air is stagnant, resulting in very low heat film coefficients. In a blast freeze the air is circulated around the freezing compartment using a fan, thereby increasing the heat film coefficient. Energy costs are slightly higher and the heat energy produced by fan also needs to be removed. Factors affecting the freezing time are the air velocity, the air temperature and the thermal properties and dimensions of the foods. Freezing times can range from several minutes for small samples to several days for large carcases of meat. One problem that may occur is surface dehydration or freezer burn. It is most likely to occur when the food is relatively warm as there will be a large water vapour pressure gradient between the surface of product and the bulk air. This can be reduced by ensuring that the cold air is saturated and by more rapid freezing methods. Heat film coefficients are low but can be increased by increasing the air velocity at the expense of increased power consumption which is proportional to the velocity raised to the power 3. These systems can be either batch or continuous.

Freezing times can be further reduced for particulate matter by fluidizing the particles using higher air velocities. This is suitable for peas, sliced beans, diced vegetables and foods of similar dimensions. The energy requirements are much higher, but the products are 'individually quick frozen'. Obviously there is a compromise between rapid freezing times which result in a greater throughput and improved quality but higher energy costs. Freezing times range between 1 min and 10 min and depend mainly on the size of the product. Fluidization is covered in more detail in section 3.12.

10.13 IMMERSION FREEZING

Cold liquids can be used for freezing rather than air, resulting in higher heat film coefficients. The lowest temperature attainable with chilled water is about 0 °C. It is suitable for chilling and is so used in batch and continuous heat exchangers. For freezing applications, the temperature is reduced by adding substances such as inorganic salts, sugars or ethylene glycol. Fig. 10.14 shows how the freezing point of a sodium chloride solution is affected by salt concentration. The presence of salt reduces the freezing point (line

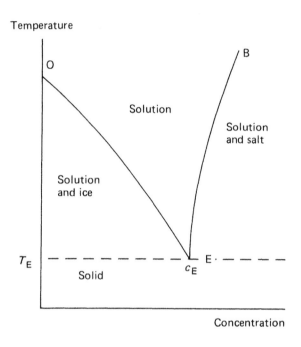

Fig. 10.14 — Eutectic diagram for sodium chloride and water: T_E, eutectic temperature; C_E, eutectic concentration.

OE); as freezing proceeds, ice separates out and is in equilibrium with the solution. As the concentration increases, a point is reached where the entire mass freezes (salt and water). This is known as the eutectic concentration and is defined as the strongest solution which will freeze without salt separating out. Fig. 10.14 represents the equilibrium conditions between the liquid and solid phases present in this mixture. Below the eutectic tempeature, no liquid will be present. Table 10.4 gives freezing point and eutectic data for some commonly used eutectics. Temperatures of -23.8 °C and -51.2 °C, respectively, can be achieved by using 40% and 60% ethylene glycol solutions. These solutions can be chilled and used with either batch or continuous heat exchangers, e.g. ice-cream freezers. In certain cases,

Table 10.4 — Freezing point and eutectic data for some inorganic salts.

Parameter	Concentration (w/w)	Sodium chloride	Calcium chloride	Magnesium chloride
Freezing point (°C)	4	−2.4	−1.8	−2.3
Freezing point (°C)	8	−5.1	−4.3	
Freezing point (°C)	12	−8.2	−7.7	−10.5
Freezing point (°C)	16	−11.9	−12.3	−17.6
Freezing point (°C)	20	−16.5	−18.3	−27.4
Freezing point (°C)	24		−25.3	—
Freezing point (°C)	28		−34.7	—
Eutectic temperature (°C)		−21.2	−51	−32.8
Eutectic composition		22.2	30	22

See Table 2.13.

products can be immersed in the solution, e.g. fish in a sodium chloride solution, but usually the product would need to be packaged before immersion, e.g. frozen chicken in a plastic bag. Eutectic bricks are used in insulated cool-boxes. They are taken from the freezer, in their frozen form, when required and placed in the box with the food. Strong sugar solutions can also be used for immersion freezing, but they may be very viscous at low temperatures. The eutectic temperature and concentration for sucrose in water are −13.95 °C and 63.6%, respectively. The eutectic point can also be regarded as the intersection of the freezing point curve OE and the solubility curve EB. Further details, including information on corrosive properties, have been given by the American Society of Heating, Refrigerating and Air-Conditioning Engineers (1981).

The freezing behaviour of foods is complicated because many solutes are present, and the eutectic temperature of most foods is below −30 °C. Sugar–water systems have been used for modelling fruit juices. Eutectic data for liquid foods would be extremely useful for understanding and predicting their behaviour during freeze concentration and freeze drying. The eutectic composition would give a good indication of the maximum concentration achievable by freeze concentration.

10.14 CRYOGENIC FREEZING

A cryogenic freezant usually refers to a fluid whose boiling point is well below the normal freezing point of the food. When the food is immersed in the fluid, evaporation of the freezant takes place on the surface of the food, thereby improving the surface heat film coefficient.

Liquid nitrogen has a boiling point of −196 °C at atmospheric pressure. When a food is immersed in liquid nitrogen, heat is rapidly transferred from the surface of the food to the liquid nitrogen and the liquid evaporates at the surface. Consequently the surface heat film coefficient is high and this, combined with a large temperature driving force, makes it a rapid method for freezing foods. However, as the size of the food increases, conduction within the food becomes the controlling mechanism and many of the advantages incurred by liquid nitrogen are lost. As liquid nitrogen is expensive to date, it is used commercially mainly with small high-quality delicate foods where the advantages of the high quality produced by rapid freezing outweigh the higher costs incurred by the process, e.g. shrimps, prawns, raspberries and strawberries. The use of liquid nitrogen becomes prohibitively high above a particle size of about 3.5 cm. In practice the food can be frozen by passing the food through the liquid on a conveyor belt although this often results in excessive thermal stress within the food, the end result being a cracked or broken product.

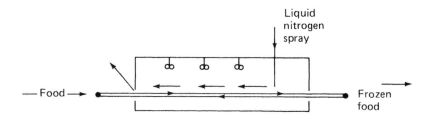

Fig. 10.15 — Liquid nitrogen tunnel freezer.

Fig. 10.15 shows the layout of a liquid nitrogen tunnel, where nitrogen is sprayed onto the food, thereby reducing thermal shock because of the short contact time with the very cold liquid. To make the best use of the energy, the spray is towards the end of the tunnel. The liquid evaporates on the surface of the food, giving up its latent heat (199.8 kJ kg^{-1}). Fans within the tunnel force the very cold nitrogen in a counter-current direction to the food, thereby pre-cooling the food before it meets the spray and utilizing the sensible heat in the cold gas (approximately 204 kJ kg^{-1} from -196 °C to 0 °C). The heat film coefficient on the surface of the product in the spray region is of the order of 1420 W m^{-2} K^{-1} whereas in the pre-cooling section it is between 17 W m^{-2} K^{-1} and 140 W m^{-2} K^{-1}.

Compared with other methods of freezing, the capital costs are low but the energy costs are high. Some properties of nitrogen are given in Table 10.5; the nitrogen consumptions when freezing foods from 20 °C to -18 °C are given in Table 10.6. In contrast with liquid nitrogen, Freon 12 (CCl_2F_2) has a boiling point of -29.8 °C at atmospheric pressure. It is debatable whether this is a true cryogenic fluid although it has been suggested that surface heat transfer coefficients are slightly higher than for liquid nitrogen freezing. This is now used for direct contact techniques. Surface heat film coefficients are high, temperature stress is reduced, and it is possible to liquefy and recover most of the Freon 12 by condensing it at a lower temperature, i.e. -40 °C. A conventional refrigeration unit is required for this. Some relevant properties of Freon 12 are given in Tables 10.5 and 11.1.

Table 10.5 — Some properties of liquid nitrogen and Freon 12.

Property	Nitrogen	Freon 12
Normal boiling point (°C)	−195.8	−29.8
Critical temperature (°C)	−150	112.2
Liquid density at normal boiling point (kg m^{-3})	809.7	1491
Gas density normal boiling point (kg m^{-3})	11.6	6.33
Specific heat of gas at 21°C (kJ kg^{-1} K^{-1})	1.04	0.63
Latent heat of vaporization (kJ kg^{-1})	199.8	165.1

Table 10.6 — Values of nitogen consumption for some products.

Product	Consumption of liquid nitrogen (kg kg^{-1})
Peas	1.01
Lean beef	0.97
Lean fish	1.19
Strawberries	1.19

Cryogenic fluids used for direct contact refrigeration must also be non-toxic and not leach any of the food components. The solubility of water in Freon 12 is 0.028% (w/w), but the solubility of oil can be as high as 5%.

Solid carbon dioxide (Cardice) is widely used for transporting frozen foods. Cardice sublimes and considerable pressure will build up, unless provision is made to allow the vapour to escape. Cardice can be deposited onto the surface of foods for crust freezing. Liquid (supercritical?) carbon dioxide is now being investigated in solvent extraction processes.

10.15 VACUUM COOLING AND FREEZING

If a food is placed in a vacuum chamber and the pressure is reduced, a point will be reached when the chamber pressure reaches the saturated water vapour pressure of the food material. At this point the water will evaporate from the surface of the food; this requires energy which is supplied by the food and will result in a localized cooling effect. If the pressure is further reduced, evaporation proceeds and further cooling results. An estimate of the amount of water removed can be obtained from the equation

heat lost by food = heat required to evaporate water

Therefore,

mass food × specific heat × temperature fall = mass evaporated × latent heat

If the pressure is reduced below about 4.6 Torr, the food will reach 0 °C and start to freeze. The latent heat term can be added to the right-hand side of the equation.

For most foods a temperature fall involving sensible heat changes results in only a small loss of water. Much larger quantities are lost as the food starts to freeze.

This method has found use in the quick cooling of horticultural products taken straight from the field. Evaporative freezing is a very useful preliminary step for freeze drying, as between 20% and 25% of the water is lost during the freezing process. It is an extremely useful method for freezing foods prior to the sublimation process in freeze drying. It is not suitable for liquids or soft fruit and vegetables.

For most foods a temperature drop of about 20 °C results in the removal of between 2% and 4% of the water. Much larger amounts are removed in order to produce a frozen food (approximately 20%). Water can be sprayed onto the food beforehand in order to assist the cooling process.

The extent of cooling or freezing can be controlled by regulating the vacuum conditions. For example, if a product temperature of 5 °C is required, the pressure is adjusted to a value corresponding to the saturated water vapour pressure at that temperature, thereby ensuring that the temperature will not fall below that value. Further details have been given by The American Society of Heating, Refrigerating and Air-Conditioning Engineers (1982).

10.16 CHILLING

Chilling involves reducing the temperature to below the ambient temperature, but above -1 °C. Microbial reaction rates are retarded, and it is often used in combination with pasteurization for extending the shelf-life of 'fresh' products. There has been a considerable increase in the sale of chilled foods in retail outlets over the last few years.

Chilling can be achieved by a variety of methods, and the unsteady-state heat transfer methods, developed earlier, can be used for estimating chilling times.

An International Institute of Refrigeration (1979) publication has given recommendations for chilled storage of perishable produce, namely fruit and vegetables, meat, poultry, eggs, fishery products, dairy products, cut flowers and seeds. Some of the typical data and correct storage conditions for a wide variety of products are given in the text together with product disorders that may occur when products are incorrectly stored. A Ministry of

Agriculture, Fishery and Food (1979) publication lists similar data for refrigerated storage of fruit and vegetables.

Recent guidelines for the handling of chilled foods have been produced by the Institute of Food Science and Technology (1982). Most meat, fish and dairy products are stored as near to 0 °C as possible. Factors affecting the quality of the final product and product defects due to incorrect chilling conditions, such as cold shortening and soft–pale exudate have been discussed by the International Institute of Refrigeration (1979).

10.17 CONTROLLED-ATMOSPHERE STORAGE AND HEAT OF RESPIRATION

Many horticultural products are stored under controlled-atmosphere storage conditions, which is a method for controlling ripening. Such products are often picked before they are ripe and they continue to respire or ripen during storage. During this process, starch is broken down to sugars and carbon dioxide, and heat is evolved. Thus, ripening is controlled by a combination of a low temperature and a controlled atmosphere. Although most temperate produce is best stored as near to -1 °C as possible, some damage may occur with some tropical fruit, e.g. bananas and pineapple. The correct conditions for a wide variety of products are detailed in Table 10.7.

The two most commonly encountered controlled-atmosphere conditions are when the sum of the concentrations of oxygen and carbon dioxide equals 21%, as in air or the sum of the two concentrations is much reduced (4–6%) (see Table 10.7).

Two extremes are as follows: (1) when oxygen is very high and the carbon dioxide is very low and (2) when carbon dioxide is added in a high concentration to air. A reduction in oxygen concentration slows down respiration but, if taken too far, products such as ethyl alcohol or acetaldehyde start to form. An increase in the carbon dioxide concentration reinforces this effect and slows down the ripening process.

Such rooms need to be as air tight as possible and need special equipment to change carbon dioxide and oxygen concentrations. In addition, on a smaller scale, controlled atmospheres can be achieved in wholesale and retail packages by methods such as gas flushing and vacuum packaging.

The heats of respiration for some fruits and vegetables are given in Table 10.8. These need to be accounted for when evaluating the refrigeration loading. Heats of respiration increase as the storage temperature increases.

Further data have been given by Polley *et al.* (1980), Mohsenin (1980) and the American Society of Heating, Refrigerating and Air Conditioning Engineers (1981). The latter publication also gives information on how the respiration rate changes with storage time for selected commodities.

10.8 SYMBOLS

a   smallest dimension (Planck's equation, section 10.7)
a thermal diffusivity

Table 10.7 — Recommended storage conditions for chilled products.

Product	Temperature (°C)	Relative humidity (%)	Storage time	Packaging material	Composition of the atmosphere[a] (%)
Apricot	−0.5	90	1–3 weeks	—	—
Banana (Grosvenor)	12	85–90	10–20 days	—	—
Cabbage (white)	0	95	6–7 months	—	—
Beef (retail cuts)	4	—	14 days	Vacuum packed	—
Liquid egg (past)	0–4	—	4–7 days	Tight packing	—
Chicken (eviscerated)	4	—	7 days	Permeable plastic	—
Cod (gutted)	0	—	11–12 days	—	—
Granny Smith apples	−1	95	6–7 months	Sealed polythene bags	2–3 CO_2 2–3 O_2
Conference pears	0	95	5–7 months		2–3 CO_2 2–3 O_2

[a] From the data of the International Institute of Refrigeration (1979).

Table 10.8 — Heats of respiration and highest freezing point for some foods.

Food	Highest freezing point (°C)	Heat (W t^{-1}) of respiration at the following temperatures	
		0 °C	10 °C
Apple	−1.1	10–12	41–61
Pears	−1.6	8–20	23–63
Potatoes	−1.3	—	20–30
Banana	−0.8	—	65–116
Celery	−0.5	21	58–81
Oranges	−0.8	9	35–40
Strawberries	−0.8	36–52	145–280
Carrots	−1.4	46	93

1 W t^{-1} = kJ day^{-1} t^{-1}, where t is the abbreviation for tonne.
Adapted from the data of the International Institute of Refrigeration.

A surface area
c specific heat
D_T decimal reduction time
E activation energy
F_0 total integrated lethal effect
f_c time for the approach temperature to pass through one logarithm cycle (cooling)
f_h time for the approach temperature to pass through one logarithm cycle (heating)
h heat film coefficient
h height
h_c can height
h_p modified heat film coefficient
k reaction velocity constant (section 10.4)
k thermal conductivity
L slab thickness
L lethality (section 10.6)
L_f latent heat of fusion
m inverse Biot number (Fig. 10.4, Fig. 10.5 and Fig. 10.6)
m mass
n an integer (section 10.3)
n x/δ (section 10.4)
N number of spores after time t
N_0 initial number of spores
dQ/dt rate of heat transfer

r_c	can radius
t	time
T	temperature
U	overall heat transfer coefficient
x	distance
Z	heat resistance parameter for spores

10.18.1 Greek symbols
δ	distance
θ	temperature
ρ	density

10.18.2 Dimensionless groups
Bi	Biot number
Fo	Fourier number

11
Properties of gases and vapours

11.1 INTRODUCTION

In many food-processing operations, it is necessary to understand the behaviour and properties of gases and vapours. In drying operations, hot air is used for removing water; heat is lost from the air whilst its moisture content increases. Similarly, during the storage of foods, there may well be an exchange of moisture between the food and the surrounding air. During heating and cooling of foods in cans or bottles, or in the production of carbonated beverages, the pressure developed in the head-space can be considerable and needs to be accounted for. Saturated and superheated vapours, e.g. steam, ammonia and Freon, are used extensively for heating and cooling processes, whilst carbon dioxide and nitrogen are both used for direct refrigeration processes.

11.2 GENERAL PROPERTIES OF GASES AND VAPOURS

It is not always easy to distinguish between gases and vapours; when vapours are superheated, there is no need to differentiate. Some of the properties which gases and vapours have in common is that they all expand rapidly, diffuse and mix well. Most of them are also transparent.

The gases and vapours present in a mixture exert pressures independently of each other. This behaviour is summarized by Dalton's law which states that

$$\left\{\begin{array}{l}\text{total pressure}\\\text{of a mixture}\end{array}\right\} = \sum \left\{\begin{array}{l}\text{partial pressure of components}\\\text{of that mixture}\end{array}\right\}$$

The partial pressure exerted by a gas in a mixture is the pressure exerted if that component alone occupied the volume of the mixture. In normal air, the total air pressure is made up of

$$\left\{\begin{array}{l}\text{total}\\\text{pressure}\end{array}\right\} = \left\{\begin{array}{l}\text{partial pressure}\\\text{of air}\end{array}\right\} + \left\{\begin{array}{l}\text{partial pressure}\\\text{of water vapour}\end{array}\right\}$$

Sec. 11.2] General properties of gases and vapours

and the partial pressure of air itself is made up of the sum of the partial pressures of its constituents, i.e. oxygen, nitrogen, carbon dioxide and the inert gases.

Avogadro stated that equal volumes of different gases, at the same temperature and pressure, contain the same number of molecules. One kilomole (1kmol) of any gas (i.e. the molecular weight expressed in kilograms) contains 6.02×10^{26} molecules and occupies a volume of $22.4 \, m^3$ at standard temperature and pressure.

For moderate temperature and pressure changes, many gases obey the ideal gas equation, which can be derived from the kinetic theory of gases. The derivation assumes that the molecules behave like elastic spheres, that they occupy a negligible volume, and that there are no forces of attraction between them..

The ideal gas equation can be written as

$$pV = RT$$

where p (N m^{-2}) is the pressure exerted by a gas, V is the molecular volume, i.e. the volume occupied by 1 kmol of gas, T (K) is the absolute temperature and R (= 8.318 kJ kmol^{-1} K^{-1}) is the ideal gas constant. This equation can be used to determine the pressure exerted by a gas under a specific set of conditions. For example, to calculate the pressure exerted by 10 kg of air (molecular weight, 29) occupying a volume of 15 m^3 at 27 °C, the molar volume V is calculated from

$$V = \frac{\text{volume}}{\text{number of kilomoles}}$$

$$= \frac{15 \, m^3}{10/29}$$

$$= 43.5 \, m^3 \, kmol^{-1}$$

Using $pV = RT$

p N m$^{-2} \times 43.5$ m^3 kmol^{-1} = 8.318×10^3 N m kmol$^{-1} \times 300$ K

Therefore,

$$p = 57.4 \times 10^3 \, N \, m^{-2} \text{ or } 57.4 \, kN \, m^{-2}$$

An alternative way of expressing and using this equation is in the form

$$\frac{p_1 V_1}{T_1} = \frac{p_2 V_2}{T_2}$$

This form of the ideal gas equation is extremely useful when a gas at one set of conditions is changed to a different set of conditions.

Many of the situations encountered in food-processing operations involve only relatively moderate temperatures and pressures or changes in these variables and the behaviour can be described by the ideal gas equation.

However, at very high temperatures and pressures, the behaviours of many gases deviate from the ideal. The molecules are packed so closely together that forces of attraction exist, and the molecules no longer occupy a negligible volume compared with the total volume of the mixture. Many equations have been proposed to account for such deviations; one of the earliest was van der Waals equation:

$$\left(p + \frac{a}{V^2}\right)(V - b) = RT$$

where a and b are constants which take into account the forces of attraction and the volume occupied by the molecules, respectively. Other equations such as the compressibility equation and the viral equations have been discussed in more detail by Perry and Chilton (1973).

11.2.1 Distinction between gases and vapours

The distinction between gases and vapours is not at all clear although by convention the term vapour refers to the gaseous state of substances that are normally liquid at room temperature, e.g. water, alcohol and most organic solvents.

In the middle of the nineteenth century, Andrews did some experiments on the compression of carbon dioxide at different temperatures and made some interesting observations. When carbon dioxide was compressed under isothermal conditions at a temperature of 50 °C, no pressure was high enough to cause the carbon dioxide to liquefy. He repeated the experiment at lower temperatures and found that, when the temperature was reduced below 31.1 °C, the carbon dioxide was liquefied; as the temperature was further reduced, the carbon dioxide was liquefied more easily. This temperature of 31.1 °C was termed the critical temperature for carbon dioxide. The critical temperature of a substance is defined as the temperature above which it is impossible to cause liquefaction by pressurizing that substance.

Gases such as air, hydrogen and nitrogen were found to have extremely low critical temperatures. Therefore, they were difficult to liquefy and were known as the permanent gases. The critical temperatures for various gases and liquids are as follows: carbon dioxide, 31.1 °C; water, 374.2 °C; oxygen, −118.8 °C; hydrogen −239.8 °C; Freon 12, 112.0 °C; ammonia, 132.4 °C. Therefore, substances with high critical temperatures, e.g. greater than 50 °C, are usually referred to as vapours, whilst those with critical temperatures below 50 °C are referred to as gases. However, the distinction is arbitrary.

To summarize, if a substance such as water or carbon dioxide is above its critical temperature, it can only exist in the gaseous state. When it is reduced below its critical temperature, it can exist as a vapour, liquid, solid or various combinations of these in equilibrium depending upon the temperature and

Sec. 11.3] Properties of saturated vapours 327

pressure. This information can be summarized in terms of the phase diagrams for carbon dioxide and water (see Fig. 8.3).

11.3 PROPERTIES OF SATURATED VAPOURS

If a closed container held at a constant temperature is completely evacuated

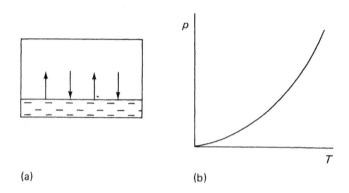

Fig. 11.1 — (a) Saturated vapour pressure of water in a sealed container; (b) relationship between saturated vapour pressure and temperature.

and then filled with water or some other liquid (Fig. 11.1), molecules from the liquid will evaporate and the pressure of the vapour will increase. Eventually the pressure exerted by the vapour reaches an equilibrium value at which the rate of evaporation equals the rate of condensation. Under these conditions, this pressure is termed the *saturated vapour pressure* of the liquid and the space is described as being saturated with the vapour of that liquid.

The saturated vapour of a liquid is independent of the volume and depends only upon the temperature. If the vessel had not been originally evacuated, the saturated water vapour pressure (at a constant temperature) would be the same. In this case the total pressure would be equal to the partial pressure of air plus the saturated vapour pressure. When the saturated vapour pressure is determined at different temperatures, it is found that the saturated vapour pressure increases as the temperature increases, usually in a non-linear fashion.

Many empirical equations hve been proposed to predict how the saturated vapour pressure changes with temperature. The best known of these is the Clausius–Clapeyron relationship:

$$\ln p_s = \frac{h_{fg}}{RT} + C$$

where p_s is the saturated vapour pressure, h_{fg} is the latent heat of vaporization, R is the gas constant and C is a constant. A substance that obeys this relationship will give a straight-line relationship when $\log p_s$ is plotted against $1/T$, the gradient being equal to $h_{fg}/2.303R$.

The saturated vapour pressure–temperature relationship is given for several substances in Table 11.1 and for water in Table 11.2.

Table 11.1 — Saturated vapour pressure data for different refrigerants.

Temperature (°C)	Saturated vapour pressure (kgf cm^{-2}) of following refrigerants			
	Freon 12	Ammonia	Carbon dioxide	Sulphur dioxide
−40	0.66	0.73	10.25	0.22
−30	1.02	1.22	14.55	0.39
−20	1.54	1.94	20.06	0.65
−10	2.23	2.96	26.99	1.03
0	3.15	4.38	35.54	1.59
10	4.31	6.27	45.95	2.35
20	5.78	8.74	58.46	3.37
30	7.58	11.90	73.34	4.71
40	9.70	15.85		6.43
Latent heat at −20 °C (kcal kg^{-1})	39.06	316.79	68.85	95.17

Table 11.2 — Saturated water vapour pressure p_{wvs} against temperature.

Temperature (°C)	p_{wvs}(Torr)	Temperature (°C)	p_{wvs}(Torr)	Temperature (°C)	p_{wvs}(Torr)
0	4.579	18	15.477	36	44.563
1	4.926	19	16.477	37	47.067
2	5.294	20	17.535	38	49.692
3	5.685	21	18.650	39	52.442
4	6.101	22	19.827	40	55.324
5	6.543	23	21.068	41	58.34
6	7.013	24	22.377	42	61.50
7	7.513	25	23.756	43	64.80
8	8.045	26	25.209	44	68.26
9	8.609	27	26.739	45	71.88
10	9.209	28	28.349	46	75.65
11	9.844	29	30.043	47	79.60
12	10.518	30	31.824	48	83.71
13	11.231	31	33.695	49	88.02
14	11.987	32	35.663	50	92.51
15	12.788	33	37.729	60	149.38
16	13.634	34	39.898	70	233.7
17	14.530	35	42.175	80	355.1

1 Torr = 133.3 Pa.
Adapted from the data of Weast (1982).

A space will be saturated with a vapour at a particular temperature if the partial pressure of the vapour is equal to the saturated vapour pressure. Situations in which air is saturated with water are assumed to exist in the

Sec. 11.4] Properties of saturated water vapour (steam tables)   329

immediate vicinity of water or food or in the head-space in a sealed can or flexible pouch.

Boiling occurs when the saturated vapour pressure of a liquid is equal to the external pressure. Therefore, saturation vapour pressure data are extremely useful for determining the relationship between boiling point and pressure.

The *thermodynamic properties* of saturated vapours and mixtures of vapour and liquid in equilibrium can be summarized in tabular form. Tables for saturated water vapour are referred to as the steam tables.

11.4 PROPERTIES OF SATURATED WATER VAPOUR (STEAM TABLES)

Table 11.3 shows the layout of the steam tables. Normally, for any such set

Table 11.3 — Properties of saturated steam.

Temperature (°C)	Pressure (absolute) (bar)	Specific volume v_g (m^3 kg^{-1})	Specific enthalpy (kJ kg^{-1})			Specific entropy (kJ kg^{-1} K^{-1})		
			h_f	h_{fg}	h_g	s_f	s_{fg}	s_g
0	0.006	206.30	0.0	2501.6	2501.6	0.0	9.16	9.16
2	0.007	179.92	8.4	2496.8	2505.2	0.03	9.07	9.10
10	0.012	106.43	42.0	2477.9	2519.9	0.15	8.75	8.90
20	0.023	57.84	83.9	2454.3	2538.2	0.30	8.37	8.67
30	0.042	32.93	125.7	2430.7	2556.4	0.44	8.02	8.45
40	0.073	19.55	167.5	2406.9	2574.4	0.57	7.69	8.26
50	0.123	12.05	209.3	2382.9	2592.2	0.70	7.37	8.08
60	0.199	7.68	251.1	2358.6	2609.7	0.83	7.10	7.93
70	0.311	5.05	293.0	2334.0	2626.9	0.95	6.80	7.75
80	0.474	3.41	334.9	2308.8	2643.8	1.08	6.54	7.62
90	0.701	2.36	376.9	2283.2	2660.1	1.19	6.29	7.40
100	1.013	1.67	419.1	2256.9	2676.0	1.31	6.05	7.36
110	1.433	1.21	461.3	2230.0	2691.3	1.42	5.82	7.24
120	1.985	0.89	503.7	2202.2	2706.0	1.53	5.60	7.13
130	2.701	0.67	546.3	2173.6	2719.9	1.63	5.39	7.03
140	3.610	0.51	589.1	2144.0	2733.1	1.74	5.19	6.93
150	4.760	0.39	632.1	2113.2	2745.4	1.84	4.99	6.83
175	8.924	0.22	741.1	2030.7	2771.8	2.09	4.53	6.62
200	15.549	0.13	852.4	1938.6	2790.9	2.33	4.10	6.43
225	25.501	0.08	966.9	1834.3	2801.2	2.56	3.68	6.25
250	39.776	0.05	1085.8	1714.7	2800.4	2.79	3.28	6.07
300	85.927	0.02	1345.1	1406.0	2751.0	3.26	2.45	5.71

Adapted from the data of Spalding and Cole (1973).

of tables a base state is chosen, and in this example the enthalpy and entropy of the saturated liquid are equal to 0.0 at a temperature of 0 °C.

The first two columns give the relationship between temperature and saturated vapour pressure; this shows how the boiling point of water changes with pressure. This is useful for predicting the boiling point of water and dilute solutions at different pressures. Many evaporation processes are performed below atmospheric pressure to reduce the boiling point and, at an operating pressure of 0.073 bar (absolute), milk will boil at approximately 40 °C. This will significantly reduce the loss of heat-sensitive components and may well improve the product quality. The rate of heat transfer is also increased.

Generally, steam used for heating purposes is saturated so that there is a fixed relationship between its pressure and temperature. Consequently, pressure gauge readings can be used to give an indication of the temperature and are often incorporated into steam pipes and retorts to augment the temperature readings. Saturated steam at 10 lbf in^{-2} (g) (1.70 bar) will give a temperature of approximately 115 °C and that at 15 lbf in^{-2} (g) (2.05 bar) will give a temperature of 121 °C. These pressures are often used for canning low-acid products.

The presence of air in steam will lower the temperature of the steam, at a fixed pressure. A mixture of saturated steam and air at a pressure of 2.05 bar containing 10 vol.% air would exert a saturated water vapour pressure of 0.9 × 2.05 = 1.85 bar and an air pressure of 0.1 × 2.05 = 0.21 bar. Using the steam tables, this would correspond to a temperature of 118 °C, resulting in a drop of 3 deg C because of the presence of the air. Even a small reduction in temperature can result in considerable under-processing of canned and bottled products. Therefore, it is very important to remove as much air as possible from steam retorts, prior to processing; this procedure is known as venting. In addition to lowering the temperature, the presence of air also reduces the rate of heat transfer because of its excellent insulating properties.

Air is removed from most canned food prior to sealing and processing. Air trapped within the can will expand and may result in excessive pressure during heating and cooling, leading to damaged seams and deformed cans. To some extent the pressure developed within the can is balanced by the external pressure, i.e. the steam pressure during heating, and the use of compressed air during cooling.

Air can be removed by a mechanical vacuum pump, hot filling, steam-flow closing and thermal exhausting. More details of these methods have been given by Hersom and Hulland (1980). The last three methods all rely on generating a high water vapour pressure in the head-space immediately prior to sealing. The following example, using hot filling of the liquid at 90 °C, will illustrate these points. It is assumed that the head-space is saturated with water vapour, that there is no change in volume of the head-space during processing and that air behaves as a perfect gas.

When a can is filled at 90 °C, the saturated vapour pressure of water is 0.701 bar. Therefore, the air pressure, which equals the atmospheric pressure minus the saturated vapour pressure, is 1.013–0.701 = 0.31 bar. When the can contents reach 120 °C, the saturated vapour pressure of water is 1.99 bar. The partial pressure of the air at 120 °C can be evaluated from $p_1/T_1 = p_2/T_2$ ($\triangle V=0$), i.e. it is given by 393 × 0.31/363 = 0.34 bar. Therefore the total pressure within the can is (1.99 + 0.34) = 2.33 bar. Note that the steam pressure outside the can would be approximately 2 bar. The largest difference in pressure may occur when the steam in the retort is condensed at the beginning of the cooling cycle. When the can contents cool down to 20

Sec. 11.4] Properties of saturated water vapour (steam tables) 331

°C, the pressure within the can is evaluated in a similar fashion and equals 0.27 bar, i.e. a vacuum is developed within the can.

For higher-temperature processes, such as UHT processing or for spray drying, steam temperatures in excess of 150 °C may be required. During UHT processing, it is important to maintain a pressure to prevent milk or fruit juice from boiling. So steam tables are useful for indicating the minimum pressure level needed to prevent this from happening.

In the specific volume, enthalpy and entropy columns, the subscripts f and g refer to the saturated liquid and the saturated vapour, respectively. The specific volume v_g column shows how the volume of unit mass of saturated vapour changes with pressure. It should be noted that water vapour occupies a very large volume at low pressures, so that in operations involving the removal of water vapour at low pressures, i.e. vacuum evaporation or freeze drying, substantial volumes need to be handled and the vacuum pumps need to be very large. Assistance is provided by removing as much water vapour by either condensing it or freezing it, the residual vapour removed by either a vacuum pump or steam ejectors.

The difference between the enthalpy h_f of the saturated liquid and the enthalpy h_g of the vapour is given by the column labelled h_{fg} and represents the latent heat value at these operating conditions. It can be seen that the latent heat is not constant but increases as the operating pressure is reduced. Consequently the amount of energy required to evaporate water increases slightly as the vacuum increases. Since most textbooks quote latent heat values for water at atmospheric pressure, considerable error may result from using this value at other pressures.

When a mass m of saturated water vapour at a pressure of 1.99 bar condenses to a saturated liquid, the enthalpy change $h_f - h_g$ is 503.7 − 2706.6 = −2202 kJ kg^{-1} and the total heat evolved is $m \times 2202$ kJ (often the sign convetion is ignored, as it is fairly obvious whether heat is evolved or absorbed). Further cooling of the saturated lquid will result in sensible heat losses.

Entropy changes also occur on evaporation or condensation and the changes associated with these processes are given by the s_{fg} value, where

$$s_{fg} = s_f - s_g$$

Since phase changes take place at a constant pressure, the following relationships apply for such phase changes:

$$s_{fg} = \frac{Q}{T} = \frac{h_{fg}}{T}$$

where T (K) is the absolute temperature. This can easily be verified using the steam tables. At 1.99 bar,

$$s_{fg} = \frac{h_{fg}}{T} = \frac{2202 \text{ (kJ kg}^{-1})}{393 \text{ (K)}} = 5.60 \text{ kJ kg}^{-1} \text{ K}^{-1}$$

In practical terms, enthalpy changes associated with phase changes provide valuable information for evaluating steam requirements and energy costs for different processing operations.

Table 11.4 — Properties of saturated Freon 12 (CCl_2F_2).

Saturation Temperature (°C)	Pressure (bar)	Specific volume v_g (m^3 kg^{-1})	Specific enthalpy (kJ kg^{-1})		Specific entropy (kJ kg^{-1} K^{-1})	
			h_f	h_g	s_f	s_g
−40	0.64	0.242	0	169.6	0	0.727
−30	1.00	0.159	8.9	174.2	0.037	0.717
−20	1.51	0.109	17.8	178.7	0.073	0.709
−10	2.19	0.077	26.9	183.2	0.108	0.702
0	3.08	0.055	36.1	187.5	0.142	0.697
10	4.23	0.041	45.4	191.7	0.175	0.692
20	5.67	0.031	54.9	195.8	0.208	0.689
30	7.45	0.024	64.6	199.6	0.240	0.685
40	9.60	0.018	74.6	203.2	0.272	0.683
50	12.22	0.014	84.9	206.5	0.304	0.680

Freezing point at 1 atm, −155 °C; critical temperature, 112 °C; critical pressure, 41.15 bar.

Table 11.4 shows similar data for Freon 12. Freon 12 boils at a temperature of −29.8 °C at atmospheric pressure. In some tables, h_{fg} and s_{fg} values are not directly quoted, as in Table 11.4.

11.5 WET VAPOURS

So far we have dealt with the properties of saturated liquids and saturated vapours; these represent the two extremes of condition for a substance at its saturation temperature. In many instances, we may well be dealing with mixtures of saturated vapour and saturated liquid. Such a mixture is termed a wet vapour. For example, saturated steam at 1.43 bar at 110 °C would be entirely in the vapour state. If, however, water droplets were present in the steam, the steam would be termed wet and can be regarded as a mixture of saturated liquid and vapour in thermal equilibrium.

Fig. 11.2 shows the temperature–time relationships for water at three different pressures, during cooling. The conditions or 'quality' of a wet vapour is given by the dryness fraction X, where

$$X = \frac{\text{mass of saturated vapour}}{\text{total mass of mixture (i.e. mass of vapour} + \text{liquid)}}$$

Dryness fraction values range from 0 for a saturated liquid to 1.0 for a saturated vapour. Thus, a mixture with a dryness fraction of 0.6 would consist of 60% vapour and 40% liquid on a mass basis.

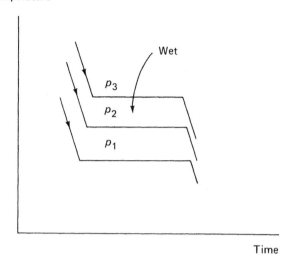

Fig. 11.2 — The relationship between temperature and time whilst cooling wet vapours at different pressures ($p_3 > p_2 > p_1$).

The specific volume, enthalpy and entropy of wet vapours can be determined from the following equations, using values from the steam tables:

$$v_{mix} = Xv_g + (1-X)v_f \text{ or } v_f + Xv_{fg}$$
$$h_{mix} = Xh_g + (1-X)h_f \text{ or } h_f + Xh_{fg}$$
$$s_{mix} = Xs_g + (1-X)s_f \text{ or } s_f + Xs_{fg}$$

Thus the specific enthalpy h_{mix} and specific entropy s_{mix} of this particular mixture at 1.43 bar are $461.3 + 0.6 \times 2230 = 1799 \text{ kJ kg}^{-1}$ and $1.421 + 0.6 \times 5.82 = 4.19 \text{ kJ kg}^{-1} \text{ K}^{-1}$, respectively.

When a wet vapour condenses, the amount of heat liberated is given by the enthalpy change and is equal to

$$h_f - h_{mix} = -Xh_{fg}$$

In this example the amount of heat liberated equals $461.3 - 1799 = 1337.7 \text{ kJ kg}^{-1}$.

It can be seen from the equation that this amount of heat liberated by unit mass of wet steam depends on the dryness fraction and that, as the steam becomes wetter, the amount of heat evolved decreases. Therefore, when using steam for heating purposes, it is a normal practice to remove as many of the entrained water droplets as possible; this is particularly important in direct steam injection processes, to avoid excessive dilution of the product. One example of such a process is the heating of milk by direct steam injection during the UHT process.

The UK heat treatment regulations (Jukes, 1984) specify the use of dry saturated steam, free from dirt and rust. This is injected into the milk which has been pre-heated to about 75 °C, resulting in rapid heating to the UHT temperature (140–150 °C). After a short period in the holding tube (1–10 s), the excess water vapour is removed by a flash cooling process, whereby it is passed into a chamber maintained at reduced pressure and starts to boil. The pressure is selected to give a boiling temperature about 2 degC higher than the temperature of the milk prior to steam injection (see steam tables). The process is summarized in Fig. 9.18. Cooling is almost instantaneous. It takes approximately 12 kg of steam to heat 100 kg of milk in this fashion.

11.6 SUPERHEATED VAPOURS

If a saturated vapour is heated at a constant pressure, its temperature will rise above its saturation temperature and will then be termed superheated. The amount of superheat is the difference between the actual temperature and the corresponding saturation temperature at the same pressure, i.e. the degree of superheat is $T - T_s$.

Water vapour at a temperature $T = 150$ °C and a pressure of 1.43 bar (saturation temperature $T_s = 110$ °C) would possess 40 degC of superheat (72 °F).

The thermodynamic properties of superheated vapours can also be represented in the form of tables. Extracts from the tables for superheated steam are presented in Table 11.5.

The tables are presented for different temperatures and pressures. The saturation temperature corresponding to each pressure is given. The value of the specific volume, the specific enthalpy or the specific entropy at any combination of temperature and pressure can be found by locating the value where the appropriate temperature column intersects the specific volume, specific enthalpy or specific entropy row and reading off the respective value. Thus, steam at 4 bar and 200 °C would have the following properties: the degree of superheat is $200 - 143.6 = 56.4$ degC (K); the specific volume is 0.53 m^3 kg^{-1}; the specific enthalpy is 2860.4 kJ kg^{-1}; the specific entropy is 7.17 kJ kg^{-1} K^{-1}.

A more detailed table can be found in the book by Spalding and Cole (1973). Superheated steam can also be regarded as dry steam, as it can be cooled down to its saturation temperature without condensation taking place. It is used in the Dole canning process for sterilizing cans, prior to aseptic filling. In this situation, high temperatures are required for achieving sterilization without actually wetting the cans.

Superheated steam is not as effective as saturated steam is as a heat

Table 11.5 — Properties of superheated steam.

Parameter	Value of the parameter at the following temperatures				
	100 °C	150 °C	200 °C	250 °C	300 °C
Pressure, 0.04 bar; saturation temperature, 29.0 °C					
specific volume ($m^3 kg^{-1}$)	43.03	48.81	54.58	60.35	66.12
specific enthalpy ($kJ kg^{-1}$)	2688.3	2783.5	2879.9	2977.6	3076.8
specific entropy ($kJ kg^{-1} K^{-1}$)	8.87	9.11	9.33	9.52	9.71
Pressure, 1.0 bar; saturation temperature, 99.6 °C					
specific volume ($m^3 kg^{-1}$)	1.69	1.94	2.17	2.41	2.64
specific enthalpy ($kJ kg^{-1}$)	2676.2	2776.3	2875.4	2974.5	3074.5
specific entropy ($kJ kg^{-1} K^{-1}$)	7.36	7.61	7.83	8.03	8.22
Pressure, 4.0 bar; saturation temperature, 143.6 °C					
specific volume ($m^3 kg^{-1}$)	—	0.47	0.53	0.60	0.654
specific enthalpy ($kJ kg^{-1}$)	—	2752.0	2860.4	2964.5	3067.2
specific entropy ($kJ kg^{-1} K^{-1}$)	—	6.93	7.17	7.38	7.57
Pressure, 15.0 bar; saturation temperature, 198.3 °C					
specific volume ($m^3 kg^{-1}$)	—	—	0.13	0.15	0.17
specific enthalpy ($kJ kg^{-1}$)	—	—	2794.7	2923.5	3038.9
specific entropy ($kJ kg^{-1} K^{-1}$)	—	—	6.45	6.71	6.9207

Adapted from the tables of Spalding and Cole (1973).

transfer fluid since it does not evolve as much heat. Therefore, in most heating applications, saturated steam is preferred. If the processing steam is slightly superheated, it will quickly lose this heat and condense at its saturation temperature.

Superheat tables for other fluids can be found in the publications by the American Society of Heating, Refrigerating and Air-Conditioning Engineers (1981) and Perry and Chilton (1973).

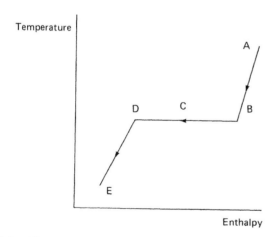

Fig. 11.3 — The temperature–enthalpy relationship for cooling a superheated vapour at a constant pressure.

If superheated steam at 4.0 bar and 200 °C is cooled at a constant pressure, the changes can be represented as in Fig. 11.3: AB, cooling of superheated vapour; B, saturated vapour; C, wet vapour; D, saturated liquid; E, liquid. Over the temperature range AB the average specific heat of water vapour is given by $\Delta h/\Delta T$.

When superheated steam at 4.0 bar, 200 °C and $h = 2860.4$ kJ kg^{-1} is cooled at a constant pressure to a saturated vapour ($T_s = 143.6$ °C; $h_g = 2737$ kJ kg^{-1}), the average specific heat is given by

$$\text{specific heat} = \frac{\text{enthalpy change}}{\text{temperature change}} = \frac{123.4}{56.4} = 2.19 \text{ kJ kg}^{-1}\text{ K}^{-1}$$

11.7 THERMODYNAMIC CHARTS

The thermodynamic properties of gases and vapours can also be represented by charts. The three major types of chart that are used are as follows.
(1) Temperature–entropy (T–S) diagrams.
(2) Pressure–enthalpy (p–H) diagrams.
(3) Enthalpy–entropy (H–S) diagrams (less common).

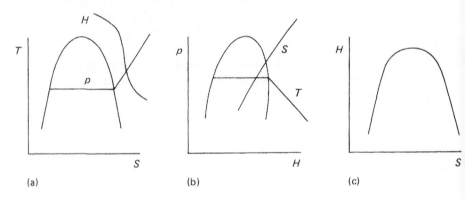

Fig. 11.4 — Types of thermodynamic chart: (a) temperature–entropy; (b) pressure–enthalpy; (c) enthalpy–entropy.

These are represented in Fig. 11.4(a), Fig. 11.4(b) and Fig. 11.4(c), respectively. All the charts have certain features in common, which will be explained in terms of the T–S diagram shown in Fig. 11.5.

Temperature and entropy are represented on the major axes; pressure and enthalpy are represented by lines of constant pressure and enthalpy respectively. If two of these four properties are known, the other two are also fixed. As with thermodynamic tables, enthalpy and entropy are measured above a defined datum, i.e. a temperature where s_f and h_f are set equal to zero; this is often -40 °C for refrigerants.

There are certain well-defined regions. The lines AB and BC represent the properties of the saturated liquid and vapour respectively. B is the point

Sec. 11.8] Representation of some thermodynamic processes 337

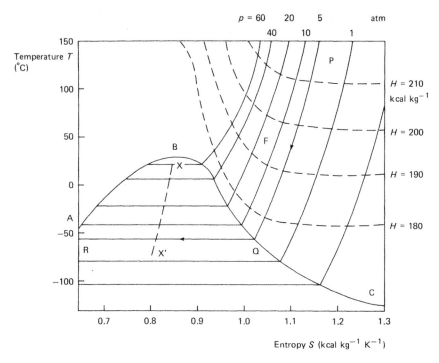

Fig. 11.5 — Temperature–entropy diagram for carbon dioxide, where p is given in atmospheres and h is given in kilocalories per kilogram.

at which vapour and liquid cannot be distinguished, i.e. the critical point; the critical temperature for carbon dioxide is 31.1 °C. The area under ABC is the two-phase or wet region. A dryness fraction of 0.5 is indicated in this region XX′. The area F above the saturated vapour line is the superheated vapour region, whilst the area to the left of the saturated liquid line is the supercooled liquid (which is not of great importance in food-processing operations).

Lines of constant pressure are shown. When superheated carbon dioxide at a pressure of 5 atm is cooled at a constant pressure, it will follow the path PQR; it will fall in temperature until it reads a point Q, corresponding to its saturation temperature; the carbon dioxide then starts to condense at a constant temperature (this is shown by QR) until condensation is complete at point R. Lines of constant enthalpy (in kilocalories per kilogram) are also represented. These diagrams are extremely useful for representing and evaluating compression and expansion processes, as explained in the section 11.8. The same principles apply to interpretation of data on p–H and T–S diagrams. The American Society of Heating, Refrigerating and Air-Conditioning Engineers (1981) have given many of these charts in more detail. Also included on some charts are lines of constant specific volume and constant dryness fraction.

11.8 DIAGRAMMATIC REPRESENTATION OF SOME THERMODYNAMIC PROCESSES

11.8.1 Compression

In any compression process, work is required to compress the gas; this is

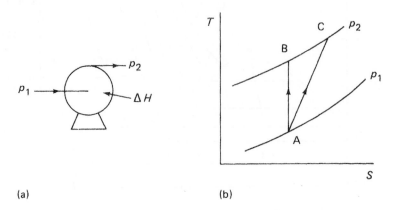

Fig. 11.6 — (a) Work of compression; (b) work of compression illustrated on a temperature–entropy diagram where AB is an isentropic compression and AC a real compression.

usually accompanied by a rise ΔT in temperature. It can be shown that the work W of compression is equal to the enthalpy change. Therefore, thermodynamic charts are useful for evaluating such temperature and enthalpy changes which occur during compression.

Isentropic compression is given by

$$\Delta S = 0$$

An isentropic compression is equivalent to a reversible adiabatic process and is illustrated in Fig. 11.6:

$$W = \Delta H_{\text{isen}} + H_B - H_A$$

An isentropic compression is an ideal compression as it represents the situation in which no heat is lost to the surroundings. In a *real compression*, heat is lost to the surroundings, and more work is required to compress the gas between the same pressures. A real compression is compared with an isentropic compression in terms of the *isentropic efficiency* IE, where

$$\text{IE} = \frac{\Delta H_{\text{isen}}}{\Delta H_{\text{real}}} \times 100$$

Therefore,

Sec. 11.8] Representation of some thermodynamic processes

$$\Delta H_{real} = \frac{\Delta H_{isen} \times 100}{IE}$$

Typical values for isentropic efficiencies range between 80% and 90%. A real compression is represented by AC (Fig. 11.6):

real enthalpy change = $H_C - H_A$

real temperature change = $T_C - T_A$

For example, carbon dioxide at 1 atm and $-50\,°C$ has an enthalpy of 178 kcal kg^{-1}. When the carbon dioxide is compressed isentropically to 10 atm, the enthalpy and temperature are 207 kcal kg^{-1} and 100 °C respectively. Therefore,

$$\Delta H_{isentropic} = 29 \text{ kcal kg}^{-1}$$

If the isentropic efficiency is 0.8, then $\Delta H_{real} = 29/0.8 = 36.3$ kcal kg^{-1}. Therefore the final enthalpy is $178 + 36.3 = 214.3$ kcal kg^{-1}. The final temperature can be determined from the chart ($p = 10$ atm; $H = 214.3$ kcal kg^{-1}) and is approximately 125 °C.

11.8.2 Cooling and condensation at a constant pressure
This process is represented on the two types of diagram (Fig. 11.7). A

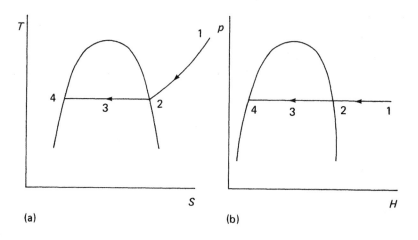

Fig. 11.7 — Cooling and condensation at a constant pressure: (a) temperature–entropy diagram; (b) pressure–enthalpy diagram.

superheated vapour 1 is cooled to a saturated vapour 2, a wet vapour 3 or a saturated liquid 4. The heat evolved (the specific enthalpy change) in these cases is $h_2 - h_1$ when it cools to a saturated vapour ($h_g - h$), $h_3 - h_1$ when it cools to a wet vapour ($h_{mix} - h$) and $h_4 - h_1$ when it cools to a saturated liquid ($h_f - h$). The specific enthalpy values associated with these changes can be determined from the charts or tables.

11.8.3 Throttling expansion

When a gas or vapour is suddenly expanded, it can be shown that there is no change in enthalpy, i.e. the process takes place under *isenthalpic conditions*. Such situations exist when a gas passes through a porous plug, e.g. cotton-wool, or an expansion valve. This process is represented for the expansion of a saturated liquid on the two types of chart in Fig. 11.8. In this case a throttling expansion results in a fall in temperature, together with the formation of a wet vapour. This is how low temperatures are produced in a refrigeration cycle. A similar expansion of a wet vapour or saturated vapour may well lead to the formation of a superheated vapour, i.e. the production of superheated steam. Throttling of a superheated vapour may also result in a cooling effect, particularly at higher pressures. The cooling effect of any throttling process can easily be determined from such charts. This type of expansion is sometimes called a Joule–Thomson expansion.

11.8.4 Evaporation at a constant pressure

When a saturated liquid or a wet vapour is allowed to evaporate at a constant pressure at a temperature below the ambient temperature, energy is withdrawn from the surroundings, and a cooling effect is observed (Fig. 11.9).

In a refrigeration system for freezing foods, the evaporation temperature chosen is between $-30\,°C$ and $-40\,°C$ whereas for chilled-water production it is about $0\,°C$. Energy is withdrawn from air, from liquids or directly from food materials. The evaporator is the working part of most indirect refrigeration units. The amount of heat absorbed by the vapour is equal to $H_2 - H_1$; this is also known as the refrigeration effect.

11.9 VAPOUR COMPRESSION REFRIGERATION CYCLE

If a working fluid, which is termed a refrigerant, is taken through the following cycle of events, a refrigeration effect can be produced. This cycle is known as a vapour compression refrigeration cycle and will be illustrated for the refrigerant Freon 12 (CCl_2F_2) working under ideal conditions (isentropic compression).

The steps involved in the cycle are shown in Fig. 11.10, and the processes are represented on the p–H and T–S diagrams.

AB The saturated vapour is taken into the compressor and compressed isentropically. The temperature rises considerably. The isentropic work of compression is equal to the enthalpy change. The discharge pressure is selected to give a condensation temperature of approximately $40\,°C$ (9.30 bar for Freon 12). The isentropic work of compression is given by

$$H_B - H_A = 218 - 170$$

$$= 48 \text{ kJ kg}^{-1}$$

BC The compressed vapour is allowed to cool and condense to a

Sec. 11.9]   Vapour compression refrigeration cycle   341

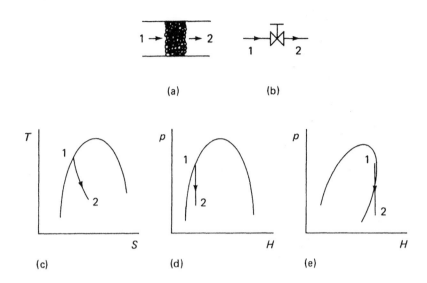

Fig. 11.8 — Throttling expansion: (a) through a porous plug; (b) through an expansion valve; (c) representation on a temperature–entropy diagram; (d) representation on a pressure–enthalpy diagram; (e) production of a superheated vapour from a wet vapour (p–H diagram).

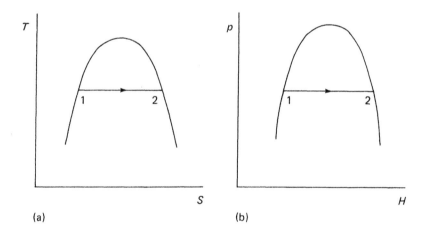

Fig. 11.9 — Evaporation at a constant pressure: (a) representation on a temperature–entropy diagram; (b) representation on a pressure–enthalpy diagram.

saturated liquid. During this process, heat is lost to the surroundings. The temperature of condensation is 40 °C; condensation is achieved using ambient air or water. The heat removed is given by

$$H_C - H_B = 75 - 218$$

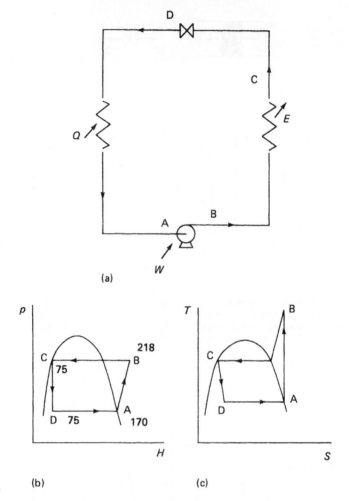

Fig. 11.10 — Vapour compression refrigeration cycle: (a) processes involved; (b) representation on a pressure–enthalpy diagram, enthalpy values are given; (c) representation on a temperature–entropy diagram.

$$= 143 \text{ kJ kg}^{-1}$$

CD   The saturated liquid is expanded under isenthalpic conditions. This results in a fall in temperature and a certain amount of evaporation. The outlet pressure is selected to reduce the temperature of the wet vapour to $-40\,°C$ (0.62 bar for Freon 12).

DA   The wet vapour is then allowed to evaporate at a constant pressure at a temperature of $-40\,°C$. Heat is withdrawn from the surroundings. The refrigeration effect is given by

$$H_A - H_D = 170 - 75$$
$$= 95 \text{ kJ kg}^{-1}$$

The efficiency of such refrigeration cycles is measured by their coefficient of performance COP where

$$\text{COP} = \frac{\text{refrigeration effect}}{\text{compressional work done}}$$

$$= \frac{H_A - H_D}{H_B - H_A}$$

$$= \frac{95}{48}$$

In this example, for an isentropic compression, the coefficient of performance is equal to 1.98.

In a similar fashion, the cycle can be evaluated for a compression with an isentropic efficiency of 80%. Since the work of compression is higher, the coefficient of performance will be reduced.

High values are desirable and indicate more economic use of energy. Ammonia is an extremely efficient refrigerant and gives a coefficient of performance between 4 and 5, over the same range of operating temperatures.

If an energy balance is performed on the refrigerant,

$$Q + W = E$$

i.e. the amount E of heat lost from the condenser is equal to the heat Q removed from the food, air or liquid during the evaporation stage plus the work W of compression. Effectively, one is using compressional work to remove energy from a source at a low temperature and to reject this to the surroundings at a higher temperature. To make the best efficient use of such a cycle, the heat rejected should be fully utilized, i.e. for heating water or space heating; when used in this way, it is also called a heat pump.

Condenser and evaporator temperatures of 40 °C and −40 °C are selected because these are of most practical use. In the condenser, air or water are used to remove the energy; typical temperatures may be between 10 °C and 30 °C, allowing a reasonable temperature driving force for heat transfer. In tropical climates, it may be necessary to have a slightly higher condensation temperature.

When freezing foods or producing low temperature environments, temperatures between −35 °C and −18 °C are commonly used. Thus an evaporation temperature of −40 °C again provides a reasonable temperature driving force for achieving these temperatures.

Once the condenser and evaporator temperatures have been fixed, this automatically sets the operating (suction and discharge) pressures for the particular refrigerant. The pressures required to condense different refrigerants are given in Table 11.1, together with critical temperatures and latent heat values at −20 °C.

11.10 INTRODUCTION TO AIR–WATER SYSTEMS

The amount of water associated with air can have a considerable bearing on dehydration processes and the storage of fresh and processed foods.

In hot-air drying processes, air acts as both a heat transfer medium and a vehicle for the removal of water. During a drying process, energy is transferred from the air to the food (providing the latent heat of evaporation), and the water vapour is transferred in the opposite direction from the food to the air. Dehydration processes are consequently referred to as simultaneous heat and mass transfer processes. It is important to maximize these processes in order to minimize drying times. Theoretically, drying rates can be increased by increasing the air temperature, increasing the air velocity and decreasing the amount of water in the air. However, there are other factors which limit the extent to which these can be increased, e.g. scorching or energy costs or diffusion processes (at low moisture contents).

11.10.1 Absolute and relative humidity

The amount of water associated with air can be measured in terms of its *absolute himidity* and *relative humidity*.

Absolute humidity is defined as the mass of water divided by the mass of dry air. It can be regarded as a measure of the moisture content of air on a dry-mass basis; it is a dimensionless quantity:

$$\text{absolute Humidity } \bar{H} = \frac{\text{mass of water}}{\text{mass of dry air}}$$

$$= \frac{\text{number of molecules water} \times \text{molecular weight of water}}{\text{number of molecules air} \times \text{molecular weight of air}}$$

$$= \frac{18 \, p_{wv}}{29(p_a - p_{wv})} \tag{11.1}$$

where p_{wv} is the water vapour pressure (consistent units) and p_a is the atmospheric (air) pressure. Thus, if the water vapour pressure is known, the absolute humidity can be determined, provided that consistent pressure units are used. Many dehydration processes take place at atmospheric pressure; so, if the water vapour pressure is expressed in atmospheres, equation (11.1) can be simplified to give

$$\bar{H} = \frac{18 p_{wv}}{29} \tag{11.2}$$

Thus, if ambient air at 20 °C exerts a water vapour pressure of 8 torr; the absolute humidity values calculated from equation (11.1) and equation (11.2) are given by

Sec. 11.10]   Introduction to air–water systems   345

$$\bar{H} = \frac{18 \times 8}{29(760 - 8)} \text{ or } \frac{18 \times 8/760}{29}$$

$$6.603 \times 10^{-3} \text{ or } 6.534 \times 10^{-3}$$

The error involved in the use of the simplified expression (equation (11.2)) is about 1%. As the water vapour pressure increases, the magnitude of the error increases. For the air in question, 0.0066 kg of water are associated with each kilogram of dry air. When air is heated or cooled, its absolute humidity remains virtually constant.

Air becomes saturated when the water vapour pressure is equal to the saturated vapour pressure at that temperature (see section 11.3). Therefore the absolute humidity H_s of saturated air is given by the expression

$$\bar{H}_s = \frac{18 p_{wvs}}{29(p_a - p_{wvs})}$$

where the subscript s denotes saturated.

The absolute humidity of saturated air can be evaluated at any temperature by use of the saturated vapour pressure data for water. It represents the maximum amount of water that air can hold at that temperature. It can be seen that air has a much greater capacity for holding moisture as its temperature increases and is one of the main reasons why hot air is used for dehydration processes.

If air is saturated with water at a particular temperature, lowering the temperature will result in condensation as the system adjusts itself to its new conditions.

The *relative humidity* RH of air is a measure of how close the air is to being saturated at a particular temperature. It is expressed as a percentage, using the following formula:

$$\text{RH} = \frac{\text{absolute humidity of air}}{\text{absolute humidity of saturated air at the same temperature}} \times 100$$

$$= \frac{\bar{H}}{\bar{H}_s} \times 100$$

Using equation (11.2), this gives

$$\text{RH} = \frac{p_{wv}}{p_{wvs}} \times 100$$

Relative humidity is more commonly used than absolute humidity. If relative humidity is to be used as an indicator of the amount of water in the environment, it should be accompanied by the temperature; otherwise it is meaningless. The absolute humidity can then be determined, as shown below.

As an example, consider air at 30 °C which has a relative humidity of 60%. From the tables the saturated vapour pressure p_{wvs} of water at 30 °C is 31.8 Torr Therefore, using

$$RH = \frac{p_{wv}}{p_{wvs}} \times 100$$

it is found that

$$p_{wv} = 19.08 \text{ Torr}$$

Thus,

$$\bar{H} = \frac{18 p_{wv}}{29(p_a - p_{wv})}$$

$$= \frac{18 \times 19.08}{29(760 - 19.08)}$$

$$= 0.016$$

Conversely, if the absolute humidity and temperature are known, the relative humidity can be evaluated.

11.10.2 Dew-point temperature

If this air is now cooled, the absolute humidity will remain the same (since no water is removed from the air), but the relative humidity will increase. As cooling proceeds, the air will become saturated and condensation will occur (when $p_{wv} = p_{wvs}$). The temperature at which condensation occurs is referred to as the *dew-point temperature*. Note that, when condensation commences, the absolute humidity of the air will start to fall.

Therefore the dew-point temperature depends solely upon the amount of moisture in the air, i.e. the absolute humidity. As the absolute humidity increases, the dew-point temperature increases. The dew-point temperature provides an extremely useful method for determining the moisture content of air; the instruments used for this method are termed dew-point hygrometers and have been described in more detail by Jones (1974b). Weast (1982) has presented tables giving the relative humidity for different values of the dew-point and dry-bulb temperatures. Alternatively the relative humidity can be calculated from the ratio of the saturated vapour pressures at the dew-point and dry bulb temperatures:

$$RH = \frac{p_{wvs} \text{ (dew-point temperature)}}{p_{wvs} \text{ (dry-bulb temperature)}} \times 100$$

For example, air with a dry-bulb temperature of 20 °C and a dew-point temperature of 12 °C will have a relative humidity given by

$$RH = \frac{10.518}{17.535} \times 100 = 60\%$$

The saturated vapour pressure data are taken from the data of Weast (1982) (see Table 11.2).

11.10.3 Wet-bulb temperature
Another important temperature measurement in hygrometry is the wet-

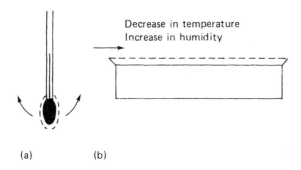

Fig. 11.11 — (a) Wet-bulb temperature; (b) evaporation from a plane water surface.

bulb temperature. This can be determined by covering the measuring element of a thermometer with a moistened wick (Fig. 11.11). The thermometer is then shaken; this causes water to evaporate from the moistened wick and the temperature recorded by the thermometer will fall because of the latent heat requirement for evaporation. Eventually, equilibrium is achieved, and the thermometer records a steady temperature. This is known as the *wet-bulb temperature* T_w. The temperature recorded by a normal thermometer is known as the *dry-bulb temperature T*.

The wet-bulb depression is the difference between the dry-bulb temperature and the wet- bulb temperature, i.e.

wet-bulb depression = $T - T_w$

The magnitude of the wet-bulb depression will depend upon the dry-bulb temperature and the amount of water in the atmosphere. As the air becomes less humid, or its temperature increases, the wet-bulb depression increases.

The wet-bulb temperature can be regarded as the temperature assumed by a free moisture surface in equilibrium with air at a fixed temperature and humidity. During the initial stages of most dehydration processes, the surface of a food behaves like a free moisture surface and the surface temperature approximates to the wet-bulb temperature.

11.10.4 Adiabatic saturation temperature
This can be examined by considering the flow of an air stream over a free moisture surface, e.g. a trough of water (Fig. 11.11(b)). Usually, it is assumed that the processes of heat transfer and evaporation take place

under adiabatic conditions, i.e. all the heat lost by the air causes evaporation. As the air flows over the moisture surface, its temperature will fall and its humidity will rise. Eventually, if the trough is long enough, the air will become saturated; the temperature of this saturated air is known as the adiabatic saturation temperature; the air and water are in equilibrium.

The rate Q' of heat transfer and the rate m' of mass transfer (i.e. the evaporation rate), across the boundary layer can be written as follows:

$$Q' = hA(T_a - T_s)$$

This causes evaporation and the evaporation rate m' is given by

$$m' = kA(p_{wvs} - p_a)$$
$$= k'A(\bar{H}_s - \bar{H}_a)$$

where k and k' are mass transfer coefficients. However,

rate of heat transfer = evaporation rate × latent heat

Therefore, $hA(T_a - T_s) = k'A(\bar{H}_s - \bar{H}_a) h_{fg}$

$$\frac{T_a - T_s}{H_s - H_a} = \frac{k' h_{fg}}{h}$$

For air–water systems the wet-bulb temperature and the adiabatic saturation temperatures are approximately equal and, for most engineering calculations, it is assumed that they are.

Thus, when heat transfer takes place under adiabatic conditions, there is a linear relationship between the fall in temperature and rise in absolute humidity. This will be examined in more detail in section 11.11.

Wet-bulb and dry-bulb measurements provide a useful means of determining the water vapour pressure, the relative humidity and the dew-point temperature of samples of air.

The relationship between the water vapour pressure p_{wv} and the saturated water vapour pressure p_{wvs} (at the wet-bulb temperature) is given by

$$p_{wv} = p_{wvs} - Ap_a(T - T_w)$$

where A (= 6.66×10^{-4} degC^{-1}) is a constant for moving air.

If air has a wet-bulb temperature and a dry-bulb temperature of 19 °C and 24 °C respectively, then p_{wvs} = 16.47 Torr and p_a = 760 Torr; so p_{wv} = 16.47 − (6.66 × 10^{-4} × 760 × 5) = 13.94 Torr. The saturated vapour pressure at the dry-bulb temperature is 22.38 Torr (Table 11.2) giving a relative humidity of (13.94/22.38) × 100 = 62.3%.

These relationships have been summarized in tabular form by Kaye and Laby (1973) for the dry-bulb temperature range 10–24 °C.

Wet-bulb hygrometers such as the sling hygrometer have been described in more detail by Jones (1974b). Additional correction factors have to be made where there is no relative motion between the air and the wet-bulb thermometer (still air).

11.11 HUMIDITY CHARTS

It is more convenient to summarize the properties of air–water systems in the form of a chart rather than in tables. Usually, charts are presented at atmospheric pressure.

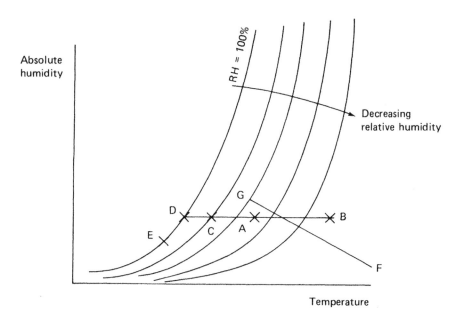

Fig. 11.12 — Simplified humidity diagram.

The chart normally takes the form of a plot of the absolute humidity against temperature (Figs. 11.12 and 11.13). Lines of constant relative humidity are inserted. The 100% saturation line is important because it can be used to give an indication of wet-bulb and dew-point temperatures. The chart can be used for interpreting humidity data. Thus, air at 42 °C and 40% relative humidity would have an absolute humidity of 0.022 (Fig. 11.12, point A).

Heating and cooling of this air can be represented by Fig. 11.12, lines AB and AC, respectively (no change in absolute humidity). Heating this air to 70 °C (point B) would decrease the relative humidity to below 10%, whereas cooling to 30 °C would increase the relative humidity to 80% (point C). If the air is further cooled, it becomes completely saturated at point D (26 °C); this represents the dew-point temperature; further cooling causes the air to follow line DE, with the occurrence of condensation.

The humidity chart can be used with dew-point data for humidity determination as follows. The dew-point temperature is located on the 100% saturation line. A horizontal line is drawn from this point to the corresponding dry-bulb temperature; the resulting point gives the state of the air. Air with a dew-point temperature of 18 °C and a dry-bulb temperature of 40 °C will have an absolute humidity of 0.014 and a relative humidity of approximately 30%. Note that the dew-point temperature alone is sufficient to fix the absolute humidity.

Also shown are adiabatic cooling lines; these are a series of straight parallel lines with a negative gradient of approximately 5.97×10^{-4} degC^{-1}. They relate how the absolute humidity increases as the air temperature decreases during a drying or humidification process. If hot air at 90 °C and an absolute humidity of 0.01 (point F) is used for drying, the relationship between absolute humidity and temperature will be described by the adiabatic cooling line going through point F. If this air leaves the drier at 60% relative humidity, it will have a dry-bulb temperature of 40 °C and an absolute humidity of 0.03 (point G). The change in absolute humidity is

$$\bar{H}_0 - \bar{H}_I = 0.03 - 0.01$$
$$\bar{H}_0 - \bar{H}_I = 0.02$$

This means that every kilogram of dry air entering picks up 0.02 kg of water. The dew-point temperature of the air leving the drier is equal to 32 °C. The wet-bulb temperature can be found from the point of interception of the adiabatic cooling line and the 100% saturation line; in this case the wet-bulb temperature is 33 °C. At higher temperatures the data are presented by plotting the logarithm of the absolute humidity against temperature. One important difference is that the adiabatic cooling lines are no longer straight lines, although they are followed in a similar way for interpreting wet-bulb temperatures.

Humidity charts are extremely useful for representing and illustrating changes that can occur during the heating and cooling of air and processes involving moisture exchange processes, such as dehydration and humidification processes, where air is used to cool water (cooling towers).

Note that freeze drying can occur at temperatures below 0.01 °C when water vapour falls below 4.6 Torr. This corresponds to an absolute humidity of 3.78×10^{-3}. Such conditions can be found on cold dry winter days.

Wet-bulb and dry-bulb readings can be used in conjunction with the psychrometric chart to determine the absolute and relative humidity of an air sample. The wet-bulb temperature is located on the 100% saturation line, and a line is drawn through this point parallel to the adiabatic cooling lines. The point of location of the dry-bulb temperature along this line represents the humidity of that air.

Sec. 11.13] Example of interpretation of charts 351

11.12 DETERMINATION OF OTHER PROPERTIES FROM HUMIDITY CHARTS

On many psychrometric charts, additional data are presented. This is briefly summarized below.

The *humid heat* c_H is the amount of heat required to raise the temperature of a mass of mixture containing unit mass of dry air by unit amount:

$$c_H = c_a + c_{vw}\bar{H}$$

where $c_a = 1.005$ (kJ kg^{-1} K^{-1}) is the specific heat of dry air, $c_{wv} = 1.88$ (kJ kg^{-1} K^{-1}) is the specific heat of water vapour and c_H is the heat required to raise $1 + H$ kg of moist vapour by 1 K. Therefore the specific heat of *unit mass* of moist air would equal

$$\frac{c_a + c_{wv}\bar{H}}{1 + \bar{H}} \qquad (11.3)$$

Humidity charts often present humid heat values plotted against absolute humidity. The *humid volume* V_H is the volume of a mass of mixture containing a unit mass of dry air.

At 273 K, 1 kg of dry air occupies 22.4/29 m³ (molecular weight of air equals 29), and H kg of water vapour occupies 22.4H/18 m³ (molecular weight of water equals 18). Therefore,

$$\text{total volume} = 22.4\left(\frac{1}{29} + \frac{\bar{H}}{18}\right)$$

At any other temperature θ and a pressure of 1 atm, $(1 + \bar{H})$ kg of moist air occupies

$$22.4\left(\frac{1}{29} + \frac{\bar{H}}{18}\right)\frac{\theta + 273}{273} \text{ m}^3$$

Therefore, unit mass of moist air will occupy

$$\frac{22.4}{1 + \bar{H}}\left(\frac{1}{29} + \frac{\bar{H}}{18}\right)\frac{\theta + 273}{273} \text{ m}^3 \qquad (11.4)$$

Humid volumes are important for converting mass flow rates to volumetric flow rates, for correctly sizing fans and blowers.

11.13 EXAMPLE OF INTERPRETATION OF CHARTS

Consider air at 60 °C and an absolute humidity of 0.02. From the chart in Fig. 11.14, the relative humidity is 15% $T_w = 33$ °C and the dew-point temperature is 25 °C. Each kilogram of dry air is associated with 0.02 kg of water vapour. The volume occupied by 1.02 kg of moist air is given by

352   **Properties of gases and vapours**   [Ch. 11

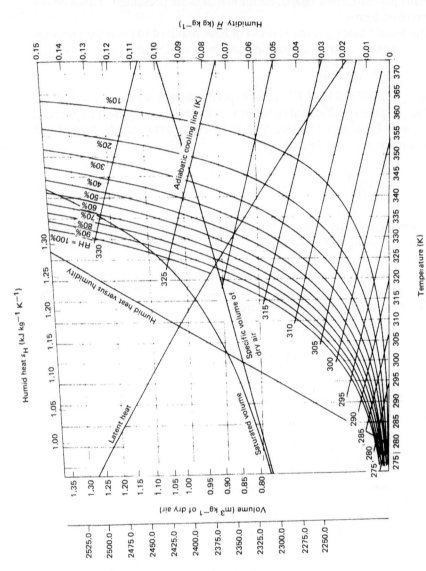

Fig. 11.13 — Humidity diagram. (From Coulson and Richardson (1977), with permission.)

Sec. 11.13] Example of interpretation of charts

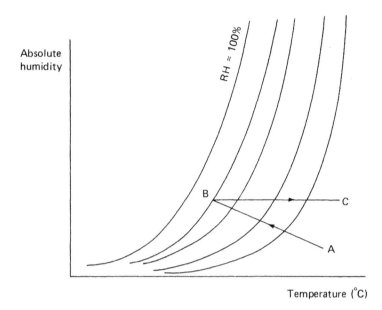

Fig. 11.14 — Representation of a drying process on the humidity chart.

$$V_H = 22.4 \left(\frac{1}{29} + \frac{0.02}{18} \right) \frac{333}{273}$$

$$= 0.972 \text{ m}^3$$

Therefore the specific volume of the moist air at 60 °C equals $0.972/1.02 = 0.953$ m^3 kg^{-1}. The specific heat of the moist air is given by equation (11.3) and is equal to $c_a + (0.02c_v)/1.02 = 1.022$ kJ kg^{-1}. If this air is passed through a drier and leaves at 80% relative humidity (Fig. 11.14, point B), then the absolute humidity is 0.03, the dew-point temperature is 32 °C and the dry-bulb temperature is 36 °C. 1 kg of dry air (i.e. 1.02 kg of moist air) will pick up $\bar{H}_0 - \bar{H} = 0.03 - 0.02 = 0.01$ kg of water vapour. Therefore it is required to remove 10 kg h^{-1} of water, the mass flow rate of dry air required is given by

$$\text{mass flow rate of dry air} = \frac{\text{mass of water removed per hour}}{\text{mass of water removed/mass (kg) of dry air}}$$

$$= \frac{10}{0.01}$$

$$= 1000 \text{ kg h}^{-1}$$

Therefore the mass of moist air is 1020 kg h^{-1} and the volume of moist air is

1020 kg h^{-1} × 0.953 m³ kg^{-1} = 972 m³ h^{-1}. The specific heat capacity of the air leaving the drier is 1.005 + (0.03 × 1.88)/1.03 = 1.030 kJ kg^{-1}. If this air is then reheated to 70 °C, the following values are obtained (Fig. 11.14, point C): dry-bulb temperature, 70 °C; wet-bulb temperature, 38 °C; dew-point temperature, 32 °C; relative humidity, 12%; absolute humidity = 0.03 and the humid volume, 22.4 (1/29 + 0.03/18) 343/273 = 1.017 m³ kg^{-1}.

11.14 MIXING OF AIR STREAMS

Often it is required to mix two air streams. For example, some of the hot air

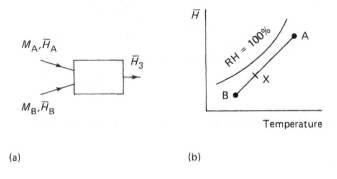

Fig. 11.15 — (a) Mixing of air streams; (b) representation on the humidity chart.

leaving a drier is recycled by mixing it with the inlet air and thus acts a means of conserving energy. The composition of the resulting mixture can be obtained from a mass balance (Fig. 11.15(a)).

If two air streams containing m_A kg and m_B kg of dry air, respectively, are mixed, the mass of dry air leaving will equal $m_A + m_B$ and its absolute humidity will equal $\bar{H}_3$.

A moisture balance gives

$$m_A \bar{H}_A + m_B \bar{H}_B = (m_A + m_b)\bar{H}_3$$

$$\bar{H}_3 = \frac{m_A \bar{H}_A + m_B \bar{H}_B}{m_A + m_B}$$

Thus, when 50 kg of air ($\bar{H} = 0.03$) is mixed with 20 kg of air ($\bar{H} = 0.01$), the resulting absolute humidity of the mixture will be

$$H_3 = \frac{50 \times 0.03 + 20 \times 0.01}{70}$$

$$= 0.024$$

If the temperatures of the two streams are known, the two points can be located on the psychrometric chart and joined by a straight line; Fig. 11.15(b), point $\times (H = 0.024)$ corresponds to the condition of the mixture. Note that the ratio of AX to XB corresponds to the ratio of the masses of the streams, i.e. m_B to m_A. This is known as the lever rule.

Some confusion may arise because absolute humidity is measured on a dry-mass basis, whereas the air actually being used is moist and used on a wet-mass basis. Dry masses can be converted to moist masses by multiplying by $1 + \bar{H}$.

A wide variety of other methods are available for determining humidity, including manometry, hair hygrometry, electric hygrometry and various chemical methods. These have been reviewed by Troller and Christian (1978).

11.15 WATER IN FOOD

The amount of water in food can be expressed on either a wet weight basis or a dry-weight basis as follows.

On a wet-mass basis,

$$\text{moisture content } m = \frac{\text{mass of water}}{\text{mass of sample}} \times 100$$

The mass of sample can be made up of water and dry matter or solids. Thus,

$$\text{moisture content } m = \frac{\text{mass of water}}{\text{mass of water} + \text{solids}} \times 100$$

On a dry-weight basis, moisture is calculated as

$$\text{moisture } M = \frac{\text{mass of water}}{\text{mass of solids}}$$

It is sometimes expressed on a percentage dry-weight basis, i.e. 100 multiplied by the moisture. It can be shown by eliminating the mass of the solids that

$$m = \frac{100M}{1 + M}$$

or

$$M = \frac{m}{100(1 - m/100)}$$

These equations are useful for converting from one form to another. The relationship between moisture and moisture content is also shown in Fig. 11.16. A food with a moisture content of 80% will have a moisture value of 4

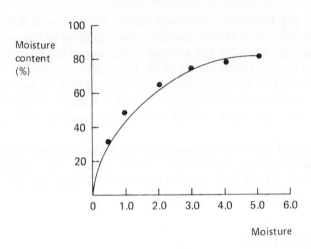

Fig. 11.16 — Relationship between moisture content (wet-weight basis) and moisture (dry-weight basis).

(or 400%, as a percentage). Imagine that half the water originally present in this food is removed during a dehydration process. If we assume that the original mass of food was 100kg, a straightforward calculation shows that 40 kg of water is removed, that the final moisture value is 2 and that the final moisture content value is 66%. It is important to realize that the moisture content is not linearly related to the amount of water removed, whereas the moisture value is. The reason for this is that the amount of dry matter remains constant during dehydration.

For example, if 250 kg of tomatoes (initial moisture content, 94%) are freeze dried, and the final mass of the product is 25 kg, determine final the moisture value and moisture content.

This is best solved by determining the mass of dry solids in the product, i.e. a solids balance. The mass of dry solids present is $0.06 \times 250 = 15$ kg. Therefore, the mass of water in the final product is 10 kg. Thus,

$$m = \frac{10}{25} \times 100$$

and

$$= 40\%$$

$$M = \frac{10}{15}$$

$$= 0.667$$

The final moisture content is 40%. Expressed as a dry-weight basis, this equals 0.667 or 66.7 on a percentage dry-weight basis. Both systems are in common use. Moisture content (wet-weight basis) is most often used in food composition tables, whereas moisture (dry-weight basis) is more often encountered with sorption isotherms and drying curves (see section 13.9). The amount of water in a food is most easily determined by taking a representative sample of the food and drying it in an oven to constant mass. During dehydration processes, considerable moisture gradients will be established within the food.

11.16 SORPTION ISOTHERMS

When a food is placed in an environment at a constant temperature and

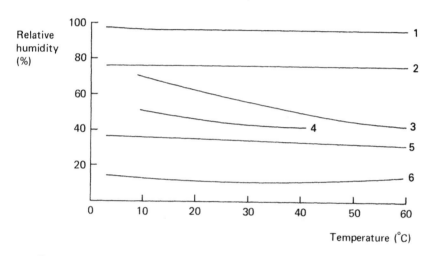

Fig. 11.17 — Relative humidity of air over some saturated solutions at different temperatures: line 1, potassium sulphate; line 2, sodium chloride; line 3, ammonium nitrate; line 4, potassium carbonate; line 5, magnesium chloride; line 6, lithium chloride.

relative humidity, it will eventually come to equilibrium with that environment. If the food was originally fresh, it will probably lose moisture to the atmosphere (known as wilting with fruits and vegetables; if the food is dehydrated, it may well gain moisture, becoming soft or soggy; at this point, mould or bacteria may start to grow). Eventually the food will reach a steady mass; the corresponding moisture content is expressed as the *equilibrium moisture content* on a dry or wet weight basis. Fig. 11.18 shows the generalized form of the relationship between equilibrium moisture content and relative humidity which is found for many food materials at a constant temperature; this is known as the sorption isotherm for that food. Sorption isotherms for most foods show this typical sigmoid shape. For convenience they can be divided into three regions: RH = 0–20%, water is tightly bound and unavailable for reactions; RH = 20–80%, water is loosely bound; RH > 80%, water in the capillaries is free for reaction (it freezes easily and has a vapour pressure approaching that of bulk water). It should be appreciated that this is very much over-simplification of the true state of water in a food system, but it gives an introduction to the concept of 'bound' and 'free' water.

11.16.1 Determination of sorption isotherms

The food, usually in the dehydrated form, is equilibrated with air at a known temperature and relative humidity. The easiest way of generating environments of constant relative humidity is to place a saturated solution of an inorganic salt in a desiccator. For example a saturated solution of ammonium sulphate at 25 °C will produce a relative humidity of 80%. A list of saturated solutions and their equilibrium relative humidities has been given by Weast (1982), Troller and Christian (1978) and British Standards Institution (1964b). The variation in equilibrium relative humidity with temperature for a range of components is shown in Fig. 11.17. Aqueous solutions of both sulphuric acid and glycerol have also been used (Table 11.6).

Table 11.6 — Glycerol solutions and sulphuric acid–water mixtures for control of relative humidity at 25 °C.

Relative humidity	Glycerol (wt. %)	Sulphuric acid (wt. %)
90	34.90	18.5
75	58.61	30.4
65	69.05	36.0
50	80.65	43.4
40	86.30	—
25	—	55.9
10	—	64.8

The food is then placed in the desiccator in the air space above the saturated solution and its mass determined at regular intervals. When it achieves a constant mass, the moisture content is determined. By selecting appropriate solutions, the equilibrium moisture content can be determined at different relative humidities. Fig. 11.18 shows typical sorption isotherms for the same food at three different temperatures. As the temperature increases, the equilibrium moisture content at any relative humidity is decreased. Thus, high temperatures or low relative humidities are required to obtain very dry food.

A distinction is made between the adsorption isotherm (where the starting material is dry) and a desorption isotherm (where the starting material is wet). The most comprehensive collection of water sorption isotherms has been given by Iglesias and Chirife (1982) where over 500

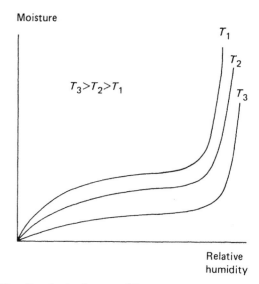

Fig. 11.18 — Sorption isotherms at different temperatures (qualitative).

isotherms for a wide variety of foods are presented diagramatically. The use of such isotherms in predicting the water activity of foods is discussed below.

11.17 WATER ACTIVITY IN FOOD

Water plays a very important part in the stability of fresh, frozen and dried foods; it acts as a solvent for chemical, microbiological and enzymatic reactions. Water activity a_w is a measure of the availability of water to participate in such reactions.

The water in a food will exert a vapour pressure. The amount of pressure will depend upon the amount of water present, the temperature and the composition of the food. Food components will lower the water vapour pressure to different extents, with salts and sugars being more effective than larger molecules such as starch and proteins. Therefore, two different foods with similar moisture contents may not necessarily exert the same vapour pressure and have the same water activity.

Water activity is the ratio of the vapour pressure exerted by the food to the saturated vapour pressure of water at the same temperature.

$$a_w = \frac{\text{vapour pressure of water exerted by food}}{\text{saturated vapour pressure of water at the same temperature}}$$

(11.5)

Values range from almost 0 for dried foods through to 1.0 for foods in which water is readily available, such as most fresh fruit, vegetables, meat, fish and milk.

The water activity values of a variety of foods are shown in Table 11.7.

Table 11.7 — Some typical water activity values for foods.

a_w	Food
0.98–1.00	Fresh vegetables, fruit, meat, fish, poultry, milk, cottage cheese
0.93–0.96	Cured meats, most cheese varieties
0.86–0.93	Salami, some dry cheeses
0.8–0.87	Flour, cakes, rice, beans, cereals, sweetened condensed milk
0.72–0.88	Intermediate moisture foods, jams, old salami
0.6–0.66	Dried fruits
0.6	Dehydrated foods

Compiled from the data of Mossel (1975), Karel (1975) and Walstra and Jenness (1984).

By inspection of equation (11.5) it can be seen that there is a relationship between water activity and relative humidity of the form

$$a_w = \frac{RH}{100} \qquad (11.6)$$

The easiest way of determining the water activity of a food is to place the food in a sealed container, and to measure the relative humidity of the air in the container, once equilibrium has been achieved. The water activity is then determined from equation (11.6). The sample should almost fill the container, so that its moisture content will not change appreciably as equilibrium is attained. Other methods for the determination of a_w have been described by Rockland and Stewart (1981) and Jowitt (1984).

It can be seen that the sorption isotherm is extremely useful because it also gives the relationship between the water activity and the water content. Thus, in a drying application, it is possible to evaluate the lowest possible moisture content attainable at specified conditions of temperature and relative humidity and the water activity of the dehydrated product.

In addition, many spoilage reactions are influenced by water activity (Rockland and Stewart, 1981). Almost all microbial activity is inhibited below 0.6, most moulds below 0.7, yeasts below 0.8 and bacteria below 0.9. More detailed information has been given by Mossel (1975). However, oxidation reactions, browning reactions and enzymatic activity will occur at low moisture contents; the rate of oxidation goes through a minimum at an a_w value of 0.4, whereas the maximum browning rate occurs at about 0.6 a_w. Although most enzymes are inactive below 0.8 a_w, enzymatic activity can be observed down to very low water activity values, e.g. lipase activity in dried products. The water activity of a food can also influence the reaction of micro-organisms to different types of sterilizing agent. For vegetative organisms a minimum rate of inactivation is found at about 0.7 a_w, whereas for bacterial spores the minimum is about 0.25 a_w (Mossel, 1975). The presence of oil has also been reported to increase the heat resistance of bacterial spores, again attributed to lowering of the water activity.

When certain solutes are added to foods, they lower the water activity by depressing the water vapour pressure. The extent of the depression can be predicted for ideal solutions by Roalt's law which states that the partial pressure of a component over a solution is the product of the vapour pressure of that component and the mole fraction of that component (see Section 1.23). Unfortunately, most food systems are too concentrated for this to occur and a_w depressions are determined experimentally. Such compounds are known as *humectants*; examples of humectants are salt, sugar and polyhydric alcohols such as glycerol and sorbitol. Growth of micro-organisms can thus be inhibited by adding such substances to foods or by formulating foods with these ingredients, i.e. salting, curing and sugar preserves. Substances with strong water binding properties are particularly useful in this respect. The water activities of foods containing various levels of sugar and salt are shown in Table 11.8.

Table 11.8 — Water activities of some foods containing salt and sugar.

Water activity	Sugar (w/w)	Salt (w/w)
1.00–0.95	40	7
0.95–0.91	55	12
0.91–0.87	65 (saturated)	15
0.80–0.75		26 (saturated)

Adapted for the data of Mossel (1975).

Considerable interest has been shown in foods with water activities in the range 0.6–0.9; such foods have been termed intermediate-moisture foods. In fact, many everyday foods fall within this range, e.g. cheese and rich fruit cakes. Intermediate-moisture foods have a much reduced moisture content and a reasonably long shelf-life but are palatable without the need to rehydrate them.

Heidelbaugh and Karel (1975) give formulations for many ready-to-eat products, such as intermediate-moisture sausage cubes, together with recipes for foods developed for the US Space Program, such as *Sky-lab* intermediate-moisture bread and *Apollo 17* beef stew. Following these recipes will result in the production of a food whose water activity is known precisely.

Also described are intermediate-moisture pet foods in which much interest has been shown in recent years. Intermediate-moisture have been discussed in more detail by Davies *et al.* (1976), Duckworth (1975) and Goldblith *et al.* (1975).

11.17.1 Hysteresis

In the determination of a sorption isotherm, different results may arise, depending on whether the test material is dry or fresh; such a phenomena is known as hysteresis. Fig. 11.19(b) shows an adsorption isotherm (the starting material is dry food) and a desorption isotherm (the starting

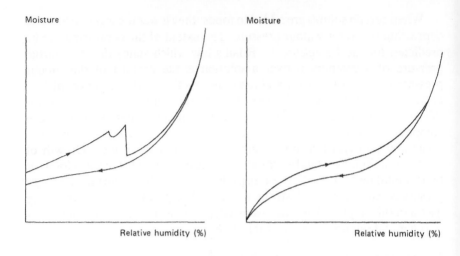

Fig. 11.19 — (a) A broken isotherm; (b) an isotherm showing hysteresis.

material is fresh or wet food) for raw chicken at 5 °C. For most foods the water content is higher when a particular a_w value is achieved by desorption than absorption. Often the greatest discrepancies occur in the intermediate-moisture food a_w range. If hysteresis does occur, the sorption isotherm most relevant to the situation should be selected. For example, for determining the equilibrium moisture content for a dehydrated food, the desoprtion isotherm should be selected. More detailed consideration of hysteresis has been given by Troller and Christian (1978).

Another interesting deviation occurs with some sugars. Crystalline sucrose has a completely different sorption isotherm from that for amorphous sugar, with the equilibrium moisture content being much lower for the crystalline form at any water activity value. The amorphous form often results when the food is dried quickly, e.g. spray drying. During storage, it may slowly revert to the crystalline form. The data for sucrose have been given by Karel (1975). If crystallization occurs during determination of the isotherm, a broken isotherm may result (Fig. 11.19(a)).

11.17.2 Frozen foods
When a food freezes, the ice which separates exerts a vapour pressure which depends only on its temperature. This is in equilibrium with the unfrozen water within the food. Therefore the water activity of the frozen food is as follows:

$$a_w = \frac{\text{vapour pressure of ice (or solution)}}{\text{saturated vapour pressure of water}}$$

The vapour pressures of ice and water at various temperatures below 0 °C are given in Table 11.9.

Water activity–moisture relationships

Table 11.9 — Water activities of frozen foods.

Temperature (°C)	Saturated vapour pressure of water (Torr)	Vapour pressure of ice (Torr)	a_w
0	4.579	4.579	1.00
−5	3.163	3.013	0.953
−10	2.149	1.950	0.907
−15	1.436	1.241	0.864
−20	0.943	0.776	0.823
−25	0.607	0.476	0.784
−30	0.383	0.286	0.75

11.18 WATER ACTIVITY–MOISTURE RELATIONSHIPS

There have been many attempts to relate the water activity of a food to its moisture, the most used of these relationships being the Brunauer–Emmett––Teller isotherm, which is as follows:

$$\frac{a_w}{M(1 - a_w)} = \frac{1}{M_1 C} + \frac{C-1}{M_1 C} a_w$$

where a_w is the water activity, the moisture M (dry-mass basis) is expressed as percentage dry mass, M_1 is the moisture value (dry-mass basis) of a monomolecular layer and C is a constant.

The Brunauer–Emmett–Teller equation is useful for estimating the monolayer value, which is equivalent to the amount of water held adsorbed on specific sites; an example of such a calculation has been given by Karel (1975).

Examples of some monolayer values are as follows (Karel, 1975): starch, 11%; gelatin, 11%; amorphous lactose, 6%; spray-dried whole milk, 3%. Karel has also given an example of how this is calculated. Iglesias and Chirife (1982) also evaluated the monolayer value for a wide variety of foods from experimental data.

Other equations have been summarized by Karel (1975) and Iglesias and Chirife (1982). Iglesias and Chirife listed nine such equations and used curve-fitting techniques to select which of the equations gives the best fit to the experimental data. For example, for mushrooms the adsorption isotherm at 20 °C is given by

$$M = 8.3477 \frac{a_w}{1 - a_w} + 2.2506$$

This applies in the a_w range 0.1–0.8.

The recent COST 90 project on water activity (Jowitt *et al.*, 1983) involved a collaborative trial to standardize procedures in 40 laboratories. As a result it was possible to recommend a suitable reference material (microcrystalline cellulose), convenient inexpensive test equipment and a 'most probable' sorption isotherm for the reference material. A procedure for using these to determine sorption isotherms for food materials and a format for recording and presenting the data have also been given by Jowitt et al. (1983).

11.19 SYMBOLS

a	van der Waals constant
a_w	water activity
b	van der Waals constant
c	specific heat
c_H	humid heat
E	energy lost from condenser
h	specific enthalpy
H	enthalpy
$\bar{H}$	absolute humidity
IE	isentropic efficiency
k	mass transfer coefficient
k'	mass transfer coefficient
m	mass
m'	mass transfer rate
m	moisture content (wet-weight basis)
M	moisture (dry-weight basis)
p	pressure
p_{wv}	water vapour pressure
p_{swv}	saturated vapour pressure
Q	heat liberated
Q'	heat transfer rate
R	gas constant
RH	relative humidity
s	specific entropy
S	entropy
T	temperature
T_s	saturation temperature
T_w	wet-bulb temperature
v	specific volume
V	volume
V_H	humid volume
W	work of compression
X	dryness fraction

11.19.1 Subscripts

a	air
f	saturated liquid
fg	from saturated liquid to sautrated vapour or vice versa
g	saturated vapour
isen	isentropic
mix	mixture
s	saturated or saturation

v   vapour
w   water or wet

12
Electrical properties

12.1 INTRODUCTION

Electricity plays an important part in many food-processing operations. Electrical energy is used for heating purposes, in the production of steam, hot water and hot air, and in some instances for supplying heat directly to the product. It is used in refrigeration applications for driving compressors and in fluid flow and transportation situations for powering pumps, fans and many types of conveyor. Consequently the electrical supply is one of the major services in food-processing operations and power ratings need to be determined for both installation purposes and running cost evaluation.

In addition, many instruments used for measurement and control purposes generate an electrical output. Examples are thermocouples and resistance probes for temperature measurement, force transducers for pressure measurement, and chemical sensors for humidity measurement. It is important to know the relationship between the electrical response of these sensors and the stimulus which they are responding to, as well as methods for accurately measuring and recording the electrical reponse.

Finally the electrical properties of foods assume importance in certain food-processing operations. These include the resistance, the conductance and the dielectric properties of food; these govern how the food will interact with electromagnetic radiation.

For these reasons, electrical properties will receive considerable attention.

12.2 ELECTRICAL UNITS

Electricity is concerned with the study of electrons. Electrostatic deals with the study of electrons at rest, whereas electricity currents are electrons in motion, the greater emphasis in this chapter being placed in electricity currents.

Terms which are commonly used in electrical work are charge, potential difference and current. A material is charged if it has either a surplus of electrons (negative) or a deficit of electrons (positive). A potential difference will exist whenever there is a difference in charge between two points (Fig. 12.1), and the charge on each will remain constant provided that they

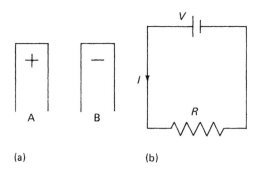

Fig. 12.1 — (a) Charged particles and current flow; (b) circuit to illustrate Ohm's law.

are isolated electrically from each other and from the earth which is usually regarded as having no charge. If A and B are connected by a material which will allow electrons to flow, electrons will flow from B to A in order to redistribute the charge (Fig. 12.1). Such a material is termed a conductor. This flow of electrons is termed the electric current and the size of the current depends upon the flow rate of electrons. The flow of a current under the inflence of a potential difference is analogous to the flow of water from the base of a tank under the influence of the head of liquid in the tank.

A battery or a cell converts chemical energy to electrical energy and produces a potential difference across two terminals. Potential difference is measured in volts and is defined later in terms of the energy dissipated by a current-carrying conductor. The convention usually used is that electricity is regarded as flowing from the positive to the negative; in reality, it is a flow of electrons in the opposite direction. Various types of cell have been described by Hughes (1977); these include a standard cell which gives a potential difference of 1.018 59 V at 20 °C and is used for calibration purposes.

Electric current is usually given the symbol I and is one of the fundamental dimensions. The unit of electric current is the ampere (A). When a current flows through a conductor, a magnetic field is produced and the ampere is defined in terms of the force between two conductors, each carrying an electric current (see section 12.5), as follows.

One ampere (1 A) is that current which when flowing in each of two infinitely long wires of negligible cross-sectional area, separated by 1 m in a vacuum, produces a force between the wires of 2×10^{-7} N for each metre of wire length. If the current flow is in the same direction, the force is one of attraction whereas, when it is in the opposite direction, the force in one of repulsion. A current of 1 A corresponds to a flow of electrons of $6.26 \times 10^{18} \text{ s}^{-1}$. Electrical current is measured by an instrument known as an ammeter.

The quantity of electricity or charge Q passing a point is measured in coulombs (C) and is obtained from the product of the current I and the time t:

$$\frac{\text{charge}}{Q \, (C)} = \frac{\text{current}}{I \, (A)} \times \frac{\text{time}}{t \, (s)}$$

One coulomb (1 C) is defined as the quantity of electricity transported in 1 s by a current of 1 A. When a current is passed through an electrolyte, chemical decomposition take place and the mass of an element liberated by 1 C of electricity is termed the electrochemical equivalent of that element.

A direct current refers to a flow of electrons in one direction only. Direct currents may be steady or variable. An alternating current occurs when the flow of electrons occurs in both directions, the most commonly encountered situation being the alternating current produced by an alternator where the current varies sinusoidally with time see (section 12.10). The first part of the chapter will be concerned with direct current (DC) applications. A rectifier is a device which converts alternating current to direct current.

The difference in electric charge between two points is known as the potential difference, or EMF, and is the driving force for the flow of electrons. The unit of potential difference is the volt (V), which is defined later in terms of the power dissipated when a current flows in a circuit (see section 12.4).

12.3 ELECTRICAL RESISTANCE AND OHM'S LAW

When a current flows through a conductor (Fig. 12.1) there is a direct relationship between the potential difference across the conductor and the size of the current, i.e. the ratio of the voltage to the current is equal to a constant, known as the resistance of that conductor. This relationship, known as Ohm's law is summarized as

$$\frac{\text{voltage } V}{\text{current } I} = \text{resistance } R$$

The unit of resistance is the ohm and is given the Greek symbol Ω. Resistances are represented diagramatically as shown in Fig. 12.2. Energy is dissipated when a current flows through a resistor (see section 12.4). Ohm's law applies to all DC applications and to alternating current (AC) circuiting containing only a resistance (see section 12.10.4).

Most metals are good conductors of electricity and have a low electrical

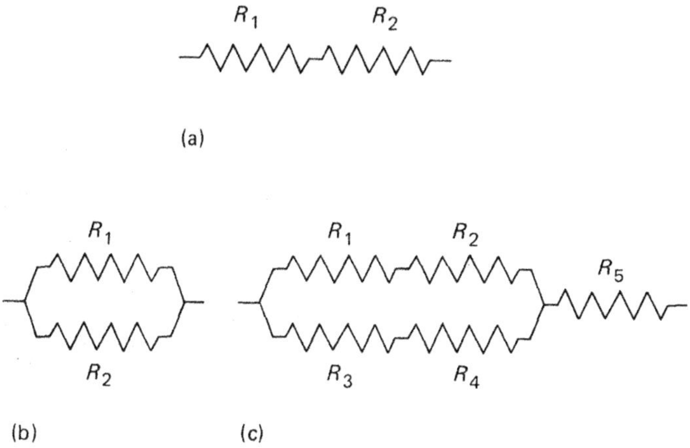

Fig. 12.2 — (a) Resistance in series; (b) resistances in parallel; (c) a combined network.

resistance. However, materials with high resistance values are known as *insulators*; materials with conduction properties midway between these extremes are known as *semiconductors*.

If a material is produced in the form of a wire, the total resistance of that wire is directly proportional to the length L and inversely proportional to the cross-sectional area a. Thus,

$$R \propto \frac{L}{a}$$

$$R = \rho_r \frac{L}{a}$$

where ρ_r is the proportionality constant known as the *resistivity*; its units are ohms per metre (Ωm^{-1}). The resistivity values for several metallic conductors and insulators are recorded in Table 12.1. Values for some food materials are discussed in more detail in section 12.7.

It can be seen that the resistivity can be related to the beahviour of electrons in the material and its temperature; the following expression can be used to predict the resistivity of a material:

$$\rho_r = \frac{m_e}{e^2 N_e t}$$

where m_e is the mass of an electron, e is the electron charge, N_e is the number of electrons per unit volume and t is the mean free time between collisions. Metals have the lowest resistivity of all materials because they contain more electrons per unit volume.

In a practical circuit, resistances can be connected in series (Fig.

Table 12.1 — Resistivity values of some conductors and insulators.

Conductor or insulator	Temperature (°C)	Resistivity ($\Omega\,m^{-1}$)
Annealed copper	20	1.72×10^{-8} [a]
Aluminium	20	2.6×10^{-8} [a]
Mild steel, 0.1 wt% carbon	20	2×10^{-7} [a]
Platinum	20	1.17×10^{-7} [a]
Carbon	20	4.6×10^{-5} [a]
Mica	27	10^{12} [b]
Glass	27	10^{10} [b]
Wood	27	$10^5 - 10^6$ [b]

[a] From the data of Say (1968).
[b] From the data of Schofield (1979).

12.2(a)), in parallel (Fig. 12.2(b)) or a combination of the two (Fig. 12.2(c)). For resistances in series, the total resistance R_T is given by

$$R_T = R_1 + R_2$$

For resistances in parallel, the total resistance R_T is given by

$$\frac{1}{R_T} = \frac{1}{R_1} + \frac{1}{R_2} + \frac{1}{R_3}$$

For complex systems the resistance of each parallel section should be evaluated before finally treating the network as a series network. In general terms the current will follow the route offering the least resistance. In this way, resistances can be added in series or parallel to measuring units to modify their working ranges.

Two other laws which are used in conjunction with Ohm's law to evaluate the flow of current in electrical circuits are known as Kirchhoff's laws.

The first law states that the current flowing towards a junction is equal to the current flowing away from the junction. For the circuit illustrated in Fig. 12.3, for any junction (B, for example)

$$I_1 = I_2 + I_3$$

Similar balances can be made at all junctions.

The second law states that in a closed circuit the algebraic sum of the products of the current and resistance of each part of the circuit is equal to the resultant EMF for the circuit.

In circuit ABCDEFGH,

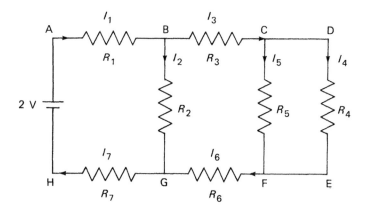

Fig. 12.3 — Circuit to illustrate Kirchhoff's laws.

$$2 = I_1R_1 + I_3R_3 + I_4R_4 + I_6R_6 + I_7R_7$$

and, in circuit BCFG where there is no resultant EMF,

$$0 = I_3R_3 + R_5R_5 + R_6R_6 - R_2I_2$$

12.3.1 Electrical conductance

The specific conductance or conductivity K is the inverse of resistivity and is preferred for comparing the resistances of liquids. Therefore,

$$K = \frac{1}{\rho_r} = \frac{L}{Ra} \quad \Omega^{-1}\mathrm{m}^{-1}$$

The reciprocal ohm (Ω^{-1}) is also known as the mho or siemen (S). Therefore the SI units of conductivity are siemens per metre ($\mathrm{S\,m^{-1}}$). However, the units of specific conductance most often encountered are mho cm^{-1}, where

$$1\,\mathrm{mho\,cm^{-1}} = 10^2\,\mathrm{S\,m^{-1}}$$

Physical chemists often use the term molar conductivity Λ which is the conductivity per mole of electrolyte. If the concentration of the solution is c ($\mathrm{mol\,m^{-3}}$), then

$$\Lambda = \frac{K}{c}$$

The units of molar conductivity of most solutions rises by about 2% degC^{-1}.

Therefore the temperature must be controlled to within about 0.05 degC to ensure that an accuracy of 0.1% is obtained for K.

12.4 ELECTRICAL ENERGY

When a current flows through a potential difference, a quantity of energy is dissipated in the form of heat. This is used to define the unit of potential difference as follows. One volt (1 V) is the difference of electric potential between two points of a conducting wire carrying a constant current of 1 A, when the power dissipated between these points is equal to 1 W ($J s^{-1}$). Instruments used for measuring potential difference are known as voltmeters.

The total amount of electrical energy E evolved when a current I flows over a potential difference V for a time t is given by

$$E(J) = V(V) \times I(A) \times t(s)$$

The electrical power rating P is given by VI (W). In this way, electrical energy is converted to thermal energy. Electrical methods are used for estimating thermal properties of foods (see sections 8.5 and 9.7).

Thus an electric bulb rated at 60 W running off a power source of 240 V would carry a current of 60/240 = 0.25 A. Energy would be dissipated at the rate of $60 J s^{-1}$.

The amount of electrical energy used by the consumer is measured in terms of kilowatt hours or units. The number of units used is equal to the product of the power rating in kilowatts and the time in hours:

$$\text{number of units used (kWh)} = \text{power rating (kW)} \times \text{time (h)}$$

Thus a lamp rated at 60 W alight for 20 h would use (60/1000) × 20 = 1.2 units. Currently the cost of electricity to the UK domestic consumer is 5.2p per unit.

The pricing system for industrial users is not so straightforward, being based on the maximum demand as well as the total number of units. Electricity costs can be reduced by trying to spread the load throughout the day, thereby ensuring that there are no excessive peaks in electrical demand.

It is perhaps interesting to note that one unit of electricity is equivalent to $1000 J s^{-1} \times 3600 s$, i.e. $3.6 \times 10^6 J$ (3.6 MJ) of energy. The latent heat of vaporization of water at atmospheric pressure is $2.257 MJ kg^{-1}$; so the evaporation of 1 kg of water vapour by direct electrical methods would require 0.63 units of electricity and would cost 3.28p (assuming no heat losses and ignoring sensible heat changes). In contrast, heating 1 kg of milk from 5 °C to 72 °C would require $1 \times 67 \times 4000 = 0.268 MJ$ and would cost 0.387p.

Other expressions for electrical power can be derived from Ohm's law:

Sec. 12.5] Magnetic effects associated with an electric current

$$P = I^2R \quad \text{or} \quad \frac{V^2}{R}$$

Electrical power is measured using a wattmeter, the most common types being those to measure AC power consumption for household supply.

12.5 MAGNETIC EFFECTS ASSOCIATED WITH AN ELECTRIC CURRENT

When a current flows through a conductor a magnetic field is produced; this can be represented by lines of force, giving the direction a compass needle would point if placed at that point. If such a conductor is placed near another current-carrying conductor, the magnetic fields interact and give rise to a force. This principle is used in the definition of electric current. The lines of force associated with the flow of current in a straight wire are shown in Fig. 12.4(a). When a current flows through a coil wrapped round an iron bar, the

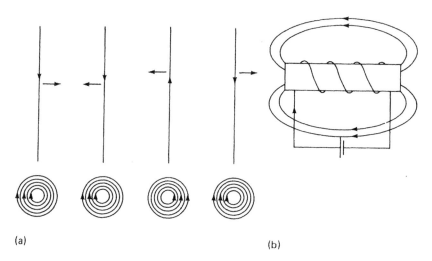

(a) (b)

Fig. 12.4 — Lines of force associated with (a) a current flow in a straight wire and (b) current flow in a wire coiled round a bar.
(Copyright © 1974 IEEE)

bar becomes magnetized. The lines of force set up are shown in Fig. 12.4(b).

The opposite effect can be observed when a permanent magnet is moved through a conductor wound in the form of a coil. As the conductor moves through the coil, an electric current in induced in the coil.

The interaction between an electric current and magnetic field has three important applications.

(1) The measurement of electrical variables, such as a voltage, current, power and resistance.

(2) The conversion of mechanical energy to electrical energy, e.g. in an alternator or generator.
(3) The conversion of electrical energy to mechanical energy, e.g. electric motors.

12.6 MEASUREMENT OF ELECTRICAL VARIABLES

When a conductor is placed in a magnetic field, the size of the force produced depends upon the current, the magnetic field strength and the length of the wire.

If the wire is arranged as shown in Fig. 12.5(a) and placed between the

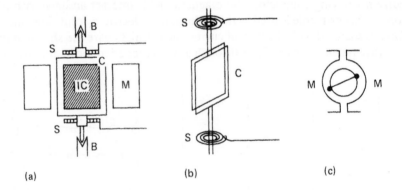

Fig. 12.5 — The moving-coil mechanism: C, coil; M, magnet; S, springs; B, bearings; IC, iron core.

poles of a permanent magnet, it can be used to measure electric current. The movement, known as the d'Arsonval movement, consists of moveable coil placed in the field of permanent magnet. The current in the coil creates a torque on the coil, which then rotates until the torque is balanced by restoring spring (Fig. 12.5(b)). A pointer is attached to coil and moves across a scale. The deflection of the pointer is directly proportional to the current flow. Such an instrument is called a moving-coil meter and is suitable only for direct current. If the direction of the current is required for example in null balance methods, it must be arranged to be a centre-zero instrument. Usually, it can only handle only a small current and must be appropriately adapted for use as an ammeter or voltmeter. For use as an ammeter, it is connected in parallel with a low resistor, known as a shunt, so that most of the current passes through the resistor. For measurement purposes an ammeter is placed in series with the load.

When used as a voltmeter, it is connected in series with a large resistance so that most of the voltage drop occurs over the resistance (Fig. 12.6). For measurement purposes a voltmeter is placed in parallel with the

Sec. 12.6] Measurement of electrical variables 375

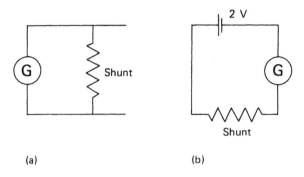

Fig. 12.6 — The use of a galvanometer G and shunt resistance (a) as an ammeter and (b) as a voltmeter.

load. By using appropriate resistances the moving-coil mechanism can be used to measure currents over the range from microamperes to kiloamperes and to measure potential differences over the range from microvolts to kilovolts.

A multimeter is a general-purpose instrument, designed to measure current, voltage, resistance and frequency over a wide range of values. For measuring alternating current and voltage, the most common type of instrument used is a moving-iron meter (see section 12.10.3).

12.6.1 Potentiometer
A potentiometer circuit can be used for comparing and measuring potential differences, the principle of action being that the unknown EMF is measured by balancing it against a known potential difference. It consists of a DC power supply, a variable resistance R_v, a potentiometer wire and sliding contact and a sensitive centre-zero galvanometer (Fig. 12.7).

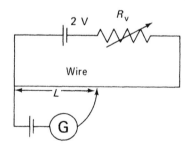

Fig. 12.7 — A potentiometer circuit: G, galvanometer.

The potentiometer wire is calibrated using a standard cell whose potential difference is accurately known by determining the position of the slide along the wire which gives a null deflection. The standard cell is replaced by the unknown voltage, and the corresponding new balance point is found. When a standard cell is used, the readings should be taken quickly to avoid removing too much current, which will affect its accuracy. The variable resistance is adjusted so that he balance lengths determined are not too short.

The unknown is $(PD)_u$ calculated from the expression

$$(PD)_u = \frac{L_v (PD)_s}{L_s}$$

where $(PD)_s$ is the potential difference of the standard cell, and L_v and L_s are the balance lengths for the unknown cell and the standard cell respectively. Accurate measurements rely on the facts that the standard cell voltage is accurately known, the potentiometer wire is uniform along its length and the galvanometer is capable of detecting very small currents, typically in the microampere range.

The potentiometer is very useful for measuring the terminal voltages of batteries. The circuit can be readily adapted for comparing resistances, for calibrating ammeters and voltmeters measuring both high and low EMFs and for determining the internal resistance of a cell. Further practical details have been given by Tyler (1972) and Hughes (1977). A commercial form of potentiometer is available where the slide wire is replaced by a number of equal resistances connected in series.

12.6.2 Electrical resistance measurement

One of the simplest ways of measuring the resistance of a material is to put a known potential difference across the material and to measure the current accurately, the resistance being calculated using Ohm's law.

The Wheatstone bridge network is an alternative method used for precision measurement of resistances in the range from $1\,\Omega$ to $1\,M\Omega$. The bridge consists of four resistance branches, a DC power supply and a centre-zero galvanometer, which are connected as shown in Fig. 12.8(a).

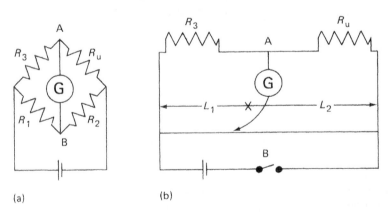

Fig. 12.8 — (a) A Wheatstone bridge network; (b) a meter bridge network: G, galvanometer.

Resistances R_1, R_2 and R_3 are known whilst R_u is the unknown resistance. The bridge is balanced when no current passes through the galvanometer when the circuit is made (null-balance method). Under these conditions the potential at A is equal to the potential at B and it can be shown, using Kirchhoff's laws, that

$$\frac{R_1}{R_2} = \frac{R_3}{R_u}$$

Therefore,

$$R_u = R_2 \frac{R_3}{R_1}$$

In a practical circuit, R_1 and R_2 may be fixed and R_3 is a variable resistance which is altered until the null-balance position is attained.

Another instrument often described is the meter bridge, which is an adaptation of the Wheatstone bridge, where R_1 and R_2 are replaced by a potentiometer wire, which is divided by a sliding contact. The balance point is found by determining the position of the sliding contact when no current flows through AB (Fig. 12.8(b)). The distances L_1 and L_2 are recorded and substituted into the following equation to evaluate the unknown resistance:

$$\frac{R_1}{R_2} = \frac{R_3}{R_u} = \frac{L_1}{L_2}$$

Note that R_3 is a standard resistance whose value is accurately known.

12.7 RESISTIVITY AND SPECIFIC CONDUCTANCE OF FOODS

Only a limited amount of attention has been paid to the determination of the resistance and conductivity of fruits and vegetables. In most cases, these investigations have been directed towards, trying to establish a relationship between the electrical property and some quality factor of the product. Mohsenin (1984) describes data for which resistance measurements have been used to measure the sugar content of water melons and to detect bruising in apple tissue. The resistivities of some foods are given in Table 12.2. The resistance was determined using an impedance bridge, and the resistivity calculated from the dimensions of the sample.

Conductance measurements have also been used to measure moisture content of materials, particularly grain products. Conductance measurements are favoured over capacitance measurements, mainly because of their simplicity and greater change with moisture content; further details have been supplied by Mohsenin (1984).

Judson-King (1971) showed how the electrical resistance of orange juice increased as the temperature decreased (as the juice was frozen). A linear

Table 12.2 — Resistivities of foods.

Food	Resistivity (Ω m)
Apple (McIntosh)	95.0
Apple (Winesap)	75.0
Potato (Alabama White)	33.3
Potato (Idaho)	45.3

Adapted from the data of Mohsenin (1984).

relationship was observed when the logarithm of the resistance was plotted against the temperature over the range from 0°C to −40°C, with the resistance levelling out at about −45°C. The increase in resistance was attributed to the transition from water to ice over this temperature range. A hysteresis effect was observed when resistance values were measured during heating and freezing, with resistance values being higher during the heating cycle. This suggests that there is a higher degree of solidification at any temperature during the heating process and that supercooling of the liquid occurred as the temperature is lowered.

Methods for measuring specific conductance have been described by Levitt (1973). Most rely on resistance measurements using a Wheatstone bridge circuit, with an alternating current to prevent electrolysis. The dimensions of the cells used for measurement are not accurately known; they need to be calibrated using a solution whose conductivity is accurately known. A potassium chloride solution is often used for this purpose at a temperature of 25°C, the procedure being described by Levitt (1973). The resistance of the cell containing the standard solution and the unknown liquid is measured and the specific conductance K' of the unknown liquid is determined from the following equation:

$$K' = \frac{KR}{R'}$$

where K (mho m^{-1}) is the specific conductance of the standard solution, R (Ω) is the resistance of the standard solution and R' (Ω) is the resistance of the unknown solution. Conductivity measurements can be used by the physical chemist for determining the solubility of sparingly soluble salts, the hydrolysis of salts, the rates of reactions in situations where changes in conductivity occur and the end-point of reactions between electrolytes.

The specific conductance of tap water is extremely variable, the major contribution being derived from impurities in the water (in particular, dissolved ions). Ordinary laboratory distilled water generally has a conductivity of $(3-5) \times 10^{-4}$ S m^{-1}. Deionized water, which is produced by replacing anions by OH$^-$ ions and cations by H$^+$ ions, has a very low specific conductance of about $10^{-5}\,\mathrm{S\,m^{-1}}$. Conductivity measurements can be used to monitor the quality of deionized water, as the specific conductance will increase as the resin becomes exhausted. Conductivity changes are not so pronounced in a water-softening process, where calcium and magnesium ions are replaced by sodium ions.

The specific conductances of a wide variety of aqueous solutions of inorganic salts have been given by Weast (1982).

Most dairy products are poor conductors of electricity. Bovine milk has conductance values ranging from $0.004\,\mathrm{mho\,cm^{-1}}$ to $0.0055\,\mathrm{mho\,cm^{-1}}$ (Lewis, 1986b); there are differences between varieties. The major contribution comes from potassium and chloride ions. It might be expected that increasing the total solids concentration would increase the specific conductance but the relationship is not so straightforward. The conductance of concentrated skim-milk was shown to increase to a maximum value of about $0.0078\,\mathrm{mho\,cm^{-1}}$ at 28% total solids, after which it decreased. This was explained by the extremely complicated salt balance between the colloidal and soluble phases.

The presence of fat tends to decrease the specific conductance. Little difference has been found between milk from different species. Milk fat has a specific conductivity of less that $10^{-14}\,\mathrm{S\,m^{-1}}$, as do other oils and fats.

It has been suggested that conductivity measurements may be useful for monitoring processes in which these changes occur. For example, in the decationization of sweet whey using Amberlite IR 120 in its hydrogen form, the conductance rose from $5.2\,\mathrm{mmho\,cm^{-1}}$ to $8.15\,\mathrm{mmho\,cm^{-1}}$. The demineralization level was 75% and the pH fell from 6.22 to 1.7. Increasing the pH to 3.0 using Amberlite IRA 47 in the OH^- ion from decreased the conductivity to $0.165\,\mathrm{mmho\,cm^{-1}}$ and gave a demineralization level of 94%.

The development of acidity which occurs during many fermentations has also been observed to increase the conductivity. Mastitic milk also has an increased conductivity because of its raised content of sodium and chloride ions. Further information on milk products has been supplied by Jenness *et al* (1974).

Conductivity measurements are also used for monitoring the concentration of sugar liquor during the concentration process which proceeds to crystallization. The specific resistance of the liquor is proportional to the viscosity, which in turn follows supersaturation closely. Therefore, there is an inverse relationship between conductivity and the degree of supersaturation. Further detail on the instrumentation has been given by Hugot (1972).

Mohsenin (1984) has presented data for the specific conductivities of several fruits and vegetables at 25 °C measured at different frequencies; a selection of these are recorded in Table 12.3.

For most materials the plot of specific conductivity against frequency has a sigmoidal shape. At a given frequency the conductivity values vary very little from one product to another. The major change in conductivity occurs as the moisture content is reduced, with reductions as high as 10^4-fold for some products during dehydration.

Table 12.3 — Specific conductivities of some fruits and vegetables.

Specific conductivity (mmho cm $^{-1}$) of the following fruits and vegetables

Frequency (Hz)	Potato (white)	Carrot	Apple	Peach
10^3	0.46	0.26	0.25	0.52
10^4	0.51	0.30	0.34	0.59
10^5	1.1	0.52	1.0	1.3
2×10^6	6.5	5.2	—	—
3×10^7	7.8	7.9	1.6	4.3

Adapted from the data of Mohsenin (1984).

The resistivity or conductance of foods is of direct interest in the resistance heating of foods; this is described in section 12.10.4.

12.8 ELECTRICAL SENSING ELEMENTS

A wide range of sensors or transducers is available; these produce an electrical response to a given stimulus, examples of which are temperature, pressure, moisture content and light intensity. When the magnitude of the electrical response is related to the magnitude of the stimulus, it allows the electrical response to be used to estimate the value of the stimulus. Electrical measurement is now widely used for temperature, pressure, humidity and other measurements. Some of these sensors in common use will now be examined in more detail.

12.8.1 Thermocouples

A thermocouple is an extremely useful instrument for measuring temperature. If two different metals, such as copper A and iron B, are joined together as shown in Fig. 12.9(a) and one of the junctions H is heated, then an EMF is developed and a current flows round the circuit. This thermoelectric effect is called the Seebeck effect after the man who discovered it and the junction is called a thermocouple. The magnitude and direction of the thermoelectric EMF depends upon the temperature difference between the hot and cold junctions and the types of wire.

In this example the current flows from copper to iron at the hot junction and from iron to copper at the cold junction. On other cases the current flow at the hot junction is from the earlier to the later metal in a series, a portion

Sec. 12.8] **Electrical sensing elements** 381

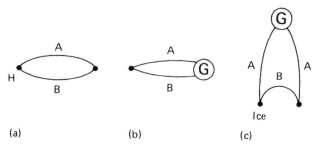

Fig. 12.9 — Thermocouples: (a) a simple thermocouple; (b) a simple thermocouple attached to a galvanometer; (c) a double-junction thermocouple with a reference junction in melting ice.

of which is as follows (Jones, 1974a, 1974b): nickel, platinum, copper, lead, tin, chromium, gold, silver, iron.

The Peltier effect is observed when a current flows across the junction. When it flows in one direction, energy is released from the junction but, when the direction is reversed, energy is absorbed from the surroundings.

The most common metal combinations used for industrial thermocouples are copper–constantan, chromel–alumel, iron–constantan and platinum–(rhodium–platinum) (British Standards Institution, 1974) (note that constantan is an alloy containing 40% nickel and 60% copper, that chromel contains 90% nickel and 10% chromium and that alumel contains 94% nickel, 2% aluminium, silicon and manganese).

The response of a thermocouple can be investigated using a thermoelectric diagram for the metals in question. This consists of a plot of microvolts per degree Celsius against temperature, using lead as the second metal in each case. Such plots for copper, iron, lead and platinum are shown in Fig. 12.10(a).

When two metals are used and the junctions are at temperatures θ_1 and θ_2, the EMF developed can be determined from this diagram and is equal to the area of the trapezium formed. The point where the lines for the two metals cross is known as the neutral temperature for that combination.

Care should be taken to avoid this region as the EMF developed starts to decrease as the temperature of the hot junction is increased above the neutral temperature, and ambiguities will creep in at temperatures close to the neutral point. Fig. 12.10(b) shows the EMF response for a copper–iron thermocouple (neutral temperature, 275 °C) when the cold junction is kept at 0 °C. The neutral temperature for all the listed industrial thermocouples is well outside their recommended ranges.

Table 12.4 shows the electrical response for the four major types of industrial thermocouple with the cold junction maintained at 0 °C. Type K and type T thermocouples are probably those most often used. The normal working range for type K is from 0 to 1100 °C and for type T from −200 °C to 400 °C, but in each case it may be modified by the construction.

If it is not convenient to keep the cold junction at 0 °C, a correction can

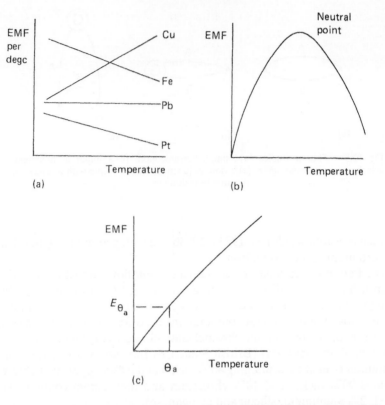

Fig. 12.10 — (a) A thermocouple diagram; (b) the relationship between the EMF and temperature; (c) cold-junction correction for a thermocouple.

be made using the law of intermediate temperatures. The EMF developed between any two temperatures θ_2 and θ_1 is determined from

$$E_{\theta_2} - E_{\theta_1}$$

If the cold junction is at the ambient temperature θ_a and the hot junction is at θ (Fig. 12.10(c)), the EMF developed $E_{\theta_a}^{\theta}$ is given by

$$E_{\theta_a}^{\theta} = E_{\theta} - E_{\theta_a}$$

The EMF developed by the same thermocouple with junctions at 0 °C and θ_a is given by

$$E_0^{\theta_a} = E_{\theta_a} - E_0$$

Therefore, if the ambient temperature is known,

Table 12.4 — EMFs for industrial thermocouples with the cold junction at 0 °C.

Temperature (°C)	EMF (mV) for the following thermocouples			
	Platinum–(platinum–rhodium), type S	Iron–constantan, type J	Chromel–alumel, type K	Copper–constantan, type T
0	0	0	0	0
50	0.297	2.584	2.02	2.021
100	0.642	5.267	4.10	4.239
150	1.024	8.004	6.13	6.631
200	1.435	10.781	8.13	9.178
250	1.866	13.561	10.16	11.862
300	2.314	16.325	12.21	14.666
350	2.777	19.089	14.29	17.580
400	3.249	21.849	16.40	20.592
450	3.730	24.607	18.51	

More detailed tables have been given by Jones (1974a).

$$E_0^\theta = E_{\theta_a}^\theta + E_0^{\theta_a}$$

$E_{\theta_a}^\theta$ is the measured EMF and $E_0^{\theta_a}$ is the correction for the ambient temperature. The sum of these terms gives the EMF developed between 0 °C and θ, allowing the tables to be used to determine the value of the unknown temperature. For example, a copper–constantan couple working with the cold junction at 18 °C and the hot junction at θ gives an EMF of 4.804 mV. The cold junction correction $E_0^{18\,°C} = 0.703\,mV$ (from tables). Therefore the total EMF developed between 0 °C and θ °C is $E_{0\,°C}^{18\,°C} + E_{18\,°C}^\theta = 4.804 + 0.703\,mV$. From the tables, θ is approximately 127 °C.

For practical measurements, the junction of a thermocouple is attached to a sensitive galvanometer, potentiometer or chart recorder and the measuring instrument acts as the cold junction. Most recording thermometers have provisions for measuring the cold-junction temperature and for automatically compensating for temperature fluctuations. Alternatively a double-junction thermocouple can be used, with one of the junctions being permanently located in melting ice, but in many cases this is not convenient.

Thermocouples have a small heat capacity and respond quickly to changes in temperature. They are useful for measuring temperatures at point locations. One source of error arises from heat conduction along the wires; these can be reduced by using thin wires. A protective sheath is occasionally used to give more mechanical strength to the instrument, but this may increase the response time.

12.8.2 Resistance thermometers

The resistance of most good conductors increases with temperature. Over a narrow range of temperatures, the relationship is linear (Fig. 12.11(a)) and

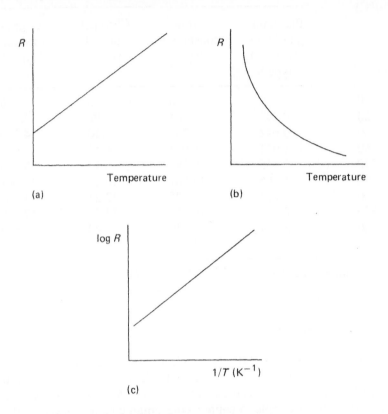

Fig. 12.11 — The relationship (a) between resistance and temperature for a good conductor, (b) between resistance and temperature for a semiconductor and (c) between log R and $1/T$ (absolute) for a semiconductor.

is given by

$$R_\theta = R_0(1 + \alpha\theta)$$

where R_θ and R_0 are the resistances at $\theta\,°C$ and $0\,°C$ respectively and α ($°C^{-1}$) is the temperature coefficient of resistance. As the temperature increases, the movement of electrons is hindered by the increased vibrations of other particles. Values for the temperature coefficient of resistance for some different metals are given in Table 12.5.

On the whole, there is little variation between the different metals. The materials usually used for industrial resistance thermometers are made of copper, nickel or platinum. The material used should conform to the

Sec. 12.8] Electrical sensing elements 385

Table 12.5 — Temperature coefficients α of resistance at 20°C.

Metal	$\alpha(\times 10^{-4}\,°C^{-1})$	Metal	$\alpha(\times 10^{-4}\,°C^{-1})$
International Standard Annealed Copper	39.3	Platinum	39
Aluminium	39.0	Nickel	50
Iron (wrought)	55.0	Lead	40
Steel (0.1 wt.% carbon)	42.0		

Adapted from the data of Say (1968).

standards laid down by the British Standards Institution (1964a) and the ratio of the resistance at the steam point to the resistance at the ice point should exceed 1.392 to comply with the requirements of the International Practical Temperature Scale. Most commercial thermometers available have a resistance of 100 Ω at 0 °C and a resistance of 138.5 Ω at 100 °C. The resistance values at some other temperatures are recorded in Table 12.6.

Table 12.6 — Variation in resistance with temperature.

Temperature (°C)	Resistance (Ω)	Temperature (°C)	Resistance (Ω)
−100	60.26	60	123.24
−50	80.30	80	130.89
−20	92.16	100	138.50
0	100.00	120	146.06
20	107.80	150	157.31
40	115.54	200	175.83
		250	194.07

A wide variety of resistance bulbs are available depending upon the required function, e.g. rapid response or surface measurement. They are very accurate and can be used over a wide temperature range (from −258 °C to 800 °C). The accuracy can be further improved by using compensating leads. Further details of the calibration and use of resistance probes have been given by Jones (1974a).

For more precise work the relationship between resistance and temperature is presented as

$$R_t = R_0(1 + \alpha\theta + \beta\theta^2)$$

12.8.3 Semiconductors (thermistors)

Semiconductor materials or thermistors show a negative temperature coefficient of resistance, with the resistance decreasing as the temperature increases. Thus, electrons are made more available as the temperature increases, because sufficient energy is provided to push them into the conduction band. More detailed explanantions have been provided by Schofield (1970).

A relationship of the following form has been observed between the resistance and the temperature (absolute):

$$R = a\exp\left(\frac{b}{\theta}\right)$$

Therefore,

$$\ln R = \ln a + \frac{b}{\theta}$$

A straight line results when the resistance is plotted against the inverse of the absolute temperature on semilogarithmic paper. Figs. 12.11(b) and 12.11(c) show the relationship between the resistance and the temperature in ordinary coordinates and semilogarithmic coordinates, respectively. The changes in resistance involved are much greater than for resistance probes for identical changes in temperature.

Thermistors are usually made from metal oxides or silicon semiconductors and for temperature measurement can be supplied in the form of beads, disks, rods or probes. They are not as accurate as resistance probes because the electrical characteristics are not so reproducible but are extremely useful for the detection of small changes in temperature and the prevention of excessive heating in electrical circuits.

12.8.4 Strain gauge transducer

When a metal is stretched, the length increases and the cross-sectional area decreases. Since the electrical resistance depends upon the length and area, the overall effect is to change the resistance of the material. This is the principle behind the strain gauge. A simple strain gauge is shown in Fig. 12.12(b). The detection element is fixed onto the material subjected to the stress. The principle of the strain gauge can be used in a pressure transducer (Fig. 12.12(b)). The force on the diapraghm is transmitted to a spring S attached to posts P. Strain gauges G are wound at the top and bottom of the posts. Deflection of the spring will cause the posts to tilt, thereby decreasing the strain on the top windings and increasing it on the bottom windings.

If the strain gauges are incorporated in a Wheatstone bridge circuit, the bridge may be unbalanced and can be used to measure the force. If correctly constructed, the output from the Wheatstone bridge can be made to be a linear function of the force applied and independent of temperature.

Pressure ranges involved are 0–0.75 bar and 0–700 bar. They are suitable for corrosive fluids with a working range from below $-40\,°C$ to $150\,°C$.

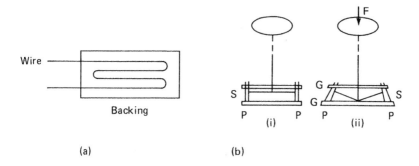

Fig. 12.12 — (a) simple strain gauge; (b) a strain gauge acting as a pressure transducer for (i) no force and (ii) on application of the force showing the spring S, the post P and the strain gauge G.

12.8.5 Humidity measurement

Electrical methods can also be used to measure the moisture content of solids, liquids and gases. Of particular interest are those for measuring the moisture content of solids and the humidity of air.

For solids, there is a change in electrical resistance with moisture content. Unfortunately the resistance is also affected by the packing density and these methods are of limited accuracy.

Hygroscopic materials will either absorb or give off moisture depending upon environmental conditions. This will result in a change in electrical resistance or capacitance. Sensors containing either calcium or lithium chloride will be subjected to such changes in resistance. Aluminium oxide sensors have also been used, and the electrical impedance changes with moisture content.

12.8.6 Other sensors

The piezoelectric effect relies on the change in EMF which is produced when a quartz crystal cut in a special way and placed between two plates is subjected to different forces. The piezoelectric effect can be used for pressure measurement and for measuring time.

Other electrical sensors can be used for flow measurement and level detection. There are also many light-dependent resistors and light-activated switches. A wide range of electrical transducers has been described in more detail by Fink (1975).

12.9 PROCESS CONTROL AND AUTOMATION

It is important to be able to measure, record and control such variables as temperatures, pressure, flow rates, levels and many others. The trend in the food industry in developed countries is towards operating food-processing plant using automation under the control of a central processing unit which is either a microprocessor or a larger computer.

Some of the benefits of automation are improved flexibility, economy, better utilization of plant, more comprehensive processing data, uniformity and a better-quality product.

On fully automated plant, there are facilities for setting routes, opening and closing valves, starting pumps and motors and controlling events such as emptying, filling and cleaning tanks, dosing operations and cleaning-in-place sequences. The central processing unit will also check that all plant items such as valves and motors are working correctly and report any failures. They can also be used for planning maintenance work by recording how long pumps and motors have been running or how many times a valve has opened or closed. More sophisticated units will also check that there are no faults in the electronic circuits of the central processing unit.

However, the effectiveness of an automatic control system is very dependent on the accuracy and reproducibility of the input signals from the measurement probes, and so these will also need checking at regular intervals. There are two types of input signal, namely digital and analogue. Digital signals are on–off signals and are used to indicate whether a motor is running or stopped or whether a valve is open or closed. Analogue signals are used for transmitting readings such as temperature, pressure or flow rate, which can take a range of values.

Furthermore, successful automatic control relies on a thorough understanding of the process, in order that the correct sequence of operations is programmed into the central processing unit.

Some of these points will be illustrated by reference to a cleaning-in-place process (Fig. 12.13) where it is required to clean tank A with either an

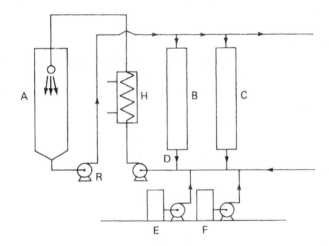

Fig. 12.13 — Application of cleaning-in-place process to a product storage tank: A, tank; B, C, detergent tanks; D, detergent pump; E, concentrated-acid detergent; F, concentrated-alkali detergent; H, heat exchanger; R, return pump.

Sec. 12.10] **Alternating current** 389

acid or an alkali detergent, which can be made up to the desired strength in either of the detergent tanks B and C by mixing soft water with the concentrated detergents from E or F. The cleaning-in-place unit must be able to make up acid or alkali detergent to a desired strength as required and to heat it to the correct temperature. There should be provision for routing the detergent to the pipes or tanks to be cleaned and for pumping it back to a collection tank for possible further use. If different detergents are to be used, it should be possible to rinse the tank with water between the detergents.

Facts about the process which should be understood are detergent strengths, operating temperatures, cleaning times and transit times through pipes. Checks should be made to ensure that different detergents do not mix with each other or that detergent does not mix with the product in other tanks. There will probably be many storage tanks, often containing different products, so that it is important that they are not allowed to mix. It may happen that tank A is only one of many and that some tanks are being filled with one product, others are being drained of another and yet others are being cleaned, all at the same time. The operations can be controlled and monitored by successfully integrating process control and automation methods. Further details on automatic control have been given by Beaverstock (1983).

12.10 ALTERNATING CURRENT

12.10.1 Introduction

In many industrial situations, it is more convenient to deal with alternating currents and voltages rather than direct current. Most of our electricity for domestic and industrial purposes is supplied in AC form.

Electrical energy is transferred at a very high voltage to mimimize energy losses. Transmission voltages vary between 132 kV up to as high as 750 kV. However, it needs to be reduced or stepped down before it can be used safely by the consumer. This is much easier to achieve for alternating current than for direct current, being accomplished by means of a transformer (see section 12.15).

12.10.2 Production and characteristics of an alternating EMF

If a coil rotates in a magnetic field, an alternating current is set up (Fig. 12.14). This can be transferred to an external circuit by brushes. The nature of the alternating current is a sine wave (Fig. 12.15) which is described as follows.

Initially, there is no current flowing, then the current starts to increase in one direction until a maximum is reached, next the current falls back to zero, followed by a change in direction, going to a maximum in the other direction, and finally the current falls back to zero.

This pattern of events is termed a cycle and the frequency of an AC source is measured in cycles per second or hertz (Hz). In Great Britain the

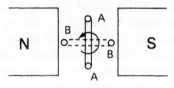

Fig. 12.14 — Production of an alternating EMF.

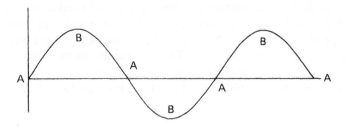

Fig. 12.15 — Sinusoidal nature of an alternating current and voltage.

frequency of AC sources is 50 Hz. Therefore this sequence of events is repeated 50 times every second. In the USA a frequency of 60 Hz is used.

One revolution of the coil produces one cycle. Therefore a generator rotating at 3000 rev min would produce an AC voltage with a frequency of 50 Hz. The size of the EMF depends upon the magnetic field strength, the number of turns in the coil, the size of the coil and the speed of rotation. The frequency can be increased by increasing the number of poled pairs. If n poled pairs are introduced, the relationship between the frequency f and rotational speed N is given by

$$f = Nn$$

The peak voltage and current obtained are referred to as the maximum voltage and current. However, in practical situations the maximum values are only attained for a very small fraction of the total time and it is much more sensible to deal with the effective or root-mean-square currents or voltages. These are the values normally recorded by AC ammeters or voltmeters, which are normally of the moving-iron type.

It can be shown mathematically that

$$I_{RMS} = 0.7071 I_{max}$$

$$= \frac{I_{max}}{\sqrt{2}}$$

and

$$V_{RMS} = 0.7071 V_{max}$$

$$= \frac{V_{max}}{\sqrt{2}}$$

The effective or root-mean-square value of an alternating current is measured in terms of the direct current that produces the same heating effect in the same resistance.

The power expended in an AC circuit containing only a resistance is similar to that for a DC circuit and is given by

$$P = I_{RMS} V_{RMS} = VI$$

The voltage and current are in phase in this type of circuit. Thus, domestic single-phase electrical equipment rated at 100 W and working at 240 V would carry a current of 0.416 A. The maximum voltage and current would be 339.5 V and 0.588 A respectively. The maximum voltage of an AC supply can be measured using a cathode-ray oscilloscope. Ohm's law can be applied to AC circuits containing only a resistance. The introduction of a capacitance and inductance complicates the procedure and is covered in section 12.11.

Such a supply is known as single phase and is the way in which electricity is supplied to individual households. Installations requiring larger electrical loads will be provided for by a three-phase supply (see section 12.16).

The process of converting an AC supply to a DC supply is known as rectification. A DC power supply will usually consist of a rectifier and a smoothing circuit to provide a reasonably constant output.

12.10.3 Voltage, current and power measurement

Alternating current and voltage are measured using a moving-iron meter. Although the current or voltage fluctuates, the moving system takes up a mean position at which the mean torque produced is proportional to the mean value of the square of the current. Therefore the instrument measures root-mean-square current or voltage and the scale is not linear. This instrument can also be used for measuring direct current and voltage.

Electrical power in an AC circuit is measured using a wattmeter, the sensing element producing a signal which is proportional to the product of the root-mean-square current and voltage. The most common application is for measuring household consumption of electricity.

There are no problems when the current and voltage are in phase, as occurs in resistance heating.

12.10.4 Electrical heating

One very interesting development of AC resistance heating is concerned with the direct conversion of electrical energy to heat by using the fluid as the conducting medium. A requirement is that the fluid is capable of conducting a current but has sufficient resistance for energy losses to occur and heat to be dissipated. The name Ohmic heating has been adopted to describe the process.

The fluid is placed in a non-conducting tube which has electrodes at each end. An alternating current at mains frequency is passed through the fluid and heat is generated within the fluid, the conversion efficiency being very efficient. The major advantages are that even heating results, that there are no temperature gradients and that none of the usual limitations due to conduction and convection arises. This is particularly useful for viscous liquids and liquids containing particulate matter. It is suitable for continuous processing and the need for a hot heat transfer surface is eliminated, thereby reducing problems arising from fouling. Liquids containing particulate matter can be processed and will not be subjected to the high shear rates found in scraped-surface heat exchangers. It is claimed that the temperatures needed for UHT processes can be easily attained and the conversion efficiency for electrical energy to heat is greater than 90%. In principle, the technique is similar to microwave heating, compared with which it is claimed to have a lower capital cost and a higher conversion efficiency.

Process vessels may be heated by attaching resistors to the vessel wall or by immersing sheathed resistors in the material to be heated. The resistors are usually made of nickel–chromium and can assume a variety of shapes, depending upon the situation. Such resistance methods are used in baking ovens, for shelf heating or in freeze drying and for heating air for hot-air drying processes. For heating fluids in tubes the heating element can be supplied as a tape or the resistance wire can be wound round a non-conducting (electrical) tube such as quartz through which the product flows. This latter application has been used for pasteurizing fluids. Electrical heating in catering operations has been described in more detail by Milson and Kirk (1980) and Kirk and Milson (1982).

12.11 AC CIRCUITS

Other electrical components play an important role in alternating circuits, these being inductors and capacitors. The symbols are shown in Fig. 12.17.

When inductances and capacitances are introduced into an AC circuit, the current and voltage may no longer be in phase. Fig. 12.16(a) shows a

Sec. 12.11] AC circuits

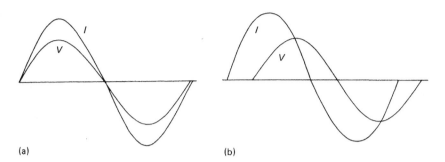

Fig. 12.16 — (a) Current and voltage in phase; (b) current and voltage out of phase with current leading the voltage by 90 °C.

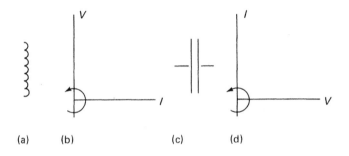

Fig. 12.17 — (a) Symbol for an inductor; (b) the current lags behind the voltage by 90° with only an inductor in the circuit; (c) symbol for a capacitor; (d) the current leads the voltage by 90° with only a capacitor in the circuit.

representation of the current and voltage in phase, and Fig. 12.16(b) shows the current and voltage out of phase; in this case the voltage lags behind the current by 90° or $\pi/2$ radians.

12.11.1 Inductance

An inductor is simply a coil of wire plays an important role in AC circuits in helping to prevent fluctuations in current.

It has been observed that the flow of an electric current gives rise to a magnetic field. When an alternating current flows through a coil of wire, a changing magnetic field is set up which links the turns of the coil and induces a current which opposes the initial current. The induced current is smaller than the initial current, and the net effect is that a resultant current flows. Therefore there is a magnetic resistance to current flow known as the inductive reactance and the property of the coil which causes this is known as the self-inductance L. The inductance value depends only on the dimensions of the coil.

A circuit has an inductance of one Henry (1 H) when the back EMF induced in it is 1 V when the current changes at the rate of 1 A s^{-1}. The henry is a very large unit of inductance.

The inductive reactance X_L depends upon the frequency of current change f as follows:

$$X_L = 2\pi f L$$

and the current flowing through an inductance is V/X_L.

Inductors (or chokes) are simply coils of wire. The inductance value can be increased by use of an iron core, which concentrates the magnetic linkage.

In a circuit containing only an inductor, the current and the voltage are out of phase. The current lags behind the voltage by 90°, and the phase angle is 90° (Fig. 12.17(b)).

12.11.2 Capacitance

Capacitors have the ability to store electric charge. Fig. 12.17(c) shows the simplest form of capacitor which consists of two parallel metal plates separated by an insulating material (dielectric). If the two plates are connected to a battery, then one plate will acquire a positive charge and the other a negative charge. A potential difference V is generated between the charge Q and the potential difference:

$$\text{capacitance } C \text{ (F)} = \frac{\text{charge } Q \text{ (C)}}{\text{voltage } V \text{ (V)}}$$

The farad (F) is an extremely large unit. It is defined as the capacitance of a capacitor between the plates of which there appears a difference of potential of 1 V when it is charged by a quantity of electricity equal to 1 C.

A capacitor will allow the passage of an alternating current but not of a direct current. Many types of capacitor are available, the most common consisting of two strips of metal foil, separated by strips of wax paper, these strips being wound spirally to form two very large surfaces near to each other. Other types of capacitor have been described by Hughes (1977).

The capacitance of a parallel plate condenser containing a vacuum can be described in terms of its dimensions; when the space between the plates is occupied by a vacuum, the capacitance is given by

$$C = \frac{\varepsilon_0 A}{d}$$

where ε_0 (8.85 × 10^{-12} F m^{-1}) is the permittivity of free space or vacuum, A (m^2) is the surface area of the plates and d (m) is the distance between the plates. For most purposes the value is the same when air fills the space.

When other materials are placed between the plates, the capacitance will increase, and the capacitance is given by

$$C = \frac{\varepsilon_0 A \varepsilon'}{d}$$

where ε' is the dielectric constant or relative permittivity for the material. Dielectrics are materials which allow more charge to be stored. Methods available for measuring capacitance can be adapted for measuring the dielectric constant.

The resistance due to the capacitor or the capacitive reactance X_C in an AC circuit of frequency f is given by

$$X_C = \frac{1}{2\pi f C}$$

and the current is V/X_C. In a capacitor the current is found to lead the voltage by 90°. When it is fully charged (maximum voltage), no more current will flow and, when fully discharged (zero voltage), the maximum charge can flow. This situation is represented in Fig. 12.17(d).

12.11.3 AC circuits containing resistors inductors and capacitors

When a resistor, a capacitor and an inductor are connected in series across an AC voltage source, the current in each of these components is the same, but the potential difference across each is different. The system adapted for solving such AC circuit problems is that the current phase is represented by OA at 90°, and the coil is rotating anticlockwise. Therefore, when an inductor is included, the current lags behind the voltage, and the voltage direction is represented by OB. When a capacitor is introduced, the current leads to voltage, which is represented by OC (Fig. 12.18(b)).

The phase angle is the angle between the current and voltage and is represented by ϕ.

The total resistance in such a circuit is measured by the impedance Z which takes into account the inductive reactance X_L, the capacitive reactance X_C and the ordinary resistance R.

The phase diagrams showing the relationship between the voltage and the current are given in Fig. 12.19. X_L and X_C can be resolved into one direction. If $X_L > X_C$, the situation is shown in Fig. 12.19(b) whereas, if $X_C > X_L$, the situation is shown in Fig. 12.19(c). In each case the resultant impedance is given by

$$Z = \sqrt{[R^2 + (X_L - X_C)^2]}$$

The phase angle is given by

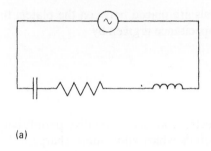

(a)

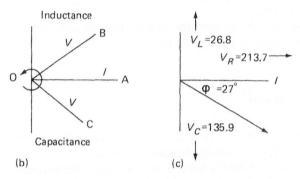

Fig. 12.18 — (a) AC circuit with a capacitor, an inductor and a resistor in series; (b) representation of current and voltage on a phase diagram; (c) voltage resolution.

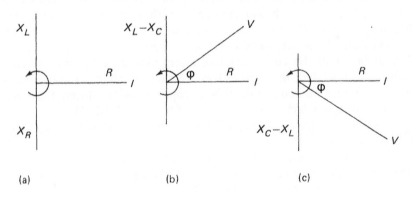

Fig. 12.19 — (a) Phase diagram for evaluating the total impedance Z; (b) for $X_L > X_C$; (c) for $X_C > X_L$.

$$\tan \phi = \frac{X_L - X_C}{R}$$

Now consider the following example. A circuit contains a resistance of $100\,\Omega$, a capacitance of $20\,\mu F$ and an inductance of $40\,mH$ in series. Calculate the phase angle, the current and the potential differences across each component when the circuit is connected to a 240 V, 50 Hz power source.

To calculate the phase angle,

$$X_L = 2\pi \times 50 \times 40 \times 10^{-3}$$
$$= 12.56\,\Omega$$

$$X_C = \frac{1}{2 \times 50\pi \times 20 \times 10^{-6}}$$
$$= 63.6\,\Omega$$

Thus,

$$X_L - X_C = 51.04\,\Omega$$

and

$$Z = \sqrt{[100^2 + (51.04)^2]}$$
$$= \sqrt{(10\,000 + 2605)}$$
$$= 112.3\,\Omega$$

Now,

$$\tan\phi = \frac{X_L - X_C}{R}$$
$$= -\frac{51.04}{100}$$
$$= -0.5104$$

and, therefore, phase angle $\phi = 27°$.

The current is given by

$$\text{current} = \frac{V}{Z}$$
$$= \frac{240}{112.3}$$
$$= 2.137\,\text{A}$$

The potential differences across the components are as follows:

$$\text{potential difference across the resistance} = 100 \times 2.137$$
$$= 213.7\,\text{V}$$

potential difference across the capacitor = 63.6 × 2.137
= 135.9 V
potential difference across the inductor = 12.56 × 2.137
= 26.8 V

The total voltage is given by resolving the voltage in the individual components (Fig. 12.18(c)). In this case the voltage lags behind the current by 27°. In this circuit the minimum impedance and maximum current occurs when

$$X_L = X_C$$

The frequency corresponding to the minimum impedance is known as the resonance frequency and is given by

$$2\pi f L = \frac{1}{2\pi f C}$$

Therefore,

$$f = \frac{1}{2\pi\sqrt{(LC)}}$$

and the resonant frequency for this circuit is

$$f = \frac{1}{2\pi(20 \times 10^{-6} \times 40 \times 10^{-3})^{0.5}}$$
$$= 178\,Hz$$

When the components are connected in parallel (Fig. 12.20(a)), the

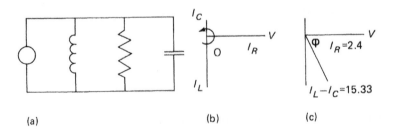

Fig. 12.20 — (a) AC circuit with a capacitor, inductor and resistance in parallel; (b) representation of the current and voltage on a phase diagram; (c) current resolution.

potential differences across each of the components are the same but the currents are different. Thus,

$$\text{current through the inductor} = \frac{V}{X_L}$$
$$= \frac{240}{2.56}$$
$$= 19.1 \text{ A}$$

$$\text{current through the resistor} = \frac{V}{R}$$
$$= \frac{240}{100}$$
$$= 2.4 \text{ A}$$

$$\text{current through the capacitor} = \frac{V}{X_C}$$
$$= \frac{240}{63.6}$$
$$= 3.77 \text{ A}$$

The total current I_r is obtained by resolving the currents (Fig. 12.20(c)), the voltage phase being represented by OV. Therefore, the total current is given by

$$I_T = \sqrt{[I_R^2 + (I_C - I_L)^2]}$$

In this example,

$$I_T = \sqrt{[2.4^2 + (3.77 - 19.1)^2]}$$

$$I_T = 15.5 \text{ A}$$

The phase angle is given by

$$\tan \phi = \frac{15.33}{2.4}$$
$$= 6.39$$

and, therefore

$$\phi = 81.10°$$

Thus the current lags behind the voltage by 81°.

12.11.4 Q factor

For units in series the currents in each of the items are the same but the voltage drops across the items are different. The voltage drop across an individual item may well be higher than that across the whole circuit. At the resonance frequency the ratio of the voltage drop across any component to that across the total circuit (i.e. source AC) is known as the Q factor. For units in parallel the voltage drops across each of the items are the same, but the currents through each of the items may be different. At the resonance frequency the Q value for a parallel system is the ratio of current flow in an item to that supplied to the parallel circuit. A Q meter can be used for measuring dielectric properties (Mohsenin, 1984).

12.11.5 Measurement of capacitance

An alternative method for comparing capacitors is to use a ballistic galvanometer; this instrument produces a deflection which is proportional to the amount of charge stored by a capacitor. The capacitor is charged using a cell and allowed to discharge through the ballistic galvanometer, the deflection being compared with that produced by a standard capacitor.

Impedance bridge methods can also be used. A standard frequency oscillator is used in a circuit similar to a Wheatstone bridge circuit, where two of the resistances have been replaced by the capacitors to be compared (Fig. 12.21). The bridge is balanced by manipulating the variable resistance

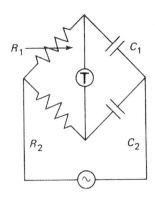

Fig. 12.21 — Impedance bridge: T, telephone detector.

R_1 until zero or a minimum sound is heard in the telephone detector. At this point,

$$\frac{R_1}{R_2} = \frac{X_{C_1}}{X_{C_2}} = \frac{2\pi f C_2}{2\pi f C_1}$$

Therefore,

$$\frac{R_1}{R_2} = \frac{C_2}{C_1} \quad \text{or} \quad R_1C_1 = R_2C_2$$

Capacitance can be measured using AC circuit theory, described in section 12.11.3. A circuit is set up using a low-voltage AC source, a standard non-inductive resistor of known resistance and the unknown capacitor in series (Fig. 12.22). If the voltage V and current I are measured for the circuit, then

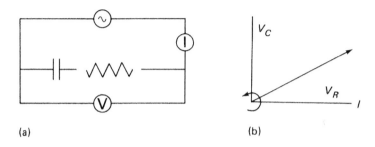

(a) (b)

Fig. 12.22 — (a) AC circuit for capacitance measurement; (b) phase diagram.

the capacitance is the only unknown:

$$\frac{V}{I} = \sqrt{\left[R^2 + \left(\frac{1}{2\pi fC}\right)^2\right]}$$

The experiment can be repeated using different known resistances and averaging the values of the capacitance obtained in each case. Tyler (1972) has described an alternative procedure in which measurements are made of the potential difference V_C across the capacitor, the potential difference V_R across the resistance and the potential difference V_T across the total circuit. The currents I_T flowing through each of the units are the same:

$$I_T = \frac{V_R}{R}$$

$$= \frac{V_C}{X_C}$$

$$= V_C 2\pi fC$$

Therefore,

$$C = \frac{V_R}{2V_C \pi fR}$$

If the capacitance is ideal (no losses), then

$$V_T = \sqrt{(V_R^2 + V_C^2)}$$

The above relationship will not apply if considerable losses occur within the capacitor, and this simple circuit theory is not then appropriate for capacitance measurement.

12.12 DIELECTRIC PROPERTIES

The dielectric properties of foods are currently receiving more attention, mainly because of dielectric and microwave heating processes. The two properties of interest are the relative dielectric constant ε' and the relative dielectric loss factor ε''. The term relative is introduced to show that values are determined relative to that of air or a vacuum, making them dimensionless. However, the term relative is often dropped. The terms permittivity and capacitivity are now recommended in preference to the term dielectric constant but at present are not often encountered.

12.12.1 Dielectric constant

The relative dielectric constant of a food is the ratio of the capacitance of the material being studied to the capacitance of air or vacuum under the conditions being studied. As the dielectric constant increases, the capacitor is capable of storing a greater quantity of energy. Therefore, methods for measuring capacitance can be adapted to measure dielectric properties.

The dielectric constant depends upon a variety of factors, such as temperature, moisture content and frequency; all these factors should be kept constant and recorded when the dielectric constant is being measured. Mohsenin (1984) has given a comprehensive review of dielectric properties of materials. Figs. 12.24 and 12.25 show how the dielectric constant is affected by temperatures and moisture content. Some of these factors are discussed in more detail in section 12.13.

12.12.2 Dielectric loss factor

The relative dielectric loss factor ε'' is a measure of the amount of energy that a component will dissipate when subjected to an alternating electrical field.

In an AC circuit containing an ideal capacitor, the current will lead the voltage by 90°. When a dielectric material is introduced into the capacitor, this angle may be reduced; the loss angle δ is a measure of this reduction and is usually expressed as a loss tangent, $\tan \delta$. This can be regarded as equivalent to introducing a resistor into the circuit in parallel with the capacitor (Fig. 12.23(a)); this will lead to a dissipation of energy. As the loss tangent increases, the amount of energy dissipated increases. This energy is

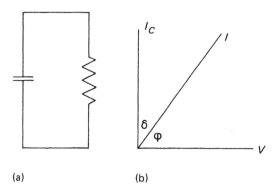

Fig. 12.23 — Representation of a non-ideal capacitor; (b) representation of the phase angle φ and the loss angle δ.

dissipated within the dielectric material. (Note that this situation also applies to a non-ideal capacitor (Hughes, 1977)).
The dielectric loss factor is realted to the dielectric constant.

$$\varepsilon'' = \varepsilon' \tan \delta$$

or

$$\tan \delta = \frac{\varepsilon''}{\varepsilon'}$$

It can be observed that the loss angle is equal to 90° minus the phase angle (Fig. 12.23(b)).

12.13 DIELECTRIC PROPERTIES OF FOODS

The dielectric loss factor is a very important property in dielectric and microwave heating processes. Mudget (1982) has reviewed equations for predicting the dielectric behaviour of water, ionic solutions and emulsions. Some values for the dielectric properties of semisolid foods, predicted on the basis of moisture and free salt levels, are recorded in Table 12.7.

The dielectric properties of some seeds (dry materials) are recorded in Table 12.8.

Materials with a high dielectric loss factor are known as lossy materials. Dielectric loss factors are recorded either directly or in terms of the dielectric constant and the loss tangent, tan δ. Some values for the dielectric properties of cereals are given in Table 12.9.

Fig. 12.24(a) and Fig. 12.4(b) show how the dielectric constant and dielectric loss factor, respectively, are affected by temperature; there is a sharp transition associated with the transition from ice to water (see section

Table 12.7 — Predicted values of the dielectric properties at microwave frequencies.

Food	Temperature (°C)	ε' 915 MHz	ε' 2450 MHz	ε'' 915 MHz	ε'' 2450 MHz
Beef	25	62	61	27	17
Beef	50	55	55	39	18
Pork	25	59	58	26	16
Potato	25	65	64	19	14
Carrot	25	73	72	20	15

Adapted from the data of Mudget (1982).

Table 12.8 — Dielectric properties of some seeds at 24.2 °C and two different frequencies.

Seed type	Moisture content (%)	ε' 10 MHz	ε' 40 MHz	ε'' 10 MHz	ε'' 40 MHz
Barley	9.2	2.9	2.8	0.24	0.34
Cabbage	5.8	2.7	2.7	0.17	0.20
Onion	7.8	2.2	2.1	0.18	0.21
Tomato	6.6	2.0	2.0	0.14	0.17

Adapted from the data of Mohsenin (1984).

Table 12.9 — Dielectric properties of some cereals.

Ceral	Temperature (°C)	Moisture (%)	ε' 1 MHz	ε' 10 MHz	tan δ 1 MHz	tan δ 10 MHz
Wheat	20	13.8	4.00	3.75	0.058	0.065
Oats (wetted)	22	13.0	2.20	2.00	0.060	0.050
Barley (wetted)	18	13.6	2.45	2.35	0.078	0.062

Adapted from the data of Mohsenin (1984).

9.21.2). In general, dielectric properties can be related to chemical composition, physical structure, frequency and temperature. Free water and dissociated salts have a high dielectric activity, whilst bound water, associated salts and colloidal solids have a low activity. Fig. 12.25 shows data for the dielectric constant and dielectric loss factor plotted against moisture

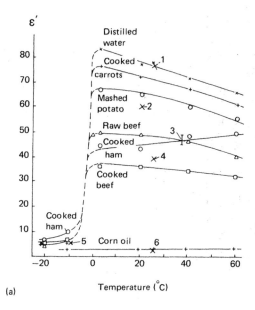
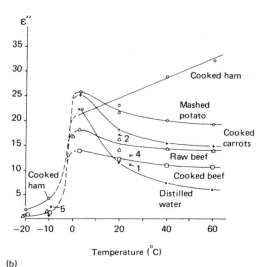

Fig. 12.24 — (a) Dielectric constant and (b) dielectric loss factor for various food materials at 2800 MHz showing the temperature dependence and sharp increase during thawing. (From Bengtsson and Ohlsson (1974), with permission.)

(Copyright © 1974 IEEE)

content for a wide variety of foods. It can be seen that the water content has a major influence on these properties, as all foods fall on a common curve.

The data in Table 12.10 show how the moisture content of a food will affect the dielectric properties and the penetration depth (see section 9.21.2).

The dielectric properties of some oils and fats are recorded in Table

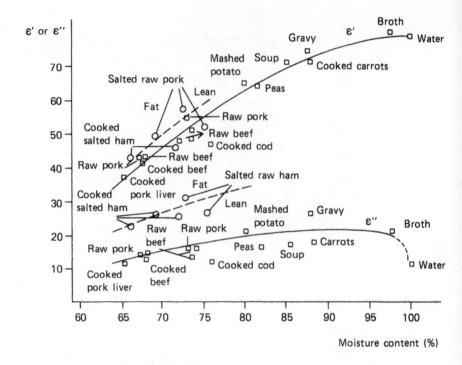

Fig. 12.25 — Dielectric properties of food at 20°C; the relationship between dielectric properties and moisture content. (From Bengtsson and Risman (1971) with permission.)

Table 12.10 — Approximate dielectric properties and penetration depths for potato samples at 25 °C and 2450 MHz.

State	ε'	tan δ	Penetration depth (cm)
High moisture	60	0.3	1.7
Intermediate moisture	10	0.2	3.0
Low moisture	4	0.1	12.3

Adapted from the data of Mudget (1982).

12.11. Compared with the corresponding values for many other foods the dielectric loss factors are very low.

Methods for measuring the dielectric properties have been discussed in more detail by Mohsenin (1984). In general, AC network theory is used to evaluate the change in phase angle when a dielectric is introduced into a

Table 12.11 — Dielectric properties of a selection of oils and fats at 77°F and two different frequencies.

Oil or fat	ε'		ε''	
	1000 MHz	3000 MHz	1000 MHz	3000 MHz
Soya bean oil	2.612	2.506	0.168	0.138
Corn oil	2.638	2.526	0.175	0.143
Lard	2.584	2.486	0.158	0.127
Tallow	2.531	2.430	0.147	0.118

Adapted from the data of Mohsenin (1984).

system containing a capacitor. The principles of microwave heating are discussed in more detail in section 9.21.

12.14 POWER FACTOR

The amount of power taken from an AC circuit is not calculated in a straightforward fashion when the current and voltage are out of phase. If the phase angle between the current and voltage is ϕ, the power factor is given by $\cos\phi$.

The average power taken from a circuit by a motor is

$$VI\cos\phi$$

This is also known as the active power. Therefore, when $\phi = 90°$, the active power is zero. The reactive power is $VI\sin\phi$ and represents power that is not available.

The power factor can also be regarded as the ratio of the average power to the apparent power:

$$\cos\phi = \frac{\text{average power}}{\text{apparent power}}$$

where the apparent power is given by the product of the effective current and voltage. The apparent power for AC equipment is rated as kilovolt amperes rather than as kilowatts to show that is the product of the effective current and voltage.

In an AC circuit a wattmeter will indicate the average power which is the quantity paid for. The apparent power can be determined by measuring the voltage and the current, and in this way the power factor can be evaluated.

The power factor for some industrial loads can be quite low, as this type of load involves a high inductive reactance. This represents inefficient utilization of energy and may be frowned upon by the electricity generating company, as they are in effect supplying the apparent power but only being paid for the average power. The power factor can be improved in a number of ways, including placing a capacitor in the circuit to decrease the phase angle or by selecting the most appropriate type of electric motor.

12.15 TRANSFORMER ACTION

A transformer is used for increasing or decreasing an AC voltage supply. The AC supply to the primary coil sets up a rapidly fluctuating magnetic field which cuts the secondary coil and induces an EMF in this coil. Fig. 12.26 shows the symbols for a transformer.

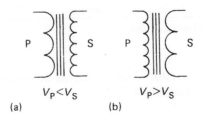

Fig. 12.26 — Transformer action: (a) a step-up transformer; (b) a step-down transformer: P, primary coils; S, secondary coils.

A step-down transformer (Fig. 12.26(b)) contains more turns in the primary coil than in the secondary coil and the voltage in the secondary coil is reduced by the factor.

$$V_S = V_P \frac{N_S}{N_P}$$

where N_P is the number of turns in the primary circuit and N_S is the number of turns in the secondary circuit. A step-up transformer (Fig. 12.26(a)) contains more turns in the secondary circuit, thereby increasing the voltage.

If a transformer is 100% efficient, then the power in the primary coil is equal to the power in the secondary coil:

$$V_P I_P = V_S I_S$$

Transformers contain no moving parts and generally work at greater than 98% efficiency. The energy which is lost is dissipated as heat, and so provision is needed to remove this in high-energy transformers.

12.16 THREE-PHASE SUPPLY

Single-phase systems are perfectly satisfactory for lighting and heating purposes. However, single-phase induction motors are not self-starting unless fitted with an auxiliary winding. By using two separate windings with currents differing in phase by a quarter of a cycle (two phase) or three windings with currents differing in phase by a third (three phase), it was found that induction motors were self-starting and had a better efficiency and power factor than a coresponding single-phase machine. Consequently, such systems are preferred for the larger motors used in the food-processing industry.

Power transmission in the UK is invariably three phase. It is generated and transported at high voltages to minimize the current and hence the transmission losses. A three-phase supply can be generated by having three coils on a rotor, spaced 120° apart, with provisions for tapping the EMF from each loop. The EMF developed by each of the phases is shown in Fig. 12.27.

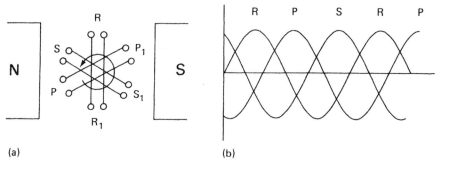

Fig. 12.27 — Three-phase supply: (a) plan view of the generation of three-phase supply; (b) characteristics of the three-phase supply: the three coils are represented by PP_1, PR_1, and SS_1.

A three-phase supply is shown in Fig. 12.28 with the three supplies and a

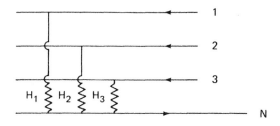

Fig. 12.28 — Supply of single-phase electricity to domestic users from a three-phase supply.

neutral return. Further details have been supplied by Hughes (1977). Household electricity is single phase and is obtained by tapping off from one of the three phases. It is generally arranged that neighbouring houses are supplied from each of the phases in turn in order to balance the load. This arrangement is shown in Fig. 12.28 for three neighbouring houses H_1, H *and* H_3.

12.17 ELECTRIC MOTORS

Electric motors convert electrical energy to mechanical energy and can be divided into DC and AC motors. DC motors are very rarely used on processing plant as DC power sources of sufficient size are not readily available. However, they may be used in the laboratory. Their real advantage is the ease at which the speed can be controlled by changing the voltage or resistance of the armature circuit or the field circuit.

AC motors can be either single phase or three phase, three phase being preferred for the larger applications. The two major types of AC motor are the synchronous and the induction motor. Three-phase induction motors are the most common for process plant applications because they are inherently simple, rugged and reliable. Fig. 12.29 shows a schematic layout of a standard squirrel cage induction motor provided with ventilation for cooling purposes. It is so called because the motor winding resembles a

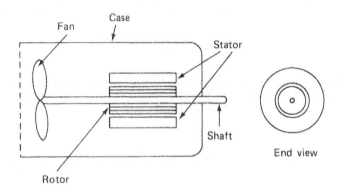

Fig. 12.29 — A simplified diagram of a squirrel cage induction motor.

squirrel cage, having longitudinal bars secured to end rings. Induction motors are characterized by having no brushes or commutator, the current in the rotor being generated by induction.

The rotor is made of laminated sheets of iron with bars of copper running parallel to the shaft of the rotor. The rotor is surrounded by magnetic coils, which when supplied with an alternating voltage appear to rotate round the rotor. The electric current induced in the rotor sets up a magnetic field and the interaction of the magnetic fields produces a turning force on the rotor.

When no load is supplied, induction motors rotate at the synchronous speed, which is governed by the number n of pole pairs and the frequency f:

$$\text{speed (rev min}^{-1}) = 60 f n$$

The synchronous speed for a 50 Hz one-pole-pair motor is 3000 rev min^{-1}. A further characteristic is that a small amount of slip occurs as a load is supplied and the slip usually increases as the load (or torque) increases. Slip is generally less than 5% and causes the rotor speed to be slightly less than the synchronous speed. Most three-phase single-pole induction motors are rated at 2880 rev min^{-1}. Other speeds commonly available are 1440 or 960 rev min^{-1}.

Three-phase induction motors are self-starting and provide a better power characteristic than single-phase motors do. A special winding is required for starting a single-phase induction motor. An induction motor is essentially a fixed-speed motor and is not easily adapted electrically for variable-speed operation. It can be done by increasing the number of poles during operation or by changing the frequency of the electrical field; frequency converters are now readily available. Alternatively, speed control can be achieved by using an external gear box or a coned-belt drive arrangement.

Further advantages of induction motors are that they are reasonably efficient and have a fairly high power factor. Fig. 12.30 shows how the speed,

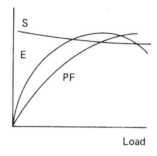

Fig. 12.30 — Characteristics of a squirrel cage induction motor; S, speed; PF, power factor; E, efficiency. S, PF and E are all plotted against load.

power factor and efficiency change as the load increases. They are also useful in applications where spark-proof equipment is required, such as solvent extraction processes.

A truly synchronous motor shows no slip, the speed remaining constant as the load increases, and is used in applications where a constant speed is essential. They are also preferred for large installations and are useful for improving the power factor of a supply when run in conjunction with induction motors. Further details on electric motors have been provided by Hughes (1977), Say (1968) and Perry and Chilton (1973).

12.18 SYMBOLS

a   cross-sectional area
A   area
c   concentration

C	capacitance
d	distance
e	electron charge
E	electrical energy
E_θ	EMF
f	frequency
I	electric current
K	specific conductance or conductivity
L	length
L	self-inductance
m_e	electron mass
N_e	number of electrons per unit volume
N	rotational speed
P	electrical power
PD	potential difference
Q	quantity of electricity or charge
R	electrical resistance
t	time
V	voltage
X_C	capacitive reactance
X_L	inductive reactance
Z	total impedance

12.18.1 Greek symbols

α	temperature coefficient of resistance
δ	loss angle
ε'	dielectric constant
ε''	dielectric loss factor
ε_0	permittivity of free space
θ	temperature
Λ	molar conductivity
ρ_r	resistivity
ϕ	phase angle between current and voltage (section 12.11)

12.18.2 Subscripts

max	maximum
P	primary
RMS	root-mean-square
S	secondary
T	total
u	unknown
v	variable

13

Diffusion and Mass Transfer

13.1 INTRODUCTION

Mass transfer operations are concerned with the transfer of matter from one stream to another. In many processes a change in phase may also be involved.

Some typical mass transfer processes are shown in Table 13.1. In most of these processes with the exception of packaging, it is desirable to maximize the transfer of material in order to speed up the process. In packaging operations the most appropriate materials are selected to minimize the transport of water vapour and oxygen or to maintain a selected atmosphere within the package.

In some other cases, mass transfer processes may be undesirable, as for example in the leaching of soluble components and essential nutrients during blanching and the subsequent storage of canned fruit and vegetables in syrups and brines. In evaporation and dehydration processes the loss of volatile components may result in a change in flavour, and hence a reduction in the overall quality. Therefore, it is important to understand the principles governing mass transfer operations, so that the conditions can be established which maximize processing rates without unduly affecting the quality of the product.

Mass transfer processes are essentially rate processes in which the mass transfer rate in influenced by a driving force and a resistance in an analogous fashion to heat transfer processes. The driving force is due mainly to concentration or partial pressure gradients for the material being transferred, and the resistance is due to the medium through which the material is being transferred and any interactions between that medium and the material. Therefore the general equation is

$$\text{mass transfer rate} = \frac{\text{driving force}}{\text{resistance}}$$

This equation will be developed for different mass transfer situations. When the transfer of gases or vapours is dealt with, driving forces are measured in terms of pressure differences but, when dissolved solids are considered, they are measured in terms of concentration differences.

The equation is usually extended to take into account the surface area:

Table 13.1 — Some mass transfer process involved in food processing.

Operation	Material transferred	Phase change	Examples and comments
Dehydration	Water	From liquid or solid to vapour	Many drying methods are available. Quality is improved by minimizing the loss of volatile components
Solvent extraction	Oil	From solid matrix to organic liquid	Extraction of oils and fats from animal plant or microbial sources
Leaching	Soluble components	From solid matrix to aqueous solvent	Tea, coffee, sugar extraction; plant protein
Sulphiting	Sulphur dioxide	From gas to solid or liquid media	Sulphiting may also be done using solutions of sodium bisulphite or metabisulphite
Smoking	Phenolic components	From vapour to solid matrix	Preservation of foods by use of antimicrobial agents
Distillation	Alcohol and other volatile components	From liquid to vapour	Recovery of alcohol from a fermentation broth
Packaging	Gases and vapours	From the external environment into the package	Prevention of microbial and oxidation reactions
Membrane processing	Water, and dissolved solutes	From liquid to liquid through a semi-permeable membrane	Reverse osmosis for concentrating liquids. Ultrafiltration for concentrating proteins
Oxygen transfer	Oxygen	From gas to liquid	Aerobic fermentation processes in which oxygen is consumed from the solution by micro-organisms
Ion exchange	Metal ions, proteins	From solution to a solid phase	Elution involves the reversal of the adsorption process

mass transfer rate = $KA \Delta c$

In most mass transfer operations the food is subjected to some form of size reduction operation to increase the surface-area-to-volume ratio and the overall mass transfer rate. The resistance K will take into account the properties of the material being transferred and those of the matrix through which it is being transferred and any interactions between them.

13.2 DIFFUSION

Diffusion is the spreading-out of a material into its surroundings. The two major types encountered are molecular diffusion and eddy diffusion. Molecular diffusion can be defined as the transport of matter on a molecular scale through a stagnant fluid or, if the fluid is in laminar flow, in a direction perpendicular to the main flow (see section 3.5).

In contrast, eddy diffusivity is concerned with mass transfer processes involving bulk fluid motion.

In practice the two types of diffusion processes are found together but it is the molecular diffusion processes that has a major influence in many processes because it is concerned with mass transfer over the boundary layer which exists in all flow situations and within the food matrix, where it is not usually possible to induce turbulence. Much of the following discussion will be concerned with molecular diffusion and diffusivity measurement in gases, liquids and solids.

13.3 FICK'S LAW

When a concentration gradient for a given component exists in one direction

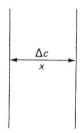

Fig. 13.1 — Diffusion across a concentration gradient.

only, its diffusion may be characterized by Fick's first law, which states that

$$N = -D \frac{dc}{dx}$$

where N (kg s^{-1} or kmol s^{-1}) is the mass or molar transfer rate, c (kg m^{-3} or kmol m^{-3}) is the mass or molar concentration, x (m) is the distance and D is the diffusion coefficient or diffusivity. In this equation the diffusivity is a measure of the resistance to the transfer of material. The equation is analogous to Fourier's equation, which describes the rate of heat transfer by conduction (see section 9.4).

At steady-state conditions, i.e. when the concentration at any point remains constant, the expression becomes

$$N = D \frac{\Delta c}{x}$$

This applies to diffusion in gases liquids and solids (Fig. 13.1).

13.4 GASEOUS DIFFUSION

For the diffusion of gases and vapours, the concentration term in Fick's Law can be replaced by the partial pressure, using the relationship for an ideal gas:

$$c_A = \frac{p_A}{RT}$$

where c_A is the molar concentration of A and, p_A partial pressure of A. Therefore the molar diffusion transfer rate becomes

$$N_A = \frac{D}{RT} \Delta p_a$$

Two cases will now be considered, one for equimolecular counter-diffusion, and the other for the diffusion of a gas through a stagnant layer of a second gas.

13.4.1 Equimolecular counter-diffusion

Equimolecular counter-diffusion occurs when the number of moles of one

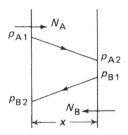

Fig. 13.2 — Equimolecular counter-diffusion.

component A moving in one direction is equal to the number of moles of component B moving in the opposite direction (Fig. 13.2), i.e.

$$N_A = -N_B$$

Sec. 13.4]   Gaseous diffusion   417

Therefore the total molar concentration remains constant across the interface. Distillation operations form a good example of this type of operation. Under these conditions the diffusion rate of component A is given by

$$N_A = \frac{D_A}{RTx}(p_{A1} - p_{A2}).$$

13.4.2 Diffusion of a gas through a stagnant layer

This occurs when a gas or vapour A moves a stagnant second gas B. In this

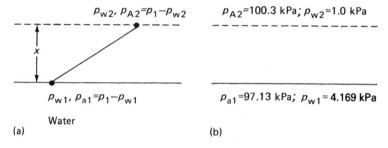

Fig. 13.3 — (a) Diffusion through a stagnant layer; (b) problem.

case, $N_B = 0$. Component B does not diffuse but maintains its concentration gradient by intermolecular friction between A and B. This hindering effect is accounted for by the addition of a term to include B. As an illustration the diffusion of water vapour through a stagnant layer of air of thickness x is represented in Fig. 13.3 and given by

$$N_A = \frac{D_w}{RTx}\frac{p_T}{p_{Am}}(p_{w1} - p_{w2}) \qquad (13.1)$$

where D_w is the diffusion coefficient of water vapour in air, p_T is the total pressure, $p_{w1} - p_{w2}$ is the water vapour pressure driving force, p_{Am} (kN m^{-2}) is the mean pressure of non-diffusing gas across the film, R ($= 8.314$ kmol kg^{-1} K^{-1}) is the gas constant and T (K) is the temperature. When the partial pressure differences across the film are small, an arithmetic mean can be used for the mean pressure of the non-diffusing gas; larger partial pressure differences require the use of a logarithmic mean.

As an example calculate the diffusion rate of water through a stagnant film of air, 3 mm thick, when the water is maintained at 29.5 °C and atmospheric pressure is 101.3 kPa. The partial pressure of water vapour in the bulk air stream is 1.00 kPa. The diffusivity of water vapour in air is 25.5×10^{-6} m^2 s^{-1} at 29.5 °C.

The partial pressure of water vapour at the surface of the water is obtained from saturated vapour pressure tables and equals 4.169 kPa (Fig. 13.3).

When an arithmetic mean is used for the mean pressue of air,

$$p_{Am} = \frac{p_{A1} + p_{A2}}{2}$$

$$= \frac{97.13 + 100.3}{2}$$

$$= 98.72 \text{ kPa}$$

When a logarithmic mean value is used,

$$p_{Am} = \frac{p_{A2} - p_{A1}}{\ln(p_{A2}/p_{A1})}$$

$$= \frac{100.3 - 97.13}{\ln(100.3/97.13)}$$

$$= \frac{3.17}{0.0321}$$

$$= 98.72 \text{ kPa}$$

In this case the two mean values are identical. Therefore from equation (13.1) the mass transfer rate is given by

$$N_A = \frac{25.5 \times 10^{-6}}{8.314 \times 298 \times 3 \times 10^{-3}} \frac{101.3}{98.72} (4.169 - 1.00)$$

$$11.15 \times 10^{-6} \text{ kmol s}^{-1} \text{ m}^{-2}$$

When the partial pressure of the diffusing material is low compared with the total pressure, the error involved in using the arithmetic mean is small. In such situations a further simplification can be made, because the mean pressure of the non-diffusing gas is approximatley equal to the total pressure. Therefore, equation (13.1) beccomes

$$N_A = \frac{D_w}{RTx} (p_{w1} - p_{w1})$$

When p_{Am} is expressed as a logarithmic mean value in terms of the diffusing component it becomes

$$p_{Am} = \frac{(p_T - p_{w2}) - (p_T - p_{w1})}{\ln[(p_T - p_{w2})/(p_T - p_{w1})]}$$

$$= \frac{p_{w1} - p_{w2}}{\ln(p_{A2}/p_{A1})}$$

Gaseous diffusion

When p_{Am} is substituted into equation (13.1), the equation becomes

$$N_A = \frac{D_w}{RTx} p_T \ln\left(\frac{p_{A2}}{p_{A1}}\right)$$

The term involving the partial pressure driving force of the diffusing component has disappeared and the mass transfer rate is expressed in terms of the partial pressure of the non-diffusing gas and the total pressure. The mass diffusivities for some common gases and vapours in air are recorded in Table 13.2. The diffusivity is also affected by temperature and pressure. At

Table 13.2 — Mass diffusivities for gases in air at 25 °C and 1 atm.

Gas	Diffusivity (m² s⁻¹)	Gas	Diffusivity (m² s⁻¹)
Ammonia	27.9×10^{-6}	Hydrogen	41.3×10^{-6}
Carbon dioxide	16.5×10^{-6}	Oxygen	20.6×10^{-6}
Ethanol	11.9×10^{-6}	Water vapour	25.5×10^{-6}

Extracted from the data of the American Soceity of Heating, Refrigerating and Air-conditioning Engineers (1981).

constant pressure, diffusivity varies approximately as $\theta^{1.75}$ and, at constant temperature, approximately as the inverse of pressure. The molecular weight of the component will also affect its diffusivity. Equations have been described for predicting diffusion coefficients from thermodynamic data, such as critical temperatures and pressure, on the basis of the kinetic theory of gases (Blackhurst *et al.*, 1974; American Society of Heating, Refrigerating and Air-Conditioning Engineers 1981). Such an approach might be useful in the absence of experimental data.

13.4.3 Experimental determination of diffusivity

One problem encountered in the experimental investigations is the elimination of convection currents. A fairly simple piece of equipment can be set up

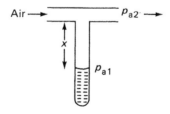

Fig. 13.4 — Equipment for the experimental determination of diffusivity.

to estimate the diffusivity of vapours. A small-diameter tube is closed at one end and filled with the liquid. A gentle stream of air is blown over the surface, the liquid and tube being maintained at the required experimental temperature (Fig. 13.4). The amount m kg of liquid evaporated in a given time t s can be determined. The distance x is also recorded.

The molar diffusion rate of component A through the non-diffusing component B is evaluated from

$$N_A = \frac{D}{RTx} \frac{p_T}{p_{Bm}} (p_{a1} - p_{a2})$$

and

$$N_A = \frac{m}{Mt}$$

where M is the molecular weight of the material. Therefore,

$$\frac{m}{Mt} = \frac{D}{RTx} \frac{p_T}{p_{Bm}} (p_{a1} - p_{a2})$$

where p_{a1} is the partial pressure of A at the interface (from saturated vapour pressure tables), p_{a2} is the partial pressure of A in air (this can be measured or assumed to be zero if the flow rate is high) and x is the mean film thickness over the duration of the experiment.

Blackhurst et al. (1974) have shown that, if the initial height is x_1 and the height after time t is x_2, then

$$D = \frac{RT\rho_L(x_2^2 - x_1^2)}{2p_T Mt \ln(p_{B1}/p_{B2})}$$

where p_{B1} is the partial pressure of air (non-diffusing gas B) in the bulk, p_{B2} is the partial pressure of air at interface $= (p_T - p_s$ where p_s is the saturated vapour pressure) and ρ_L is the liquid density.

13.5 DIFFUSIVITY IN LIQUIDS

The diffusivities of components in a liquid are several powers of ten (orders of magnitude) smaller than the diffusivity in gases. The diffusivity of a component in a liquid is again related to the rate of mass transfer by Fick's law:

$$N = -D \frac{dc}{dx}$$

Methods for measuring the diffusivities of solutes in liquids have been reviewed by Loncin and Merson (1979). In essence these methods involve measuring the change in concentration when a system essentially at equilibrium is subjected to a step change in concentration and using unsteady-state mass transfer equations. The diffusivity of a gas in a liquid is measured by making a laminar jet of the liquid pass through a gas and determining the rate of absorption.

Agar solutions have been used to form gels in the form of spheres and these have been used to measure diffusivities of components through the gel. The results can be extrapolated back to zero agar concentration to find the diffusivity of the component in water. In general the diffusivity of a component in a gel is reduced slightly (about 20%) but certainly not as much as would be expected from the resulting increase in viscosity. Diffusivities for components in gelatin gels are quoted by Schwartzberg and Chao (1982).

As a general approximation the diffusivity in a dilute liquid phase is approximately inversely proportional to the mean radius of the diffusing particles. Increasing the temperature significantly increases the diffusivity. The Stokes–Einstein equation predicts that diffusivity is related to temperature θ and viscosity μ as follows:

$$D = k \frac{\theta}{\mu}$$

Therefore, increasing the temperature is effective in itself and because it reduces the viscosity. Blackhurst et al. (1974) have described how a modified form of this equation can be used to predict liquid diffusivities.

For most food solutions, particularly sugar solutions, and vegetable oil in hexane, the diffusion coefficient decreases as the solute concentration increases, again because of the corresponding increase in viscosity. Some diffusivity values for gases and liquids in water are shown in Table 13.3. More detailed tables have been provided for organic acids, amino acids, proteins and enzymes by Schwartzberg and Chao (1982) and for other components by Coulson and Richardson (1977).

The addition of solutes can decrease the diffusivities of volatile aroma components; maltodextran has been found to have a significant effect.

13.6 SOLID DIFFUSION

Diffusion in a solid matrix is more complex than diffusion in a liquid or gas because, although the product may appear to be diffusing within the solid matrix, it may actually be diffusing through liquid contained within that matrix or through the gas phase in a porous solid. Therefore, diffusivities in solids are poorly known.

Structure-insensitive diffusion has been used to described a situation in which the solute is dissolved to form a homogeneous solid solution, with diffusion then taking place through concentration gradients within the solution.

In leaching operations the leachable solids are contained in a framework

Table 13.3 — Diffusivities of components in water (dilute solutions).

Substance	Temperature (°C)	Diffusivity ($m^2 \, s^{-1}$)
Oxygen	25	2.07×10^{-9}
Nitrogen	25	1.90×10^{-9}
Carbon dioxide	25	1.98×10^{-9}
Glucose	25	0.67×10^{-9}
Sucrose	25	0.52×10^{-9}
Glycerol	25	0.94×10^{-9}
Acetic acid	25	1.24×10^{-9}
Soluble starch	—	0.10×10^{-9}
Ethanol	25	1.28×10^{-9}
Sodium chloride		1.611×10^{-9}

Adapted from data in Loncin and Merson (1979).

of insoluble solids (known as the marc), the occluded solution and, in some cases, sparingly soluble solids or solute precursors (Schwartzberg and Chao, 1982).

Diffusion occurs primarily within the occluded solution but the marc restricts the diffusion process and will strongly affect the rate of diffusion. Therefore, when material is being leached from a dry solid, the solid must first imbibe the solvent, to dissolve the external solute, before diffusion begins. However, diffusion processes are normally based in terms of the solute concentration in the solid rather than the occluded liquid, as these are easier to measure.

Schwartzberg and Chao (1982) have reviewed solid diffusion coefficients for a wide variety of systems. Some examples are given in Table 13.4. These workers have provided a comprehensive review of solid diffusivities obtained during leaching. Often, agreement between different research

Table 13.4 — Diffusivities of solids in foods.

Food	Solute	Solvent	Temperature (°C)	Diffusivity $\times 10^{10} \, m^2 \, s^{-1}$	Comments
Apple slices	Sugars	Water	75	11.8 ± 0.9	
Sugar beet	Sucrose	Water	75	7.2	
Roast ground coffee	Coffee solubles	Water	70–85	0.8–1.1	
Roasted whole beans	Coffee solubles	Water	97–100	3.05	
Pickled cucumbers	NaCl	Water	25	5.3–11	Affected by size
Soya bean flakes	Oil	Hexane	69	1.13–0.13	Decreases during extraction

Taken from the data of Schwartzberg and Chao (1982).

workers is poor. In some cases the diffusivities of components are not much higher than those found in liquids, suggesting that the marc offers little resistance to material transfer; in other cases the solid diffusivity values are significantly lower, suggesting that the marc offers a greater resistance. It should also be remembered that foods have a complex structure, with many components contained within a cellular structure, in which each cell in turn has its own membrane. This structure may have been also disrupted by size reduction operations such as crushing, grinding or pressing.

A second type of solid diffusion is referred to as structure-sensitive diffusion. This refers to diffusion in porous or granular solids which permit the flow of a liquid or gas through the void volume or capillaries within the solids. Various mechanisms may be involved, such as diffusion in the solid itself, diffusion in the gas-filled pores, capillary flow resulting from gradients in surface pressure and convective flow resulting from differences in total pressure.

It is usually more convenient to work out an overall resistance or diffusivity without necessarily needing to evaluate the contribution of the individual factors. Again, Fick's law may be used to determine the value of the resistance, which in gas or vapour transport may be known as an overall permeability coefficient.

13.7 TWO-FILM THEORY

In most mass transfer problems, there is often more than one resistance to consider. Such a situation will be illustrated by reference to a gas dissolving in a liquid. The gas stream may be broken into a stream of bubbles or may be in contact with a thin film of liquid, the objective in both cases being to increase the interfacial area. These two situations are illustrated in Fig. 13.5 with the gas phase and liquid phase boundary layers.

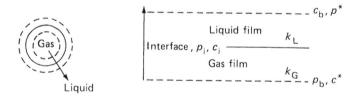

Fig. 13.5 — Two-film theory illustrate by the transfer of material from a gas bubble into solution.

The gas must diffuse through the gas phase boundary layer to the interface, where it dissolves; the dissolved gas then diffuses through the liquid boundary layer to the bulk of the liquid. All concentration gradients exist over the boundary layer and it is assumed that there is an equilibrium situation at the interface beween the diffusing component in the gaseous and liquid phase.

The rate N_G of gas transfer over the gas film is given by

$$N_G = k_G A(p_b - p_i) \quad (13.2)$$

and the rate N_L of mass transfer through the liquid film is given by

$$N_L = k_L A(c_i - c_b) \quad (13.3)$$

where A is the interfacial area. The equilibrium between the concentration of a gas in a liquid and its partial pressure is described by Henry's law, where

$$p = Hc$$

or

$$c = \frac{p}{H}$$

where c is the concentration, p is the partial pressure and H ($=p/c$) is Henry's law constant. For soluble gases such as ammonia and carbon dioxide, the Henry's law constant is high, whereas it is much lower for sparingly soluble gases such as air, oxygen and nitrogen.

At steady state, $N_G = N_L$. Therefore,

$$k_L(c_i - c_b) = k_G(p_b - p_i)$$

Therefore the liquid film coefficient can be regarded as being related to the liquid diffusivity divided by the film thickness, i.e. D_L/x_L. Similarly the gas film coefficient can be regarded as being related to D_G/x_G. It is much more convenient to use film coefficients because the film thicknesses are not known.

The expression can be rearranged to give

$$\frac{k_L}{k_G} = \frac{p_b - p_i}{c_i - c_b}$$

If $k_L \gg k_G$, the mass transfer process is gas film controlled, and the partial pressure difference over the gas phase is much greater than the concentration difference over the liquid film. If $k_G \gg k_L$, then the transfer process is liquid film controlled, and the concentration gradient over the liquid film is much greater than that over the gas film.

13.7.1 Overall mass transfer coefficients
Overall mass transfer coefficients are often used and are defined on the basis of the gas phase, i.e. K_G, or on the basis of the liquid phase, i.e. K_L, in terms of the overall concentration gradient.

Sec. 13.7] Two-film theory 425

With reference to Fig. 13.5,

$$N = K_G A(p_L - p^*)$$

where $p^*(= c_L H)$ is the partial pressure of gas in equilibrium with bulk liquid concentration. Therefore,

$$N = K_G A(p_b - c_b H) \tag{13.4}$$

In a similar fashion, N can be defined in terms of an overall liquid phase coefficient K_L where

$$N = K_L A(c^* - c_b)$$

where $c^*(= p_b/H)$ is the concentration of gas in equilibrium with bulk liquid partial pressure. Therefore,

$$N = K_L A \left(\frac{p_b}{H - c_b}\right) \tag{13.5}$$

13.7.2 Relationship between overall mass transfer coefficients and film coefficients

The relationship between the overall gas mass transfer coefficient and the individual film coefficients will now be investigated, for unit surface area $(A = 1)$:

$$\frac{1}{K_G} = \frac{p_b - p^*}{N}$$

and $p_b - p^*$ can be written as $(p_b - p_i) + (p_i - p^*)$. Therefore,

$$\frac{1}{K_G} = \frac{p_b - p_i}{N} + \frac{p_i - p^*}{N}$$

From the definition of k_G (equation (13.2)) and the knowledge that $p_i - p^* = H(c_i - c_b)$, then

$$\frac{1}{K_G} = \frac{1}{k_G} + \frac{H(c_i - c_b)}{N} \tag{13.6}$$

Substituting $(c_L - c_b)/N = 1/k_L$ (from equation (13.3)) into equation (13.6) gives

$$\frac{1}{K_G} = \frac{1}{k_G} + \frac{H}{k_L}$$

In a similar fashion, it can be shown that

$$\frac{1}{K_L} = \frac{1}{Hk_G} + \frac{1}{k_L}$$

Therefore, the overall mass transfer coefficients can be evaluated from the film coefficients and substituted into either equation (13.4) or equation (13.5) to evaluate the actual mass transfer rates. For a further discussion on applications to distillation, gas absorption or extraction processes. The reader is referred to the work of Coulson and Richardson (1978), Loncin and Merson (1979) and Blackhurst *et al.* (1974).

An alternative theory used to describe the mass transfer is the *penetration theory* in which it is assumed that eddies in the fluid bring an element of the fluid to the interface, where it is exposed to the second phase for a definite interval of time, after which it is then mixed in with the bulk again. It is assumed that equilibrium is attained between the element and the second phase whilst it is at the interface. Therefore, mass transfer takes place through the agency of these eddies. These theories have been described in more detail by Coulson and Richardson (1977).

In all cases of mass transfer across a phase boundary, it is important to ensure that there is a large contact area between the phases and to promote turbulence in both the phases.

In continuous processes the two streams usually flow in a counter-current direction. In a steady-state process the composition of each component will change as it passes through the equipment, but conditions at any point do not change with time. Mass transfer theory is applied to determine the time required to complete a particular extraction for a batch process or the size of equipment required to achieve a separation in continuous processes. Overall mass transfer coefficients are extremely useful for expressing mass transfer rates in terms of the concentration gradients existing in such equipment.

13.7.3 Determination of mass film coefficients

Mass transfer film coefficients can be evaluated using dimensional analysis, in a manner analogous to heat film coefficients; such correlations take into account the physical properties of the fluid and have a general application. The following dimensionless groups are of importance:

Sherwood number $Sh = \dfrac{kL}{D}$, where L is a characteristic length

(this corresponds to the Nusselt number in heat transfer problems).

Schmidt number $Sc = \dfrac{\mu}{\rho D}$

(this corresponds to the Prandtl number). Other groups that are used are as follows:

Peclet number $Pe = Re\ Sc$

Stanton number $St = \dfrac{Sh}{Re\ Sc}$

$j_D = St\ Sc^{2/3}$

where j_D is a mass transfer factor. Correlations have been given for a wide variety of flow situations (using these groups) in Loncin and Merson)1979) and Milson and Kirk (1980).

Two examples are as follows:

(1) Inside circular pipes, $Sh = 0.023 Re^{0.82}\ Sc^{0.33}$ for Re values of 4000–60 000 and Sc values of 0.6–3000.
(2) For flow parallel to a flat plate,

$j_D = 0.664 Re^{-0.5}$ $Re < 80\ 000$

$j_D = 0.036 Re^{-0.2}$ $Re > 500\ 000$

13.8 UNSTEADY-STATE MASS TRANSFER

Fick's second law of diffusion can be used to solve unsteady-state mass transfer problems. When a concentration gradient exists in one direction only, Fick's second law states that

$$\frac{dc}{dt} = D \frac{d^2 c}{dx^2}$$

It describes how the concentration changes with time t and position in the food. Methods of solving this equation are similar to those for unsteady-state heat transfer problems and have been discussed for shapes such as an infinite slab, an infinite cylinder and a sphere by Loncin and Merson (1979). They involve subjecting a material at a uniform concentration to a sudden increase in concentration at its surface and noting how the concentration changes with time at different positions within the food, the charts or equations used being analogous to those used for unsteady-state heat transfer (Milson and Kirk, 1980; Jackson and Lamb, 1981; Loncin and

Merson, 1979). Unsteady-state measurements can also be used for estimating diffusion coefficients (Schwartzberg and Chao, 1982).

The unsteady-state equation can also be solved in terms of the penetration theory, where

$$N_A = 2(c_{Ai} - c_{Ab})\sqrt{\left(\frac{D}{\pi t_e}\right)}$$

where N_A is the average transfer rate of A, c_{Ai} is the concentration of component A at the interface (the equilibrium value), c_{Ab} is the concentration of component A in the bulk and t_e is the time of exposure at the surface. It can be seen that the mass transfer rate increases as the time of exposure decreases, i.e. as turbulence increases. One problem is that it is not easy to assign a value to t_e for industrial mass transfer equipment.

13.9 SIMULTANEOUS HEAT AND MASS TRANSFER

In many processes, heat and mass transfer occur at the same time. One example, discussed in section 11.10, is the evaporation of water from a free moisture surface. Energy is supplied from the air stream to provide latent heat energy for evaporation. Eventually the system comes to equilibrium and at steady state the rate of heat transfer to the surface and the rate of mass transfer away from the surface balance.

It is assumed that the air is saturated with water vapour at the surface and the temperature at the surface equilibrates at the wet-bulb temperature. The rate of heat transfer can be a major resistance in operations where latent heat values are high, particularly in dehydration processes.

13.9.1 Hot-air drying

During hot-air drying, if the drying rate is measured at constant conditions and plotted against time (Fig. 13.6), there are two distinct periods, namely a constant-rate period AB and falling-rate period BC; the moisture corresponding to the transition is known as the critical moisture.

During the constant-drying-rate period, the controlling resistance is due to the stagnant air film. The surface of the food behaves like a free moisture surface and the surface temperature approximates to the wet-bulb temperature. The drying rate is controlled by the rate of heat transfer and can be increased by increasing either the temperature driving force or the air velocity. In the situation where air blows over the surface of a food, the heat film coefficient h is proportional to the mass flow rate of air, raised to the power 0.8. The rate Q' of heat transfer and the rate m' of mass transfer are given by (see section 11.10.4)

$$Q' = hA(T_a - T_s)$$
$$m' = kA(p_s - p_a)$$

Sec. 13.9] Simultaneous heat and mass transfer 429

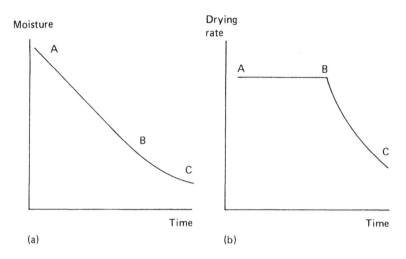

Fig. 13.6 — Drying curves: (a) moisture against time; (b) drying rate against time.

Both the heat film coefficient and the mass transfer coefficient increase as turbulence increases. The vast majority of the water in the food is removed during the constant-drying-rate period. During the constant drying rate, the rate of evaporation is given by

$$\frac{Q}{h_{fg}} = A(T_a - T_s)$$

where h_{fg} is the latent heat of vaporization. Correlations for the heat film coefficient have been given by Jackson and Lamb (1981), Perry and Chiton (1973) and Loncin and Merson (1979). During this period, moisture is moving from the centre of the food to the surface, mostly by capillary flow; it also carries soluble constituents which can be deposited onto the surface and may result in the phenomenon known as case hardening.

Jackson and Lamb (1981) quote that the critical moisture content for foodstuffs is quite high and may range between 240% and 770% for a variety of fruits and vegetables, between 300% and 450% for fish muscle, 400% for gelatin and 300% for peas (dry weight basis). In contrast, Earle (1983) states that the critical moisture content occurs at a moisture content in equilibrium with air, at a relative humidity of between 58% and 68%. This can be determined from the desorption isotherm for that food. Karel (1975) has presented an equation to predict the critical moisture content in terms of the mass transfer and heat transfer characteristics. The break point in the drying curve may also be determined experimentally.

At the critical moisture content the surface of the food begins to dry out and the dry layer advances into the food. Consequently the surface temperature rises and the drying rate becomes to a greater extent independent of the surface conditions and more strongly influenced by the movement of water within the food.

Most texts on drying deal with the drying of hygroscopic and non-hygroscopic materials. A hygroscopic material is one whose partial pressure p_{wv} of water vapour is dependent upon the moisture-content. If the moisture content m is above a limiting moisture content value m_h, then the partial pressue of water vapour equals the saturated water vapour pressure, i.e. if $m > m_h$ then $p_{wv} = p_{wvs}$. However, when the moisture content m falls below the limiting value m_h, then the partial pressure of water vapour falls below the saturated water vapour pressure, i.e. if $m < m_h$, then $p_{wvs} > p_{wv} > 0$ and the relationship between water vapour pressure and moisture content is given by the sorption isotherm (see section 11.16).

A non-hygroscopic material, however, exerts the same water vapour pressure at all moisture contents.

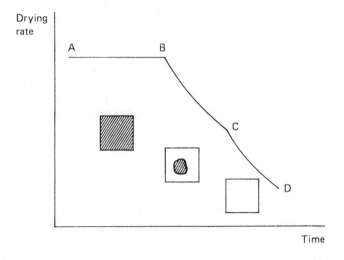

Fig. 13.7 — Drying curves and moisture gradients within food during drying; AB, constant-rate period; BC, first falling-rate period, CD, second falling-rate period; □, region where $p_{wv} \leqslant p_{wvs}$.

The drying curve for a hygroscopic material shows three distinct periods, these being a constant-rate period and two falling-rate periods (Fig. 13.7). At the end of the constant-rate period the surface of the food starts to dry out, because water can no longer be supplied from the interior at a sufficient rate to satisfy the surface evaporation requirements.

During the first falling-rate period BC, the plane of evaporation or saturation moves towards the centre of the food. Water is evaporated at the plane of evaporation and diffuses through the dry solid and through the air film into the bulk of the air; this period ends when the plane of saturation

Sec. 13.9] Simultaneous heat and mass transfer

reaches the centre, and the partial pressure of water at the centre falls below the saturated water vapour pressure.

The second falling-rate period CD then occurs when the partial pressure of water is below the saturated vapour pressure throughout the food. In this situation the drying occurs by desorption.

Karel (1975) has derived an equation to determine the length of the falling-rate period t_F for constant external drying conditions:

$$t_F = \frac{(m_c - m_e)\rho_s L}{p_{wvs} - p_{wv}} \ln\left(\frac{m_c - m_e}{m_F - m_e}\right)$$

where ρ_s is the bulk density of the solid, L is the slab thickness, m is the moisture content, (kg of water per kg of solid) the subscript c means critical, the subscript e means equilibrium and the subscript F means final), p_{wv} is the partial water vapour pressure, p_{wvs} is the saturated water vapour pressure and t_F is the falling-rate period, i.e. the time to fall from the critical moisture content m_c to the final moisture content m_F. Therefore the total drying time is obtained by summing the constant-rate-drying period and the falling-rate period.

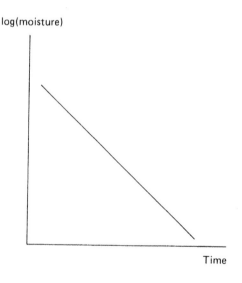

Fig. 13.8 — Semilogarithmic plot of the moisture content against time during the falling-rate period.

Drying time data during the falling-rate period are most conveniently plotted in semilogarithmic coordinates, where a straight-line relationship is generally found (Fig. 13.8). This type of experimental data can be relatively easily obtained for a selected food on laboratory or pilot-plant drying equipment and used for optimizing or scaling up the drying process.

A wide variety of hot-air drying systems are in use, including overdraught, through-draught, fluidized-bed and pneumatic driers. These can be either batch or continuous operations. In continuous operations the air and the food may move in the same direction (co-current) or in opposite directions (counter-current). Co-current operations give high initial drying rates because higher temperatures can be used, whereas counter-current processes allow the food to be dried to a lower equilibrium moisture content because the food leaving the drier is in contact with the dry air entering. An industrial drier may employ co-current drying in the initial drying stage and counter-current drying as a finishing process. Hot-air drying processes have been described in more detail by Her Majesty's Stationery Office (1946), Williams-Gardner (1971) and Hall (1979).

13.9.2 Spray drying

One special form of hot-air drying is the spray drier (Fig. 13.9). Hot air at

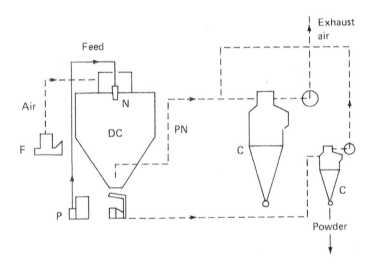

Fig. 13.9 — Diagram of a spray drier with cyclone separators: DC, drying chamber; C, cyclone; F, fan; N, nozzle or atomizer; P, pump; PN, pneumatic conveyance.

150–300 °C is blown into the drying chamber and comes into contact with the liquid feed which has been broken into a fine spray by a centrifugal atomizing device; alternatively, two-fluid or pressure nozzles can be used. Drying takes place extremely quickly and the dry powder is conveyed by the air to a system of cyclones where the fine particles are separated from the air stream. The product feed rate is controlled to give an outlet air temperature of between 90 °C and 100 °C. If it falls much below 90 °C, the resulting product is too wet, whereas energy is wasted if it exceeds 100 °C. The corresponding wet-bulb temperature is between 40 °C and 50 °C. Powder temperatures will not exceed the wet-bulb temperature provided that the system as designed to avoid a long hold-up of dry powder in the conveying system. Methods for improving the efficiency of driers are discussed in section 7.12. Spray drying is widely used for milk and cheese whey, egg, coffee and other beverages and powdered potato.

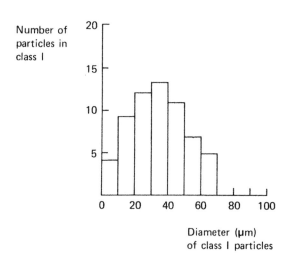

Fig. 13.10 — Particle size distribution for a spray.

The type of spray produced is characterized by the distribution of drop sizes. The sauter mean particle diameter d_{Sm} is the diameter of the particle having the same surface-area-to-volume ratio as the entire spray (Fig. 13.10). It is calculated from the equation

$$d_{\text{Sm}} = \frac{\Sigma \, n_\text{I} d_\text{I}^3}{\Sigma \, n_\text{I} d_\text{I}^{23}}$$

where n_I is the number of particles in class I and d_I is the mean diameter of class I particles.

The drying time for a single droplet can be approximated by

$$t = \frac{r^2 \rho_\text{L} h_{\text{fg}}}{3h \Delta \theta} \left(\frac{m_\text{I} - m_\text{F}}{1 + m_\text{I}} \right)$$

where r is the radius of the drop, ρ_L is the liquid density, h_{pg} is the latent heat, $\Delta \theta$ is the temperature difference, m_I is the initial moisture content and m_F is the final moisture content.

(The moisture contents were obtained from the moistures (see section 11.15) which were determined on a dry-weight basis). Heat film coefficients for the flow past spherical droplets have been given by Loncin and Merson (1979). Spray drying has been convered in more detail by Masters (1972) and quality aspects for spray-dried dairy products have been reviewed by the Society for Dairy Technology (1980).

13.9.3 Freeze drying (lyophilization)

Freeze drying is a two-stage process. In the first stage the food is frozen, a number of techniques being available. The size of the ice crystals will be

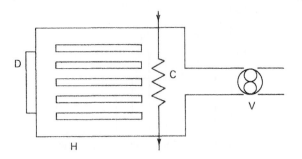

Fig. 13.11 — Schematic diagram of a freeze drier: D, door; H, heaters; C, condenser, V, vacuum pump.

affected by the rate of freezing, smaller crystals being produced at faster freezing rates. The size of the ice crystals will affect the subsequent drying rate as smaller ice crystals will produce a smaller pore size, which will decrease the permeability of water vapour. Evaporative cooling methods can also be used for freezing most solid foods (see section 10.15); this is of benefit because there is a significant reduction in the moisture content. The food is then placed into a vacuum chamber and the pressure is quickly reduced to below the triple-point pressure. If the water vapour pressure is maintained below 4.6 torr, sublimation will occur, as long as sufficient energy is supplied to provide the latent heat of sublimation. The phase diagram for water is shown in Fig. 8.3. A schematic layout for a freeze drier is shown in Fig. 13.11. The metal trays containing the frozen food are placed onto heating shelves which are heated by electricity or by circulating hot water. Water vapour is removed by a combination refrigerated condenser and a vacuum pump, the two systems being required because of the very high specific volume of water vapour at low pressures. The reverse process to sublimation occurs and water freezes out on the surface of the condenser, residual water vapour being removed by the vacuum pump. If the water vapour pressure rises above 4.6 torr, the ice will melt rather than sublime, and liquid phase drying will occur.

Fig. 13.12 shows a cross-section through a slab of food part of the way through the freeze-drying process. The surface AB is in contact with the heating surface. If the slab is bisected two distinct regions will be observed, a dry layer and a frozen layer, with a sharp interface between them. As drying proceeds, the interface moves from the surface towards the centre of the food.

The heat and mass transfer processes taking place are as follows:

(1) Heat transfer from the heater through the dry layer.
(2) Sublimination at the interface.
(3) Diffusion of water vapour through the dry layer to the surface.
(4) Movement of water vapour from the food surface to the condenser.

The heat transfer process is inherently slow, because of the extremely low thermal conductivity of the dried food, and most freeze-drying processes are heat transfer controlled.

The rate of heat transfer is given by

$$Q = \frac{kA}{x}(\theta_{surf} - \theta_i)$$

where θ_{surf} is the surface temperature and θ_i the interface temperature. Q can be increased by increasing the surface temperature θ_{surf}. However, this is limited by the temperature which the surface of the food will tolerate without scorching and by the need to avoid melting at the interface.

Note that, at equilibrium, the rate of sublimation equals the rate of mass transfer.

The mass transfer rate $- dm/dt$ is given by

$$\frac{-dm}{dt} = \frac{Ab}{x}(p_{wi} - p_{wsurf})$$

where p_{wi} is the partial pressure of water at the interface, p_{wsurf} is the partial pressure of water at the solid surface and b is the permeability of dry material.

At equilibrium the rate of heat transfer and the rate of mass transfer are related by the latent heat of sublimation.

Karel (1975) evaluated the drying time in terms of the heat transfer properties or the mass transfer properties as follows:

$$t = \frac{x^2 \rho (m_I - m_F) \Delta H_{subl}}{8k(\theta_{surf} - \theta_i)}$$

$$= \frac{x^2 \rho (m_I - m_F)}{8b(p_{wi} - p_{wsurf})}$$

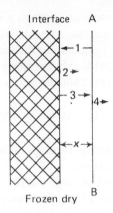

Fig. 13.12 — Cross-section through food part of the way through a freeze-drying processes; AB, heating surface; I, ice–dry food interface; 1, heat transfer; 2, sublimation; 3, mass transfer through the solid; 4, vapour removal from the chamber.

where x is the slab thickness, ρ is the bulk density of the solid, m_I and m_F are the initial and final moisture contents, ΔH_{subl} is the latent heat of sublimation, k is the thermal conductivity of the dry later, and θ_{surf} and θ_i are the surface and interfacial temperatures. The interfacial temperature remains constant throughout the drying period.

These equations are important in that they show how the various factors affect the drying time. It is important to know that the drying time is proportional to the square of the thickness; therefore, doubling the thickness will increase the drying time by a factor of 4.

It should also be possible to decrease the drying time by putting heat in through the frozen layer. This has been discussed in more detail by Judson-King (1971) and Mellor (1978). The main mechanism of drying is by conduction, and contact between the food and the heating plates is improved by applying a slight pressure; this is known as accelerated freeze drying. The history of the development of the accelerated freeze-drying process has been documented (Her Majesty's Stationery Office 1961) and provides interesting reading.

Some of the advantages of freeze drying are improved product quality due to the mild heat treatment, better volatile retention due to selective diffusion of water vapour through the dry layer and no case hardening or shrinkage. The major disadvantages are the high capital and running costs, which has meant that the major commercial successes for freeze-dried products have been restricted to relatively expensive products, for which the consumer is willing to pay a higher price for the superior-quality product produced. Examples are coffee, shrimps, prawns, chickens and mushrooms and some dehydrated complete meals. Microbiological cultures are also often preserved by freeze drying.

The selective retention of volatile components during freeze drying has been discussed by Mellor (1978).

Other methods of drying not discussed here are roller drying, osmotic drying and microwave or dielectric drying. Traditional drying methods such as solar drying are still widely used worldwide.

13.10 PACKAGING MATERIALS

One of the main functions of packaging material is to provide adequate barrier properties. These include reducing the amount of light that enters the product, preventing the entry of micro-organisms and other environmental contaminants and reducing the transmission of water vapour, oxygen or other gases, as the situation demands. Metal packaging materials in the form of foil or cans provide a complete barrier, as does glass, except for its light transmission properties. More recently a wide variety of plastics have been developed in the form of films or rigid containers, and one important property of these materials is their permeability, particularly when used in films.

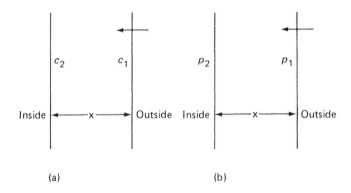

Fig. 13.13 — Mass transfer through packaging materials: (a) concentration gradients; (b) partial pressure gradients.

Consider the flow of a gas or vapour through a membrane under steady-state conditions (Fig. 13.13). Then, using Fick's law,

$$J = DA \frac{c_1 - c_2}{x}$$

where J is the volumetric flow rate or mass flow rate, D is the diffusivity of the material, A is the surface area, c_1 is the concentration of the component on the outside of the film, c_2 is the concentration of the component on the inside of the film and x is the film thickness. However, it is difficult to determine the concentration of the gas dissolved in the membrane material;

so it is assumed that it is in equilibrium with the gas in contact with it. Henry's law gives the relationship between solubility and partial pressure:

$$c = Hp$$

Therefore,

$$J = DA \frac{H(p_1 - p_2)}{x}$$

and the product of DH is the permeability B of the package to that particular gas.

The equation states that the transfer rate of gas is proportional to the permeability, the surface area and the partial pressure driving force and is inversely proportional to the thickness

$$B = \frac{Jx}{A(p_1 - p_2)}$$

Many types of units are used in practice, two examples being given in Table 13.5. It is important to ensure that consistent units are used when evaluating

Table 13.5 — Two examples of the systems of units used.

Parameter and symbol		Units in one system	Units in the other system
Flux	J	cm³ s^{-1}	kmol day^{-1}
Thickness	x	mil	cm
Area	A	m²	m²
Pressure	p_1-p_2	atm	Torr
Permeability	B	cm³ mil s^{-1} m^{-2} atm^{-1}	kmol cm day^{-1} m^{-2} Torr^{-1}

Note that 1 mil = 0.001 in = 25.4 μm.

the diffusion rate. The permeability properties of some of the more important films are given in Table 13.6.

Materials differ widely in their permeabilities towards oxygen, carbon dioxide and water vapour.

Composite packaging materials are now widely available and total resistance can be dealt with in a similar way to thermal resistances. The most usual form is to have the resistances in series (Fig. 13.14).

At steady state, the overall rate of mass transfer of a component can be expressed in trms of an overall permeability B_T and total pressure difference, i.e.

Table 13.6 — Permeability of some gases and vapours through a variety of film materials.

	Permeability (ml m^{-2} MPa^{-1} day^{-1}) for the following gases and vapours			
	Nitrogen (30 °C)	Oxygen (30 °C)	Carbon dioxide (30 °C)	Water vapour (25 °C) relative humidity, 90%)
Poly(vinylidene chloride) (Saran)	0.7	0.35	1.9	94
Polychlorotrifluoroethylene	0.20	0.66	4.8	19
Polyester (Mylar A)	0.33	1.47	10	8700
Polyamide (Nylon 6)	0.67	2.5	10	47,000
Poly(vinyl chloride) (unplasticized)	2.7	8.0	6.7	10,000
Cellulose acetate (P912)	19	52	450	500,000
Polyethylene ($\rho = 0.945$–0.960 g cm^{-3})	18	71	230	860
Polyethylene ($\rho = 0.922$ g cm^{-3})	120	360	2300	5300
Polystyrene	19	73	590	80,000
Polypropylene ($\rho = 0.910$ g cm^{-3})		150	610	480

All permeabilities are calculated for a film 25 µm (1 mil) thick.
Adapted from the data of Paine and Paine (1983).

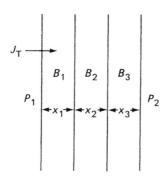

Fig. 13.14 — Composite packaging materials.

$$J_T = B_T A_T (p_1 - p_2)$$

where

$$B_T = \frac{1}{x_1/B_1 + x_2/B_2 + x_3/B_3 + \ldots}$$

B_T takes into account both the individual permeabilities and their thicknesses.

Two light-weight packages which have been developed comparatively recently are the flexible pouch and the aseptic package. The flexible pouch is a laminate consisting of plastic and metal films and is used for in-container sterilization processes. It gives a much larger surface-area-to-volume ratio than a tin can of similar capacity, resulting in faster heating and cooling times, shorter overall processing times and a better-quality product. The aseptic package, exemplified by the Tetrabrik or Combiblok is a four- or five-ply laminate, consisting of plastic film, metal film and paper. It is used for storing long-life UHT products which are produced using a continuous heat exchanger and then put into the package under aseptic conditions. The packaging material is sterilized using hydrogen peroxide. These products are regarded as being commercially sterile, having a shelf-life of 6 months at ambient temperature.

Other properties of flexible packaging that need to be accounted for include suitability for use at low temperatures (e.g. frozen storage) and high temperatures (e.g. retorting) compatability with solvents (e.g. water, acids, alkalies, oils and fats and organic materials), cost, light transmission properties, mechanical properties (e.g. tensile strength and shear stress impact strength) and many others.

Also of considerable interest is the diffusion of material from the package into the food, examples being lead solder in cans and plasticizers from the flexible packaging material. Packaging materials have been described in greater detail by Paine and Paine (1983), Palling (1980), Hersom and Hulland (1980) and Sacharow and Griffin (1980).

13.11 MEMBRANE PROCESSES

Polymeric materials are used to fabricate semipermeable membranes for use in processes such as reverse osmosis (hyperfiltration), ultrafiltration, dialysis and electrodialysis. The membranes are permeable to some components but not to others, discrimination being based on the molecular shape, size and charge of the component.

In both *reverse osmosis* and *ultrafiltration*, the liquid is placed on one side of the membrane and subjected to a pressure; the material which passes through the membrane is known as the permeate. The major differences between ultrafiltration and reverse osmosis are summarized in Table 13.7.

The flow rate of solvent through the membrane can be expressed as

$$J_{solvent} = K_{solvent} (\Delta p - \Pi)$$

where Δp is the mean pressure difference across the membrane, Π is the osmotic pressure of the solution and $K_{solvent}$ is the resistance to solvent flow. For dilute solutions the osmotic pressure can be estimated using the Van't Hoff equation

$$\Pi = \frac{RTc}{M}$$

Table 13.7 — Distinction between reverse osmosis and ultrafiltration.

	Reverse osmosis	Ultrafiltration
Operating pressure	Greater than 50 bar	Between 1 and 10 bar
Membrane characteristics	Small pores; 100% rejection of all components	Larger pore size; low rejection of low-molecular-weight solutes; high rejection of proteins and other high-molecular-weight components
Nature of permeate	Water, perhaps with traces of low-molecular weight components	Low-molecular-weight components, present at similar concentration to that in the feed
Selection mechanism	Diffusion activated	Sieving process

where Π is the osmotic pressure, c is the concentration and M is the molecular weight. (SI units throughout). An alternative expression for the permeate rate is

$$J = \frac{(\Delta p - \Pi)}{R_M + R_F + R_P}$$

where R_M is the resistance due to the membrane R_F is the resistance due to the fouling layer and R_P is the resistance due to concentration polarization. Fouling may occur because of the deposition of particulate matter onto the surface of the membrane. It is minimized by centrifuging or filtering the feed prior to processing. Concentration polarization occurs as a result of a build-up rejected material at the membrane surface and offers an additional resistance. It can be reduced by inducing turbulence in the feed. The permeate flux rate is usually increased by increasing the pressure, increasing the flow rate and increasing the temperature.

In practice the performance is characterized by its rejection R, where

$$R = \frac{c - c_{permeate}}{c}$$

where c is the concentration of the component in the feed and $c_{permeate}$ is the concentration in the permeate. Rejection values can be obtained for any component by analysing samples of the feed and permeate (Fig. 13.15(a)).

The rejection characteristics of reverse osmosis and ultrafiltration membranes are briefly described in Table 13.7. Lewis (1982) has shown how rejection data can be used to evaluate the yield of a component and has reviewed rejection data for ultrafiltration membranes used for concentrating proteins.

The solute flux through the membrane J_{solute} is given by a similar equation to that for solvent flux:

$$J_{solute} = K_{solute}\, c$$

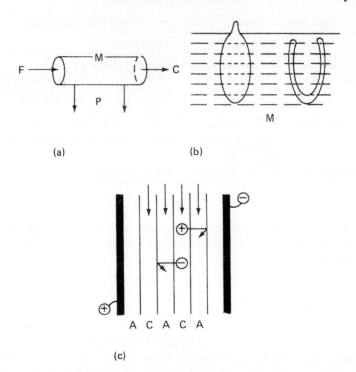

Fig. 13.15 (a) Flow through a tubular membrane (reverse osmosis or ultrafiltration), showing the membrane M, the feed F, the concentrate C and the permeate P; (b) dialysis processes, showing the membrane M; (c) the movement of charged particles in electrodialysis showing the anion-permeable membranes A and the cation-permeable membranes C.

where K_{solute} is the resistance to solute flow. Therefore the concentration of solute in the permeate will equal

$$\frac{\text{solute flux}}{\text{solvent flux}} = \frac{K_{solute} c}{K_{solvent}(\Delta p - \Pi)}$$

$$= c_{permeate}$$

Therefore,

$$\text{rejection} = \frac{c - c_{permeate}}{c}$$

$$= 1 - \frac{K_{solute}}{K_{solvent}(\Delta p - \Pi)}$$

Thus, it should be possible to predict the membrane rejection values in terms of the mass transfer characteristic $K_{solvent}$ of the solvent, the mass transfer characteristic K_{solute} of the solute the operating pressure Δp and the osmotic pressure Π of the feed. However, mass transfer theory is not sufficiently developed to allow this to be done with any degree of confidence for complex food systems.

Reverse osmosis is used for concentrating liquids such as milk, cheese whey, fruit juices and other beverages. Operating temperatures are low and there is no phase change and a smaller loss of volatiles compared with evaporation. More recently, it has been used for removing alcohol from beer and lager.

Ultrafiltration is used for concentrating high-molecular-weight components under mild operating conditions. This is useful for proteins including enzymes, polysaccharides and fermentation products. Membrane techniques have been reviewed in more detail by Glover et al. (1978), Lewis (1982) and the International Dairy Federation (1979).

However, *dialysis* is not pressure activated and is a slower process than ultrafiltration. The solution to be dialysed is placed inside a dialysis membrane or sack, which is then sealed. The membrane is then surrounded by the dialysis fluid, which is usually pure water, and components which will permeate through the membrane will diffuse from the solution into the water under the influence of the concentration gradient that has been established (Fig. 13.15(b)). The major use of dialysis has been in the laboratory for cleaning up or purifying extracts prior to analysis; usually the components removed are of low molecular weight. One drawback results from having to change the dialysis fluid at regular intervals to maintain a reasonable concentration gradient throughout the process.

A further development of dialysis is known as *electrodialysis*. It involves the use of an electric field and specially tailored charged membranes which are arranged in stacks (Fig. 13.15(c)). Two types of membranes are used: cation permeable C; anion permeable A. The material to be treated is passed through every second channel, alternating with water. Under the influence of the electric field, cations such as Na^+ and K^+ move towards the cathode and anions such as Cl^- and SO_4^{2-} move towards the anode, the overall result being the removal of low-molecular-weight charged particles from the solution into the water. The layout of the plant and the movement of the ions is illustrated in Fig. 3.15(c). Each ion passes through one membrane before being repelled by the next and the water is converted to a dilute salt solution. Therefore the process is similar to demineralization, using ion exchange resins, in its overall effects and has been used commercially for demineralizing cheese whey. A comparison of the economics of the two processes has been reported by Houldsworth (1980).

13.12 SYMBOLS

A   surface area
b   permeability of dried material
B   permeability of packaging
c   concentrations

d	diameter
d_{Sm}	Sauter mean particle diameter
D	diffusion coefficient or diffusivity
h	film heat coefficient
h_{fg}	latent heat of vaporization
H	Henry's law constant
ΔH_{subl}	latent heat of sublimation
j_D	mass transfer factor
J	volumetric flux
k	thermal conductivity
k_G	gas film coefficient
k_L	liquid film coefficient
K	mass transfer coefficient
K_G	overall gas mass transfer coefficient
K_L	overall liquid mass transfer coefficient
L	length (slab thickness)
m	mass
m'	dm/dt, mass transfer rate
m	moisture content
M	molecular weight
n	number of particles
N	mass or molar transfer rate
P	partial pressure
p_T	total pressure
P_s	saturation vapour pressure
Q'	heat transfer rate
r	radius of drop
R	rejection
R	gas constant
R_F	resistance due to the fouling layer
R_M	resistance due to the membrane
R_p	resistance due to the concentration polarization
t	time
t_e	exposure time at surface
T	temperature
x	distance

13.12.1 Greek symbols

θ	temperature
μ	viscosity
Π	osmotic pressure
ρ	density

13.12.2 Dimensionless groups

Pe	Peclet number
Re	Reynolds number
Sc	Schmidt number

Sec. 13.2] Symbols 445

Sh Sherwood number
St Stanton number

13.12.3 Subscripts
a   air
b   bulk
c   critical
e   equilibrium
F   final
G   gas
i   interface
I   initial
L   liquid
s   saturated or saturation
S   solid
T   total
v   vapour
w   water

Bibliography and references

Adrian, J., 1976. Gums and hydrocolloids in nutrition. *World Rev. Nutr. Diet.*, **25**, 189.
Akers, R. J. (Ed.), 1976, *Foams*. Academic Press, London.
American Society of Heating, Refrigerating and Air-Conditioning Engineers, 1981, *ASHRAE Handbook, Fundamentals*. ASHRAE, Atlanta, Georgia.
American Society of Heating, Refrigerating and Air-Conditioning Engineers, 1982, *ASHRAE Handbook, Applications*. ASHRAE, Atlanta, Georgia.
Arbuckle, W. S., 1977. *Ice-cream*, 3rd edition. AVI, Westport, Connecticut.
Arthey, V. D., 1975. *Quality of Horticultural Products*. Butterworths, London.
Baird, D. G., 1983. Food dough rheology. In M. Peleg and E. B. Bagley (Eds.), *Physical Properties of Food*. AVI, Westport, Connecticut.
Barfoed, H. C., 1983, Detergents. In T. Godfrey and J. Reichelt (Eds.), *Industrial Enzymology, The Application of Enzymes in Industry*. Macmillan, Byfleet, Surrey.
Batty, J. C., and Folkman, S. L., 1983. *Food Engineering Fundamentals*. John Wiley, New York.
Beaverstock, M. C., 1983. Process control. In I. Saguy (Ed.) *Computer Aided Techniques In Food Technology*. Marcel Dekker, New York.
Becher, P., 1965. *Emulsions, Theory and Practice*. Rheinhold, New York.
Bender, A. E., 1978. *Food Processing and Nutrition*. Academic Press, London.
Berger, K. G., 1976, Ice-cream. In S. Friberg (Ed.) *Food Emulsions*. Marcel Dekker, New York.
Bengtsson, N. E. and Ohlsson, T., 1974. Microwave heating in the food industry. *IEEE Trans.*, **62** (1), 44.

Bibliography and references

Bengtsson, N. E., and Risman, P. O. 1971. Dielectric properties of foods at 3 GHz, as determined by a cavity perturbation technique, II, Measurement of food materials. *J. Microwave Power*, **6** (2), 107.

Biliaderis, C. G., 1983, Differential scanning calorimetery in food research — a review. *Food Chem.*, **10** (4), 239.

Blackhurst, J. R., Harker, J. H., and Porter, J. E., 1974. *Problems In Heat and Mass Transfer.* Edward Arnold, London.

Blanshard, J. M. V., and Mitchell, J. R. (Eds.), 1979. *Polysaccharides in Foods.* Butterworths, London.

Birch, G. G., Brennan, J. G., and Parker, K. J. (Eds.), 1977, *Sensory Properties of Foods.* Applied Science, Barking, Essex.

Bourne, M. C., 1975. Texture measurement in vegetables. In C. K. Rha (Ed.), *Theory, Determination and Control of Physical Properties of Food Materials.* D. Reidel, Dordrecht.

Bourne, M. C., 1978. Texture profile analysis. *Food Technol.*, **7**, 62.

Bourne, M. C., 1982. *Food Texture and Viscosity.* Academic Press, New York.

Bourne, M. C., 1983. Physical properties and structure of horticultural products. In M. Peleg, and E. B. Bagley (Eds.), *Physical Properties of Foods.* AVI, Westport, Connecticut.

Brennan, J. G., 1984. Texture perception and measurement. In J. R. Piggott, (Ed.), *Sensory Analysis of Foods.* Elsevier Applied Science, London.

Brennan, J. G., Butters, J. R., Cowell, N. D., and Lilly, A. E. V., 1976. *Food Engineering Operations.* Applied Science, London.

British Standards Institution, 1937. *British Standard* **734**.

British Standards Institution, 1959. *British Standard* **734**.

British Standards Institution, 1964a. *British Standard* **1904**.

British Standards Institution, 1964b. *British Standard* **3718**.

British Standards Institution, 1974. *British Standard* **4937**.

British Standards Institution, 1975. *British Standard Glossary of Rheological Terms, British Standard* **5168**.

Campbell, A. M., Penfield, M. P., and Griswold, R. M., 1979. *The Experimental Study of Food.* Houghton Mifflin, Boston, Massachusetts.

Calderbank, P. H., 1967. Mass transfer in fermentation equipment. In N. Blakebrough (Ed.), *Biochemical and Biological Engineering Science*, Vol. 1. Academic Press, New York.

Charm, S. E., 1978. *The Fundamentals of Food Engineering.* AVI, Westport, Connecticut.

Cheow, C. S., and Jackson, A. T., 1982. Circulation cleaning of a plate heat exchanger fouled by tomato juice. *J. Food Technol.*, **17**, 417.

Cleland, A. C., and Earle, R. L., 1982. Freezing time prediction for foods — a simplified procedure. *Int. J. Refrig.*, **5** (3), 134.

Collison, R., 1968. Swelling and gelation of starch. In J. A. Radley, *Starch and its Derivatives*, 4th edition. Chapman and Hall, London.

Considine, D. M., and Considine, G. D. (Eds.), 1982. *Foods and Food Production Encyclopedia.* Van Nostrand Reinhold, New York.

Coulson, J. M., and Richardson, J. F., 1977. *Chemical Engineering*, Vol. 1. Pergamon Press, Oxford.
Coulson, J. M., and Richardson, J. F., 1978. *Chemical Engineering*, Vol. 2. Pergamon Press, Oxford.
Coulson, J. M., and Richardson, J. F., 1979. *Chemical Engineering*, Vol. 3. Pergamon Press, Oxford.
Davidson, S., Passmore, R., Brock, J. F., and Truswell, A. S., 1979. *Human Nutrition and Dietetics*. Churchill Livingstone, Edinburgh.
Davies, R., Birch, G. G., and Parker, K. J. (Eds.), 1976, *Intermediate Moisture Foods*. Applied Science, London.
Deman, J. M., 1976. Texture of fats and fat products. In J. M. Deman et al. (Eds.), *Rheology and Texture in Food Quality*. AVI, Westport, Connecticut.
Deman, J. M., Dobbs, J. E., and Sherman, P., 1979. Spreadability of butter and margarine. In P. Sherman (Ed.), *Food Texture and Rheology*. Academic Press, London.
Desrosier, N. W., and Tressler, D. K. (Eds.), 1977. *Fundamentals of Food Freezing*. AVI, Westport, Connecticut.
Dickinson, E., and Stainsby, G., 1982. *Colloids In Foods*. Applied Science, London.
Dinsdale, A., and Moore, F., 1962. *Viscosity and its Measurement*. Institute of Physics and Physical Society, Chapman and Hall, London.
Duckworth, R. B. (Ed.), 1975. *Water Relations of Foods*. Academic Press, London.
Earle, R. L., 1983. *Unit Operations in Food Processing*. Pergamon Press, Oxford.
Ede, A. J., 1949, *Mod. Refrig.*, **52**, 52.
Elias, P. S., and Cohen, A. J. (Eds.) *Recent Advances in Food Irradiation*. Elsevier Biomedical Press, Amsterdam.
Ellis, P., 1982, Private communication.
Farrall, A. W., 1979, *Food Engineering Systems*, Vol. 2, *Utilities*. AVI, Westport, Connecticut.
Fennema, O. R. (Ed.), 1975a. *Principles of Food Science*, Part 2, *Physcial Principles of Food Preservation*. Marcel Dekker, New York.
Fennema, O. R., 1975b, Freezing preservation. In O. R. Fennema (Ed.), *Principles of Food Science*, Part 2, *Physical Principles of Food Preservation*. Marcel Dekker, New York.
Fennema, O. R. (Ed.), 1976. *Principles of Food Science*, Part 1, *Food Chemistry*. Marcel Dekker, New York.
Fennema, O. R., Powrie, J. D., and Morth, E. H., 1973. *Low Temperature Preservation of Foods and Living Matter*. Marcel Dekker, New York.
Fink, O. G. (Ed.), 1975, *Electronic Engineer's Handbook*. McGraw-Hill, New York.
Finney, E. E., Jr., 1973. Elementary concepts of rheology relevant to food texture studies. In A. Kramer and A. S. Szczesniak (Eds.), *Texture Measurement of Foods*. D. Reidel, Dordrecht.
Flink, J. M., 1983. Structure and structure transitions in dried carbohydrate

materials. In M. Peleg and E. B. Bagley (Eds.), *Physical Properties of Foods.* AVI, Westport, Connecticut.
Food Technology, 1979, Overview, energy analysis in food process operations. *Food Technol.*, 3, 51.
Fox, P. F. (Ed.). 1982. *Developments in Dairy Chemistry,* Vol. 1. Applied Science, London.
Friberg, S. (Ed.), 1976. *Food Emulsions.* Marcel Dekker, New York.
Glicksman, M. (Ed.), 1969. *Gum Technology in the Food Industry.* Academic Press, New York.
Glicksman, M. (Ed.), 1982, *Food Hydrocolloids,* Vol. 1. CRC Press, Boca Raton, Florida.
Glover, F. A., Skudder, P. J., Stothart, P. H., and Evans, E. W., 1978. Reviews of the progress of dairy science: reverse osmosis and ultrafiltration in dairying. *J. Dairy Res.*, 45, 291.
Godfrey, T., and Reichelt, J. (Eds.), 1983. *Industrial Enzymology — The Application of Enzymes in Industry.* Macmillan, Byfleet, Surrey.
Goldblith, S. A., Rey, L., and Rothmayr, W. W. (Eds.), 1975. *Freeze Drying and Advanced Food Technology.* Academic Press, London.
Graf, E., and Bauer, H., 1976. Milk and milk products. In S. Friberg (Ed.), *Food Emulsions.* Marcel Dekker, New York.
Graham, D. E., and Phillips, M. C., 1976. The conformation of proteins at the air-water interface and their role in stabilizing foams. In R. J. Akers, *Foams.* Academic Press, London.
Graham, H. D. (Ed.), 1977, *Food Colloids.* AVI, Westport, Connecticut.
Gruenwedel, D. W., and Whitaker, J. R. (Eds.), 1984, *Food Analysis — Principles and Techniques,* Vol. II, *Psyicho chemical Techniques.* Marcel Dekker, New York.
Hall, C. W., 1979. *Dictionary of Drying.* Marcel Dekker, New York.
Hall, C. W., Farrall, A. W., and Rippen, A. L. (Eds.), 1971, *Encyclopedia of Food Engineering.* AVI, Westport, Connecticut.
Hawthorn, J., 1981, *Foundations of Food Science.* W. H. Freeman, Oxford.
Heidelbaugh, H. D., and Karel, M., 1975. Intermediate moisture food technology, In S. A. Goldblith *et al.* (Eds.), *Freeze Drying and Advanced Food Technology.* Academic Press, London.
Heimenz, P. C., 1977. *Principles of Colloid and Surface Chemistry.* Marcel Dekker, New York.
Henderson, S. M., and Perry, R. L., 1955. *Agricultural Process Engineering.* John Wiley, New York.
Her Majesty's Stationery Office, 1946. *Vegetable Dehydration.* Ministry of Food. HMSO, London.
Her Majesty's Stationery Office, 1961. *The Accelerated Freeze Drying Method of Food Preservation.* HMSO, London.
Hermansson, A. M., 1979. Texturization of proteins. In P. Sherman (Ed.), *Food Texture and Rheology.* Academic Press, London.
Hersom, A. C., and Hulland, E. D., 1980. *Canned Foods.* Churchill Livingstone, Edinburgh.

Holland, C. R., and McCann, J. B., 1980. Heat recovery in spray drying systems. *J. Food Technol.*, 15 (1) 9.
Holland, F. A., 1973. *Fluid Flow for Chemical Engineers.* Edward Arnold, London.
Holmes, Z. A., and Woodburn, M., 1981. Heat transfer and temperature of foods during preparation. CRC Crit. Rev. Food Sci. Nut., **14** (3) 231.
Houldsworth, D. W., 1980. Demineralization of whey by means of ion exchange and electrodialysis. *J. Soc. Dairy Technol.*, **33**, 45.
Howell, N. K., and Lawrie, R. A. 1984. Functional aspects of blood plasma proteins, II, Gelling properties. *J. Food Technol.*, **19** (3), 289.
Howgate, P., 1977. Aspects of fish texture. In G. G. Birch, J. G. Brennan and K. J. Parker (Eds.), *Sensory Properties of Foods.* Applied Science, London.
Hughes, E., 1977. *Electrical Technology.* Longmans, London.
Hugot, E., 1972. *Handbook of Cane Sugar Engineering.* Elsevier, Amsterdam.
Hyde, K. A., and Rothwell, J., 1973, *Ice-cream.* Churchill Livingstone, Edinburgh.
Iglesias, H. A., and Chirife, J. 1982. *Handbook of Food Isotherms: Water Sorption Parameters for Food and Food Components.* Academic Press, New York.
Institute of Food Science and Technology, 1982. *Guidelines for Handling for Chilled Foods.* IFST, London.
International Dairy Federation, 1977. *Energy Conservation in the Dairy Industry, Annual Bulletin, IDF Doc.* **102**.
International Dairy Federation, 1979. *Equipment Available for Membrane Processing, IDF Doc.* **115**.
International Dairy Federation, 1980. *General Code of Hygienic Practice for the Dairy Industry, IDF Doc.* **123**.
International Dairy Federation, 1981. *Evaluation of the Firmness of Butter, IDF Doc.* **135**.
International Institute of Refrigeration, 1979, *Recommendations for Chilled Storage of Perishable Produce.* IIR, Paris.
Jackson, A. T., and Lamb, J., 1981. *Calculations in Food and Chemical Engineering.* Macmillan, London.
Jackson, J. M., and Shinn, B. M., 1979. *Fundamentals of Food Canning Technology.* AVI, Westport, Connecticut.
Jellinek, G., 1985. *Sensory Evaluation of Food.* Ellis Horwood, Chichester, West Sussex.
Jenness, R., Shipe, W. F., Jr., and Sherbon, J. W., 1974. Physical properties of milk. In B. H. Webb, A. H. Johnson and J. A. Alford (Eds.), *Fundamentals of Dairy Chemistry.* AVI, Westport, Connecticut.
Johnson, A. H., and Peterson, M. S., 1974. *Encyclopedia of Food Technology.* AVI, Westport, Connecticut.
Johnson, J. F., Martin, J. R., and Porter, R. S., 1975. Determination of viscosity of food systems. In C. Rha (Ed.), *Theory, Determination and Control of Physical Properties of Food Materials.* D. Reidel, Dordrecht.

Jones, E. B., 1974a. *Instrument Technology*, Vol. 1. Newnes Butterworths, London.
Jones, E. B., 1974b. *Instrument Technology*, Vol. 2. Newnes Butterworths, London.
Josephson, E. S., and Peterson, M. S. (Eds.), 1982. *Preservation of Food by Ionizing Radiation*, Vol. 1. CRC Press, Boca Raton, Florida.
Josephson, E. S., and Peterson, M. S. (Eds.), 1983a. *Preservation of Food by Ionizing Radiation*, Vol. 2. CRC Press, Boca Raton, Florida.
Josephson, E. S., and Peterson, M. S. (Eds.), 1983b. *Preservation of Food by Ionizing Radiation*, Vol. 3. CRC press, Boca Raton, Florida.
Journal of Food Protection, 1980. Articles on microwave heating. *J. Food. Prot.*, **43**, 617.
Jowitt, R. (Ed.), 1980. *Hygienic Design and Operation of Food Plant*. Ellis Horwood, Chichester.
Jowitt, R. (Ed.), 1984. *Extrusion Cooking Technology*. Elsevier Applied Science, London.
Jowitt, R., Escher, F., Hallstrom, B., Meffert, H. F. Th., Spiess, W., and Vos, G. (Eds.), 1983. *Physical Properties of Foods*. Applied Science, London.
Judson-King, C., 1971. *Freeze Drying of Foods*. Butterworths, London.
Jukes, D. J., 1984. *Food Legislation of the UK — A Concise Guide*. Butterworths, London.
Jul, M., 1984. *The Quality of Frozen Foods*. Academic Press, London.
Kalab, M., 1983. Electron microscopy of foods. In M. Peleg and E. B. Bagley (Eds.), *Physical Properties of Foods*. AVI, Westport, Connecticut.
Karel, M., 1975. Dehydration of foods. In O. R. Fennema (Ed.), *Principles of Food Science*, Part 2, *Physical Principles of Food Preservation*. Marcel Dekker, New York.
Kaye, G. W. C., and Laby, T. H., 1973. *Tables of Physical and Chemical Constants*. Longmans, London.
Keeney, P. G., and Kroger, M., 1974. Frozen dairy products. In B. H. Webb, A. H. Johnson and J. A. Alford (Eds.), *Fundamentals of Dairy Chemistry*. AVI, Westport, Connecticut.
Kessler, H. G., 1981. *Food Engineering and Dairy Technology*. A. Kessler, Freising.
Kinsella, J. E., 1976. Functional properties in food — a survey. *CRC Rev. Food Sci. Nut.*, **7**, 219.
Kirk, D., and Milson, A., 1982. *Services, Heating and Equipment for Home Economists*. Ellis Horwood, Chichester, West Sussex.
Klein, H. A., 1974, *The World of Measurement*. Simon and Schuster, New York.
Kleinert, J., 1976. Rheology of chocolate. In J. M. Deman *et al.* (Eds.), *Rheology and Texture in Food Quality*. AVI, Westport, Connecticut.
Kramer, A., and Szczesniak, A. S. (Ed.), 1973, *Texture Measurements of Foods*. D. Reidel, Dordrecht.
Krog, N., and Lauridsen, J. B., 1976. Food emulsifiers and their associa-

tions with water. In S. Friberg (Ed.), *Food Emulsions*. Marcel Dekker, New York.
Lamb, J. 1976. Influence of water on the physical properties of foods. *Chem. Ind.*, **24**, 1046.
Launay, B., and Lisch, J. M., 1979. Protein dopes. In P. Sherman (Ed.), *Food Texture and Rheology*. Academic Press, London.
Leach, G., 1976, *Energy and Food Production*. IPC Science and Technology Press, London.
Lee, C. H., and Rha, C. K., 1979. Rheological properties of proteins in solution. In P. Sherman (Ed.), *Food Texture and Rheology*. Academic Press, London.
Levenspiel, O., 1972. *Chemical Reaction Engineering*. John Wiley, New York.
Levitt, B. P., 1973. *Findlay's Practical Physical Chemistry*. Longman's, London.
Lewis, G. E. D., 1973. *Metric and Other Conversion Tables*. Longmans, London.
Lewis, M. J., 1982. Concentration of proteins by utlrafiltration. In B. J. F. Hudson (Ed.), *Developments of Food Proteins*, Vol. 1. Applied Science, London.
Lewis, M. J., 1986a. Advances in the heat treatment of milk. In R. K. Robinson (Ed.), *Modern Dairy Technology*, Vol. 1. Elsevier Applied Science, London.
Lewis, M. J., 1986b. Physical properties of milk and milk products. In R. K. Robinson (Ed.), *Modern Dairy Technology*, Vol. 2. Elsevier Applied Science, London.
Ley, F. L., 1984. Food irradiation — background and potential. *Engineering Aspects of Food Irradiation, Food Engineering Forum*.
Linko, P., Malkki, Y., Olkku, J., and Larinkari, J. (Eds.), 1980. *Food Process Engineering*, Vol. 1, *Food Processing Systems*. Applied Science, London.
Loncin, M., and Merson, R. L., 1979. *Food Engineering — Principles and Selected Applications*. Academic Press, New York.
Luh, B. S., and Woodroof, J. G., 1975. *Commercial Vegetable Processing*. AVI, Westport, Connecticut.
Lund, D. B., 1975. Heat processing. In O. R. Fennema (Ed.), *Principles of Food Science*, Part 2, *Physical Principles of Food Preservation*. Marcel Dekker, New York.
Lund, D. B., 1983. Application of differential scanning calorimetry in foods. In M. Peleg and E. B. Bagley (Eds.), *Physical Properties of Foods*. AVI, Westport, Connecticut.
Masters, K., 1972. *Spray Drying*, Leonard Hill, London.
McKenna, B. M. (Ed.), 1984a. *Engineering and Food*, Vol. 1, *Engineering Sciences in the Food Industry*. Elsevier Applied Science, London.
McKenna, B. M. (Ed.), 1984b. *Engineering and Food*, Vol. 2, *Processing Applications*. Elsevier Applied Science, London.

Mehta, R. S., 1980. Milk processed at ultrahigh temperatures — a review. *J. Food Prot.*, **43**, 212.
Mellor, J. D., 1978. *Fundamentals of Freeze Drying*. Academic Press, London.
Mettler, A. E., 1980. *Chemical and Physical Aspects of Powder Quality in Milk and Whey Powders*. Society of Dairy Technology, Wembley, Middlesex.
Miles, C. A., Van Beek, G., and Veerkamp, C. H., 1983. Calculation of thermophysical properties of foods. In R. Jowitt et al. (Eds.), *Physical Properties of Foods*. Applied Science, London.
Miller, D. S., and Payne, P. R., 1959. A ballistic bomb calorimeter. *Br. J. Nutr.*, 13.
Milson, A., and Kirk, D., 1980. *Principles of Design and Operation of Catering Equipment*. Ellis Horwood, Chichester, West Sussex.
Mohsenin, N. N., 1970. *Physical Properties of Plant and Animal Materials*, Vol. 1, *Structure, Physcial Characteristics and Mechanical Properties*. Gordon and Breach, London.
Mohsenin, N. N., 1980. *Thermal Properties of Foods and Agricultural Materials*. Gordon and Breach, London.
Mohsenin, N. N., 1984. *Electromagnetic Radiation Properties of Foods and Agricultural Products*. Gordon and Breach, New York.
Moore, E., 1967. *Detergents, Unilever Educational Booklet*. Unilever.
Mossel, D. A. A., 1975. Water and micro-organisms in foods — a synthesis. In R. B. Duckworth (Ed.), *Water Relations of Foods*. Academic Press, London.
Mudget, R. E., 1982. Electrical properties of foods in microwave processing. *Food Technol.*, **36**, (2) 109.
Muller, H. G., 1973. *An Introduction to Food Rheology*. Heinemann, London.
Muller, H. G., and Tobin, G., 1980. *Nutritition and Food Processing*. AVI, Westport, Connecticut.
Nelkon, M., 1970. *Heat*. Blackie, London.
Norrish, R. S., 1967. *Selected Tables of Physical Properties of Sugar Solutions, Scientific and Technical Surveys Publ.* **51**. Scientific and Technical Surveys, Leatherhead, Surrey.
Ohlsson, T., 1983. The measurement of thermal properties. In R. Jowitt (Eds.), Physical Properties of Foods. Applied Science, London.
Paine, F. A., and Paine, H. Y., 1983. *A Handbook of Food Packaging*. Leonard Hill, Glasgow.
Palling, S. J. (Ed.), 1980. *Developments in food packaging*, Vol. 1. Applied Science, London.
Pancoast, H. M., and Junk, W. R., 1980. *Handbook of Sugars*. AVI, Westport, Connecticut.
Paul, A. A., and Southgate, D. A. T., 1978. *McCance and Widdowsons: The Composition of Foods*. HMSO, London.
Paul, P. C., and Palmer, H. H., 1972. *Food Theory and Applications*. John Wiley, New York.

Peleg, M., 1983. Physical characteristics of food powders. In M. Peleg and E. B. Bagley, (Eds.), *Physical Properties of Foods*. AVI, Westport, Connecticut.
Peleg, M., and Bagley, E. B. (Eds.), 1983. *Physical Properties of Foods*. AVI, Westport, Connecticut.
Perkin, A. G., and Burton, H., 1970. The control of the water content of milk during ultrahigh temperature sterilization by a steam injection method. *J. Soc. Dairy Technol.*, **23** (3) 147.
Perry, R. H., and Chilton, C. H. (Eds.), 1973, *Chemical Engineers' Handbook*. McGraw-Hill, New York.
Pflug, I. J., and Esselen, W. B., 1979. Heat sterilization of canned food. In J. M. Jackson and B. M. Shinn (Eds.), *Fundamentals of Food Canning Technology*. AVI, Westport, Connecticut.
Piggott, J. R. (Ed.), 1984, *Sensory Analysis of Foods*. Elsevier Applied Science, London.
Polley, S. L., Snyder, O. P., and Kotnour, P., 1980. A compilation of thermal properties of foods. *Food Technol.*, **11**, 76.
Pomeranz, Y., and Meloan, C. E., 1978. *Food Analysis: Theory and Practice*. AVI, Westport, Connecticut.
Porter, J. W. G., 1975. *Milk and Dairy Products*. Oxford University Press, Oxford.
Potter, J. R., 1971. *Chemical Engineering — An Introduction*. Butterworths, London.
Powrie, W. D. and Tung, M. A., 1976. In O. R. Fennema (Ed.). *Principles of Food Science, Part I, Food Chemistry*, Marcel Dekker, New York.
Prentice, J. H., 1954. An instrument for estimating the spreadability of butter. *Lab. Pract.*, **3**, 186.
Prentice, J. H., 1979. Recent developments in the rheology of dairy products and some present-day problems. In P. Sherman (Ed.), *Food Texture and Rheology*. Academic Press, London.
Prentice, J. H., 1984. *Measurements in the Rheology of foodstuffs*. Elsevier Applied Science, London.
Radley, J. A., 1976. *Examination and Analysis of Starch and Starch Products*. Applied Science, London.
Reuter, H., 1984, UHT plants for milk — state of technological development. In B. M. McKenna (Ed.), *Engineering and Food*, Vol. 2. Elsevier Applied Science, London.
Rha, C. K., 1975a. Thermal properties of food materials. In C. Rha (Ed.), *Theory, Determination and Control of Physical Properties of Food Materials*, D. Reidel, Dordrecht.
Rha, C. K. (Ed.), 1975b. *Theory, Determination and Control of Physical Properties of Food Materials*. D. Reidel, Dordrecht.
Robinson, R. K. (Ed.), 1981a. *The Microbiology of Milk Products*, Vol. I. Applied Science, London.
Robinson, R. K. (Ed.), 1981b. *The Microbiology of Milk Products*, Vol. II. Applied Science, London.
Rockland, L. B., and Stewart, G. F. (Eds.), 1981, *Water Activity: Influence on Food Quality*. Academic Press, New York.

Rosen, M. J., 1978. *Surfactants and Interfacial Phenomena*. John Wiley, New York.
Sacharow, S., and Griffin, R. C., Jr., 1980, *Principles of Food Packaging*. AVI, Westport, Connecticut.
Salunkhe, D. K., 1974. *Storage, Processing and Nutritional Quality of Fruits and Vegatables*. CRC Press, Cleveland, Ohio.
Say, M. G. (Ed.), 1968. *The Electrical Engineer's Reference Book*. Hamlyn, Feltham, Middlesex.
Schofield, W., 1970. *Physics for ONC Engineers*. McGraw-Hill, Maidenhead, Berkshire.
Schwartzberg, H. G., 1977. Energy requirements for liquid food concentrations (overview). *Food Technol.*, 3, 1967.
Schwartzberg, H. G., and Chao, R. Y., 1982. Solute diffusivities in leaching processes. *Food Technol.*, 38, (2) 73.
Schwartzberg, H. G., 1983. Expression related properties. In M. Peleg, and E. B. Bagley (Eds.), *Physical Properties of Foods*. AVI, Westport, Connecticut.
Schweingruber, P., Escher, F., and Solms, J., 1979. Instrumental measurement of texture of instant mashed potato. In P. Sherman (Ed.), *Food Texture and Rheology*. Academic Press, London.
Schweitzer, P. A. (Ed.), 1979. *Handbook of Separation Techniques for Chemical Engineers*. McGraw-Hill, New York.
Shaw, D., 1970. *Introduction to Colloidal and Surface Chemistry*, 2nd edition. Butterworths, London.
Sherman, P. (Ed.), 1968. *Emulsion Science*. Academic Press, London.
Sherman, P., 1975. Factors influencing the instrumental and sensory evaluation of food emulsions. In C. Rha (Ed.), *Theory, Determination and Control of Physical Properties of Food Materials*. D. Reidel, Dordrecht.
Sherman, P. (Ed.), 1979, *Food Texture and Rheology*. Academic Press, London.
Sherman, P., 1980. Emulsion rheology and surface properties. In P. Linko, Y. Malkki, J. Olkku and J. Larinkari (Eds.), *Food Processing Systems*, Applied Science, London.
Singh, R. P., 1982. Thermal diffusivity in food processing. *Food Technol.*, 2, 87.
Singh, R. P., 1984, Energy management in the food industry (review paper). In B. M. McKenna (Ed.), *Engineering and Food*, Elsevier Applied Science, London.
Singh, R. P., and Heldman, D. R., 1984. *Introduction to Food Engineering*. Academic Press, Orlando, Florida.
Society of Dairy Techology, 1975. *Cream Processing Manual*. SDT, Wembley, Middlesex.
Society of Dairy Technology, 1980. *Milk and Whey Powders*. SDT, Wembley, Middlesex.
Society of Dairy Technology, 1982, Articles on energy economy. *J. Soc. Dairy Technol.*, 35, 3.

Society of Dairy Technology, 1983. *Pasteurizing Plant Manual.* SDT, Wembley, Middlesex.
Souci, S. W., Fachmann, W., and Kraut, H., 1981. *Food Composition and Nutrition Tables.* Wissenschaftliche, Stuttgart.
Spalding, D. B., and Cole, E. H., 1973. *Engineering Thermodynamics.* Edward Arnold, London.
Spencer, G. L., and Meade, G. P., 1957. *Cane Sugar Handbook.* John Wiley, New York.
Spicer, A. (Ed.), 1974. *Advances in Preconcentration and Dehydration of Foods.* Applied Science, London.
Stanley, D. W., 1976. The texture of meat and its measurement. In J. M. Deman et al. (Ed.), *Rheology and Texture in Food Quality.* AVI, Westport, Connecticut.
Stanley, D. W., 1983. Relation of structure to physical properties of animal material. In M. Peleg and E. B. Bagley (Eds.), *Physical Properties of Foods.* AVI, Westport, Connecticut.
Stumbo, C. R., 1973. Thermobacteriology. In *Food Processing.* Academic Press, New York.
Stumbo, C. R., Purohit, K. S., Ramokrishnan, T. V., Evans, D. A., and Francis, F. J., 1983a. *Handbook of Lethality Guides for Low-acid Canned Foods,* Vol. 1, *Conduction Heating.* CRC Press, Boca Raton, Florida.
Stumbo, C. R., Purohit, K. S., Ramokrishnan, T. V., Evans, D. A., and Francis, F. J., 1983b. *Handbook of Lethality Guides for Low-acid Canned Foods,* Vol. 2, *Convection Heating.* CRC Press, Boca Raton, Florida.
Swern, D. (Ed.), 1964, *Bailey's Industrial Oil and Fat Products.* Interscience, New York.
Sydenham, P. H. (Ed.), 1983. *Handbook of Measurement Science,* Vol. 2, *Practical Fundamentals.* John Wiley, Chichester, West Sussex.
Szczesniak, A. C., 1979. Classification of mouth feel characteristics of beverages. In P. Sherman (Ed.), *Food Texture and Rheology,* Academic Press, London.
Tamplin, T. C., 1980. Cleaning-in-place technology, detergents and sanitizers. In R. Jowitt (Ed.), *Hygienic Design and Operation of Food Plant.* Ellis Horwood, Chichester, West Sussex.
Taneya, S., Izutsu, T., and Sone, T., 1979. Dynamic viscoelasticity of natural cheese and processed cheese. In P. Sherman (Ed.), *Food Texture and Rheology.* Academic Press, London.
Tannenbaum, S. R., 1979. *Nutritional and Safety Aspects of Food Processing.* Marcel Dekker, New York.
Taranto, M. V., 1983. Structural and textual characteristics of baked goods. In M. Peleg and E. B. Bagley (Eds.), *Physical Properties of Foods,* AVI, Westport, Connecticut.
Taylor, R. J., 1961. *Surface Activity, Unilever Educational Booklet.* Unilever.

Thijssen, H. A. C., 1975. Freeze concentration. In S. A. Goldblith, L. Rey and W. W. Rothmayr (Eds.), *Freeze Drying and Advanced Food Technology*. Academic Press, London.
Thompson, D. R., 1982. The challenge in predicting nutrient changes during food processing. *Food Technol.*, **2**, 97.
Thomson, D. M. H., 1984. Flavour perception. *Br. Nutr. Found. Nutr. Bull.*, **9**, 69.
Thorne, S. (Ed.), 1981. *Developments in Food Preservation*, Vol. 1. Applied Science Publishers, Barking, Essex.
Thorne, S. (Ed.), 1983, *Developments in Food Preservation*, Vol. 2. Applied Science Publishers, Barking, Essex.
Timperley, D. A., and Lawson, G. B., 1980. Test rigs for evaluation of hygiene in plant design. In R. Jowitt (Ed.), *Hygienic Design and Operation of Food Plant*. Ellis Horwood, Chichester, West Sussex.
Toledo, R. T., 1980. *Fundamentals of Food Process Engineering*. AVI, Westport, Connecticut.
Troller, J. A., 1983. *Sanitation in Food Processing*. Academic Press, New York.
Troller, J. A., and Christian, J. H. B., 1978. *Water Activity and Food*. Academic Press, New York.
Tscheuschner, H. D. and Wunsche, D., 1979. Rheological properties of chocolate mass and the influence of some factors. In P. Sherman (Ed.), *Food Texture and Rheology*. Academic Press, London.
Tschubik, I. A., and Maslow, A. M., 1973, *Warmephysikalische Konstanten von Lebensmitteln und Halfabrikalen*. Fachbuchverlag, Leipzig.
Turnbull, A. H., Barton, R. S., and Riviere, J. C., 1962. An Introduction to Vacuum Technique. George Newnes, London.
Tyler, F., 1972. A Laboratory Manual of Physics, SI Version. Edward Arnold, London.
Uhl, U. W., and Gray, J. B., 1966. *Mixing, Theory and Practice*, Vol. 1. Academic Press, New York.
Uhl, U. W., and Gray, J. B., 1967. *Mixing, Theory and Practice*, Vol. 2. Academic Press, New York.
Unklesbay, N., and Unklesbay, K., 1982. *Energy Management in Food Service*. Ellis Horwood, Chichester, West Sussex.
Voisey, P. W., 1975. Instrumentation for determination of mechanical properties of foods. In C. K. Rha (Ed.), *Theory, Determination and Control of Physical Properties of Food Materials*. D. Reidel, Dordrecht.
von Kempf, W., and Kalender, G., 1972. Possible standardization of viscosity measuring procedures for a comparative valuation of the rheological properties and behaviour of starches. *Die Starke*, **7**, 220.
Walker, A. F., 1984. The nutritional quality of food. In G. G. Birch and K. J. Parker (Eds.), *Food and Health: Science and Technology*, Elsevier Applied Science, London.
Walstra, P., and Jenness, R., 1984. *Dairy Chemistry and Physics*. John Wiley, New York.

Warn, J. R. W., 1969. *Concise Chemical Thermodynamics*. Van Nostrand Reinhold, New York.
Weast, R. C. (Ed.), 1982. *Handbook of Physics and Chemistry*, 63rd edition. CRC Press, Cleveland, Ohio.
Webb, B. H. Johnson, A. H., and Alford, J. A. (Eds.), 1974. *Fundamentals of Dairy Chemistry*. AVI, Westport, Connecticut.
Whistler, G. (Ed.), 1975. Industrial Gums. Academic Press, New York.
Wilkinson, W. L., 1960. *Non-Newtonian Fluids, Fluid Mechanics and Heat Transfer*. Pergamon Press, London.
Williams-Gardner, A., 1971. *Industrial Drying*. George Goodwin, London.
Wiseman, A. (Ed.), 1982. *Principles of Biotechnology*. Blackie, Glasgow.
Wright, D. J., 1982. Application of scanning calorimetry to the study of protein behaviour in foods. In B. J. F. Hudson (Ed.), *Developments in Food Proteins*, Vol. 1. Applied Science, London.
Zangger, R. R., 1979. Viscoelasticity of doughs. In P. Sherman (Ed.), *Food Texture and Rheology*, Academic Press, London.

Index

absolute humidity, 344
absorptivity, 281
AC circuits, 392–402
AC circuits, evaluation, 395–399
AC circuits, introduction, 392
acceleration, 35
activation energy, 304
adiabatic cooling lines, 350
adiabatic processes, 206
adiabatic saturation temperature, 347
air–water systems, properties, 344–355
alternating current, characteristic, 389
alternating current, introduction, 389
ammeters, 374
amount of substance (mol), 28
Archimedes principle, 74
area, 29
aseptic operations, 102
Atwater factors, 215
automation, 387

basic friction factor, 83
batch heating, 269
Baume reading, 61
bed friction factor, 90
Bernoulli's equation, 80, 92, 94, 99
Bingham plastic, 115, 158
Biot number, 268, 296
Biot number, inverse, 297
black body, 280
Bond's Law, 165
boundary layer, 75
Bourdon gauge, 71
Brabender amylograph, 155
Brabender system, 154
Brabender units, 154
Brix scale, 61
Brookfield viscometer, 129
Brunauer–Emmett–Teller isotherm, 363
bubble methods (surface tension), 176
bubble size, 197
bulk density, 55

cereals, 57
fruit, 56
milk powder, 57
particulates, 56
vegetables, 56
bulk modulus, 148
Burgess model, 158

calorific value of food, 214
Cannon–Fenske viscometer, 124
capacitance, 394
capacitance, measurement, 400
capacitive reactance, 395
capillary flow viscometers, 123
capillary rise, 174
carbon dioxide, extraction, 318
carbon dioxide, phase diagram, 230
carbon dioxide, T-S diagram, 337
cardice, 318
Carmen–Kozeny equation, 90
Carnot cycle, 211
Carnot efficiency, 25, 212
Casson fluid, 115
cavitation, 99
centrifugal pumps, 97
chilling, 319
Clausius–Clapeyron equation, 327
cleaning-in-place, 388
closed systems, 205
co-current drying, 432
co-current flow, 265
coefficient of discharge, 93
coefficient of performance, 212, 343
cold-air freezing, 314
colloidal systems, 167
commercial sterility, 304, 309
compression, 338
concentration polarization, 441
concentration, conversions, 47
concentric cylinder viscometer, 128
condensation, 339
conductance, electrical, 371

Index

conduction, thermal, 246
conduction, composite wall, 250
conduction, thick-walled tube, 251
cone and plate viscometer, 129
consistency index, 119
constant rate period (drying), 428
continuity equation, 79
continuous heat exchangers, 271
continuous stirred tank reactor, 105
controlled-atmosphere storage, 320
convection, introduction, 258
convection factors, units and dimensions, 46
cooling times, 293
COST 90, 363
counter-current drying, 432
counter-current flow, 265
creep compliance, 150
creep function, 150
critical micelle concentration, 172, 189
critical moisture content, 429
critical temperature, 326
cryogenic freezing, 316

d'Arsonval movement, 374
Dalton's Law, 324
Decimal reduction time, 303
degrees of superheat, 334
density, 32, 51
  bulk, 55
  composition, 54
  fluids, 52
  foods, 55
  fruit juices, 64
  gases and vapours, 65
  liquid values, 62
  liquids, 58
  milk, 64
  overrun, 66
  solids, 52
  temperature effects, 53
  bottles, 59
derived units, 28
detergency, 188, 389
detergent formulation, 191
dew-point temperature, 346
dialysis, 443
dielectric constant, 279, 402
dielectric constant, foods, 403–407
dielectric heating, 282
dielectric loss factor, 285, 402
dielectric loss factor, foods, 403–407
dielectric loss factor, values, 286
dielectric properties of foods, 403–407
dielectric properties, measurement, 406
differential scanning calorimeter, 240
differential thermal analysis, 240
diffusion
  eddy, 415
  gaseous, 416
  introduction, 413

molecular, 415
problem, 418
solids, 421
diffusion coefficient, 415
diffusion through stagnant layer, 417
diffusivity of components in water, 422
diffusivity values, gases, 419
diffusivity, determination, 419
diffusivity, liquids, 420
diffusivity, solids in foods, 422
dilatant fluid, 115
dilatation, 243
dimensional analysis, 45, 261, 426
dimensional groups, 45
direct pull method (surface tension), 179
direct steam injection, 272
dispersion of liquids in gases, 196
drag coefficient, 88
drop weight method (surface tension), 178
dry-bulb temperature, 347
drying time, 431, 433, 435
drying, hot-air, 428
drying, spray, 432
dryness function, 332
du Nouy ring, 180
dynamic viscosity, see viscosity

Einstein's equation, 202
electric charge, 368
electric current, 27, 367
electric motors, 410
electrical energy, 372
electrical heating, 392
electrical measurement, 374
electrical measurement, AC, 391
electrical methods (specific heat), 229
electrical properties, introduction, 366
electrical units, 366
electrodialysis, 443
electromagnetic radiation, 278
emissivity, 281
emissivity values, 281
emulsifying agents, 185
emulsion stability, 187
emulsions, 184
endothermic reaction, 208
energy, 45
  conservation, 202
  conservation processes, 216–218
  conversion, 202
  transfer processes, 201
  units, conversion, 204
  value of food, 214
enthalpy, see also specific enthalpy, 207
enthalpy–composition data, 234
entrainment velocity, 91
entropy, see also specific entropy, 208
equilibrium moisture content, 357, 431
equimolecular counter-diffusion, 416
eutectic mixtures, 315

Index

eutectic temperatures, 233
evaporation (refrigeration), 340
evaporator design, 275
exothermic reaction, 208
expression (pressing), 165
extensometer, 155
extruder, FIRA-NIRD model, 159
extrusion cookers, 101

falling rate period, 428
falling sphere viscometer, 125
fats, melting characteristic data, 239
fats, percentage crystallization, 237
f_c determination, 306
f_h determination, 306
Fick's first law, 415, 437
Fick's second law, 427
film heat coefficient (see heat film coefficient)
first law of thermodynamics, 207
flow control, 100
flow nozzle, 93
flow through packed beds, 86, 89
fluid flow measurement, 92
fluidization, 92
fluidization velocity, 92
Fo evaluation, 304
foaming, 191
force, 36
force, centrifugal, 38
force, units, 37
fouling, 274
Fourier number, 296
freeze-drying, 434
freezing point, solutions, 63, 316
freezing points, foods, 322
freezing processes, 232
freezing times, 309
freon 12, properties, 317
friction chart, 83, 84
frictional loss, other fitting, 85, 87
frictional loss, straight pipe, 82
frictional loss, total system, 86
frictional losses, 82
frozen food, storage, 237
frozen water, percentage, 236
full radiator, 280
fundamental units, 18

gas film coefficient, 424
gas film control, 424
gases, general properties, 324
gases and vapours, distinction, 326
gelation, 155
General Foods texturometer, 161
Gibbs free energy, 184
Grashof number, 261
grinding, 162

gross energy, 214

hard materials, 144
heat and mass transfer, simultaneous, 428
heat balances (see also energy balances), 213
heat exchanger, design, 268
heat exchanger, duty, 266
heat exchanger, types, 270–274
heat film coefficient, 259
  evaluation, 261
  values
heat of respiration, 320
heat penetration, 306
heat pump, 344
heat transfer
  control, 435
  block, 299
  finite cylinder, 299
  infinite cylinder, 297, 301
  infinite slab, 297, 300
  mechanisms, introduction, 246
  sphere, 297, 302
heating times, 293
heats of respiration, values, 322
Henry's law, 424
HLB values, 185
homogenizer, 101
Hookean solid, 157
hot air drying, 428–432
HTST pasteurizer, 271
humectants, 361
humid heat, 351
humid volume, 351
humidity chart, 349, 352, 353, 354
  interpretation, 349–355
humidity measurement, charts, 345–349
humidity sensor, electrical, 387
hydrometer scales, 60
  conversions, 61, 62
hydrometers, 60
hygrometers, 348
hysteresis (sorption isotherms), 361

ideal elastic material, 153
ideal gas equation, 325, 65
ideal solids and liquids, 109
ideal viscous material, 153
immersion freezing, 315
inductance, 393
induction motors, 410
inductive reactance, 394
infrared radiation, 282
Instron testing machine, 161
insulators, 369
interfacial tension, 180
  values, 181, 183
intermediate moisture foods, 361

irradiation, 287
  applications, 289
  units, 287
irreversible processes, 206
isenthalpic, 208, 340
isentropic, 209
isentropic efficiency, 338
isothermal processes, 206

Jaegers method, 177
Joule–Thomson expansion, 340

Kelvin—Voigt model, 157
Kick's Law, 164
kinematic viscosity (see viscosity)
kinetic energy, 203
Kirchhoff's Laws, 370

laminar flow (see streamline flow)
latent heat, 229
latent heat of vaporization, 331
latent heat values, fusion, 233
length, 22
lethality tables, 305
limiting resistance, 266
liquid film coefficient, 424
liquid film control, 424
liquid nitrogen, 317, 318
log mean temperature difference, 265
loss angle, 402
loss modulus, 153
luminous intensity, 27
lyophilization, (see freeze-drying)

McLeod gauge, 73
magnetic effects of current flow, 373
manometers, 70
Marangoni effect, 193
mass, 18
mass balances, 19
mass transfer
  coefficient, 197
  factor, 427
  film coefficients, determination, 426
  introduction, 413
  processes, 414
membrane processes, 440
metabolic rate, units, 215
metabolizable energy, 214
meter bridge, 377
method of cooling, 227
method of mixtures, 226
microwave heating, 282
microwaves, power absorption, 285
model systems, (rheological), 157

moisture content, 234
  dry weight basis, 355
  wet weight basis, 355
moisture contents, conversion, 355
molar conductivity, 371
momentum, 34
multimeter, 375

Newton's law of cooling, 293
Newtonian fluids, 110, 157
non-Newtonian fluids, 114
Nusselt number, 47, 261

Ohm's Law, 368
Ohmic heating, 392
oils, percentage crystallization, 237
open systems, 206
orifice plate, 93
osmotic pressure, 440
Ostwald viscometer, 124
overall heat transfer coefficient, values, 264
overall mass transfer coefficient, 424
overall permeability coefficient, 423
overrun, frozen desserts, 67
overrun, measurement, 67
oxygen transfer rate, 197

packaging materials, composites, 439
packaging, flexible, 437
pasteurisation, 271
Peclet number, 427
Peltier effect, 381
penetration depth, 284
penetration theory, 426, 428
penetrometer, 158
peristaltic pump, 101
permeability units, 438
permeability, values for films, 439
permittivity of free space, 394
phase diagrams, 230
piezoelectric effect, 387
pipe roughness, 85, 86
piston pump, 97
pitot tube, 94
Planck's equation, 279
Planck's equation (freezing), 310
plastic fluids, 117
plate freezers, 313
plate heat exchanger, 272, 273
plug flow, 104
Poiseuille's equation, 121
Poisson's ratio, 147
porosity, 57
porosity, milk powder, 57
positive displacement pumps, 96
potential energy, 203
potentiometer, 375
powder, flowability, 106
power, 43

power factor, 407
power law equation, 119
power law fluids, 120
power law index, 119
power law, plastic fluids, 121
power, pumping, 44, 81
Prandtl number, 47, 261
prefixes, 28
pressure, 38
pressure gauges, 71
pressure measurement, 70
pressures, processing operation values, 41
process control, 387
pseudoplastic fluid, 115
psychrometric chart (see humidity chart)
pump characteristics, 98
pump selection, 96

Q factor, 400
quality (wet vapour), 332

radiation, introduction, 277
Ramsay–Shields equation, 173
reaction velocity constant, 304
refrigeration methods, 312–322
regeneration efficiency, 270
rejection, 441
relative humidity, 355
relaxation time, 150
residence time, average, 102
residence time, distribution, 103
residence time, minimum, 104
resistance, 368
resistance measurement, 376
resistance networks, 369
resistance of semiconductors, 385
resistance thermometers, 384
resistivity, 369
resistivity values, foods, 378
resistivity, values, 370
reverse Carnot cycle, 212
reverse osmosis, 440
reverse-flow viscometer, 124
reversible processes, 206
Reynolds number, 47, 75, 261
Reynolds number, flow in packed beds, 90
Reynolds number, mixing, 78
Reynolds number, particle, 88
rheograms, 115
rheology, 36
rheology, introduction, 137
rheopectic fluids, 118
Rittinger's Law, 164
root mean square values, 391
rotameter, 95
rotary gear pump, 97
rotational viscometers, 127–130

saturated vapour pressure, 327
   refrigerants, 328
   water, 328, 329
saturated vapours, properties, 327
sauter mean particle diameter, 433
Schmidt number, 427
second law of thermodynamics, 208
Seebeck effect, 380
semiconductors, 369
sensible heat changes, 221
sensors, electrical, 380–387
shape factor, 89
shear modulus, 147
shear rate, 109, 123
shear stress, 109, 123
shear stress, wall, 82
Sherwood number, 426
single-spindle viscometer, 129
size reduction, 162
soft materials, 144
solar heating, 280
solubility, 194
sorption isotherms, 357
sorption isotherms, broken isotherms, 362
sorption isotherms, determination, 358
specific conductance, 371
specific conductance, measurement, 378
specific conductivity, foods, 378
specific enthalpy, 208, 331
specific entropy, 209, 331
specific gravity, 32, 58
specific gravity, solutions, 62, 63
specific heat, 220
   apparent, 240
   compositional factors, 222
   gasses and vapours, 224
   measurement, 226–229
   solutions, 63
   values, 223
specific volume, 331
spray drying, 432
spray production, 197
spreading coefficient, 182
St Venant slider, 157
stabilization, 195
Stanton number, 427
steady-state heat transfer, 247
steam tables, interpretation, 329
steam tables, saturated, 329
Stefan's constant, 280
Stefan's Law, 280
Stokes Law, 89, 126
Stokes–Einstein equation, 421
storage conditions, recommended, 321
storage modulus, 153
strain, 143
strain gauge transducer, 386
streamline flow, 75, 77, 121
streamline flow, power law fluids, 123
stress, 143

464 Index

sublimation, 434
superheated vapour, 231, 334
superheated vapours, thermodynamic tables, 335
surface activity, 171
surface area to volume ratio, 31, 196
surface energy, 170
surface pressure, 172
surface tension
  values, 171
  introduction, 169
  measurement, 174–180
  temperature effects, 173
  units, 170
surroundings, 205
suspended level viscometer, 124
synchronous motor, 411
systems, 205

temperature, 22
temperature coefficient of resistance, 384
temperature conversion, 23
temperatures used in processing operations, 26
terminal velocity, 91
texture, perception, 137
texture, sensory assessment, 138
texture evaluation, foods, 163
texture evaluation, instrumental, 139, 162
thawing time, 309
thermal conductivity, 248
  compositional factors, 254
  foods, 252
  measurement, 256
  particulates, 258
  pressure effects, 255
  temperture effects, 255
  values, 249
thermal diffusivity, 257, 294, 308
thermal energy, 203
thermal processing, 301
thermal properties, introduction, 200
thermal units, 203
thermistors, 385
thermocouples, 380–384
thermodynamic charts, 336–344
thermodynamic charts, applications, 338–344
thermodynamic properties, introduction, 200
thermodynamic tables, 329–336
thermometers, 25, 380–386
Third law of thermodynamics, 213
thixotropic fluids, 118
three-phase supply, 408
throttling expansion, 340
time, 22
time-dependent fluids, 117
time-independent fluids, 115
transformer action, 408

turbulent flow, 75, 78
Twaddell hydrometer, 61
two-film theory, 423

UHT processes, 274, 306
ultrafiltration, 440
unsteady state mass transfer, 427
unsteady-state heat transfer
  introduction 247
  equations, 292
  charts, 295–302

vacuum, canned food, 330
vacuum, conversion, 41
vacuum cooling and freezing, 318
vacuum measurement, 40, 73
vacuum production, 102
van der Waal's equation, 326
Van't Hoff equation, 440
vapour compression refrigeration, 340–344
vapours, general properties, 324
variable area meters, 95
variable head meters, 92
velocity
  angular, 34
  average, 33, 103
  maximum, 77, 104
venturi meter, 93
viscoelastic behaviour, 149
viscometer, selection, 130
viscosity
  dairy products, 131
  data, 130–135
  hydrocolloids, 133
  intrinsic, 113
  introduction, 108
  kinematic, 112
  measurement, 121–130
  milk, 131
  oils and fats, 132
  proteins, 134
  reduced, 113
  relative, 113
  sensory aspects, 135
  specific, 113
  sugar solutions, 133
  temperature effects, 111
  units and conversion, 111
  values, Newtonian Fluids, 111
viscous flow (*see* streamline flow)
voidage volumes, 89
voltmeter, 374
volume, 31
volumetric flow rate, 31

water, behaviour during freezing, 232
water, phase diagram, 230

water activity, 359
   food values, 360
   frozen foods, 362
water activity–moisture relationships, 363
water in food, 355
Weber number, 197
Weissenberg effect, 152
wet vapours, 332
wet vapours, thermodynamic properties, 333
wet-bulb temperature, 347
wettability, 194
Wheatstone bridge, 376
work
   of adhesion, 182
   of adhesion values, 183
   of cohesion, 182
   compressional, 42, 338
   mechanical, 42
   units, 42
worm pump (mono pump), 97

Young's equation, 188
Young's modulus, 145

Z values, 303